ELEMENTS OF PROJECTIVE GEOMETRY

ELEMENTS OF PROJECTIVE GEOMETRY

LUIGI CREMONA

GYAN

Published by

Gyan Publishing House
5, Ansari Road
Daryaganj, New Delhi-110002
Phone: 011-47034999, 9811692060
E-mail: books@gyanbooks.com

Distribution Network
gyanbooks.com
India, USA, Canada, UK, Australia

ISBN: 978-93-7414-427-5 (PB)

First Published, 1893

2nd Impression 2026

Printed at: Gyan Press, Delhi.

ELEMENTS OF PROJECTIVE GEOMETRY
Author: LUIGI CREMONA

ELEMENTS

OF

PROJECTIVE GEOMETRY

CREMONA

ELEMENTS

OF

PROJECTIVE GEOMETRY

BY

LUIGI CREMONA

LL.D. EDIN., FOR. MEMB. R. S. LOND., HON. F.R.S. EDIN.
HON. MEMB. CAMB. PHIL. SOC.
PROFESSOR OF MATHEMATICS IN THE UNIVERSITY OF ROME

TRANSLATED BY

CHARLES LEUDESDORF, M.A.

FELLOW OF PEMBROKE COLLEGE, OXFORD

SECOND EDITION

Oxford

AT THE CLARENDON PRESS

1893

AUTHOR'S PREFACE TO THE FIRST EDITION*

Amplissima et pulcherrima scientia figurarum. At quam est inepte sortita nomen Geometriæ!—NICOD. FRISCHLINUS, *Dialog.* I.

Perspectivæ methodus, quâ nec inter inventas nec inter inventu possibiles ulla compendiosior esse videtur . . . —B. PASCAL, *Lit. ad Acad. Paris.*, 1654.

Da veniam scriptis, quorum non gloria nobis
Causa, sed utilitas officiumque fuit.—OVID, *ex Pont.*, iii. 9. 55.

THIS book is not intended for those whose high mission it is to advance the progress of science; they would find in it nothing new, neither as regards principles, nor as regards methods. The propositions are all old; in fact, not a few of them owe their origin to mathematicians of the most remote antiquity. They may be traced back to EUCLID (285 B.C.), to APOLLONIUS of Perga (247 B.C.), to PAPPUS of Alexandria (4th century after Christ); to DESARGUES of Lyons (1593–1662); to PASCAL (1623–1662); to DE LA HIRE (1640–1718); to NEWTON (1642–1727); to MACLAURIN (1698–1746); to J. H. LAMBERT (1728–1777), &c. The theories and methods which make of these propositions a homogeneous and harmonious whole it is usual to call *modern*, because they have been discovered or perfected by mathematicians of an age nearer to ours, such as CARNOT, BRIANCHON, PONCELET, MÖBIUS, STEINER, CHASLES, STAUDT, &c.; whose works were published in the earlier half of the present century.

Various names have been given to this subject of which we are about to develop the fundamental principles. I prefer

* With the consent of the Author, only such part of the preface to the original Italian edition (1872) is here reproduced as may be of interest to the English reader.

not to adopt that of *Higher Geometry* (*Géométrie supérieure*, *höhere Geometrie*), because that to which the title 'higher' at one time seemed appropriate, may to-day have become very elementary; nor that of *Modern Geometry* (*neuere Geometrie*), which in like manner expresses a merely relative idea; and is moreover open to the objection that although the methods may be regarded as modern, yet the matter is to a great extent old. Nor does the title *Geometry of position* (*Geometrie der Lage*) as used by STAUDT * seem to me a suitable one, since it excludes the consideration of the metrical properties of figures. I have chosen the name of *Projective Geometry* †, as expressing the true nature of the methods, which are based essentially on central projection or perspective. And one reason which has determined this choice is that the great PONCELET, the chief creator of the modern methods, gave to his immortal book the title of *Traité des propriétés projectives des figures* (1822).

In developing the subject I have not followed exclusively any one author, but have borrowed from all what seemed useful for my purpose, that namely of writing a book which should be thoroughly elementary, and accessible even to those whose knowledge does not extend beyond the mere elements of ordinary geometry. I might, after the manner of STAUDT, have taken for granted no previous notions at all; but in that case my work would have become too extensive, and would no longer have been suitable for students who have read the usual elements of mathematics. Yet the whole of what such students have probably read is not necessary in order to understand my book; it is sufficient that they should know the chief propositions relating to the circle and to similar triangles.

It is, I think, desirable that theoretical instruction in

* Equivalent to the *Descriptive Geometry* of CAYLEY (*Sixth memoir on quantics*, Phil. Trans. of the Royal Society of London, 1859; p. 90). The name *Géométrie de position* as used by CARNOT corresponds to an idea quite different from that which I wished to express in the title of my book. I leave out of consideration other names, such as *Géométrie segmentaire* and *Organische Geometrie*, as referring to ideas which are too limited, in my opinion.

† See KLEIN, *Ueber die sogenannte nicht-Euklidische Geometrie* (Göttinger Nachrichten, Aug. 30, 1871).

geometry should have the help afforded it by the practical constructing and drawing of figures. I have accordingly laid more stress on *descriptive* properties than on *metrical* ones; and have followed rather the methods of the *Geometrie der Lage* of STAUDT than those of the *Géométrie supérieure* of CHASLES *. It has not however been my wish entirely to exclude metrical properties, for to do this would have been detrimental to other practical objects of teaching †. I have therefore introduced into the book the important notion of the *anharmonic ratio*, which has enabled me, with the help of the few above-mentioned propositions of the ordinary geometry, to establish easily the most useful metrical properties, which are either consequences of the projective properties, or are closely related to them.

I have made use of *central projection* in order to establish the idea of *infinitely distant elements*; and, following the example of STEINER and of STAUDT, I have placed the law of *duality* quite at the beginning of the book, as being a logical fact which arises immediately and naturally from the possibility of constructing space by taking either the point or the plane as element. The enunciations and proofs which correspond to one another by virtue of this law have often been placed in parallel columns; occasionally however this arrangement has been departed from, in order to give to students the opportunity of practising themselves in deducing from a theorem its correlative. Professor REYE remarks, with justice, in the preface to his book, that Geometry affords nothing so stirring to a beginner, nothing so likely to stimulate him to original work, as the principle of duality; and for this reason it is very important to make him acquainted with it as soon as possible, and to accustom him to employ it with confidence.

The masterly treatises of PONCELET, STEINER, CHASLES, and

* Cf. REYE, *Geometrie der Lage* (Hannover, 1866; 2nd edition, 1877), p. xi of the preface.

† Cf. ZECH, *Die höhere Geometrie in ihrer Anwendung auf Kegelschnitte und Flächen zweiter Ordnung* (Stuttgart, 1857), preface.

STAUDT * are those to which I must acknowledge myself most indebted; not only because all who devote themselves to Geometry commence with the study of these works, but also because I have taken from them, besides the substance of the methods, the proofs of many theorems and the solutions of many problems. But along with these I have had occasion also to consult the works of APOLLONIUS, PAPPUS, DESARGUES, DE LA HIRE, NEWTON, MACLAURIN, LAMBERT, CARNOT, BRIANCHON, MÖBIUS, BELLAVITIS, &c.; and the later ones of ZECH, GASKIN, WITZSCHEL, TOWNSEND, REYE, POUDRA, FIEDLER, &c.

In order not to increase the difficulties, already very considerable, of my undertaking, I have relieved myself from the responsibility of quoting in all cases the sources from which I have drawn, or the original discoverers of the various propositions or theories. I trust then that I may be excused if sometimes the source quoted is not the original one †, or if occasionally the reference is found to be wanting entirely. In giving references, my desire has been chiefly to call the attention of the student to the names of the great geometers and the titles of their works, which have become classical. The association with certain great theorems of the illustrious names of EUCLID, APOLLONIUS, PAPPUS, DESARGUES, PASCAL, NEWTON, CARNOT, &c. will not be without advantage in assisting the mind to retain the results themselves, and in exciting that scientific curiosity which so often contributes to enlarge our knowledge.

Another object which I have had in view in giving references is to correct the first impressions of those to whom the name *Projective Geometry* has a suspicious air of novelty. Such

* PONCELET, *Traité des propriétés projectives des figures* (Paris, 1822). STEINER, *Systematische Entwickelung der Abhängigkeit geometrischer Gestalten von einander, &c.* (Berlin, 1832). CHASLES, *Traité de Géométrie supérieure* (Paris, 1852); *Traité des sections coniques* (Paris, 1865). STAUDT, *Geometrie der Lage* (Nürnberg, 1847).

† In quoting an author I have almost always cited such of his treatises as are of considerable extent and generally known, although his discoveries may have been originally announced elsewhere. For example, the researches of CHASLES in the theory of conics date from a period in most cases anterior to the year 1830; those of STAUDT began in 1831; &c.

persons I desire to convince that the subjects are to a great extent of venerable antiquity, matured in the minds of the greatest thinkers, and now reduced to that form of extreme simplicity which GERGONNE considered as the mark of perfection in a scientific theory *. In my analysis I shall follow the order in which the various subjects are arranged in the book.

The conception of *elements lying at an infinite distance* is due to the celebrated mathematician DESARGUES; who more than two centuries ago explicitly considered parallel straight lines as meeting in an infinitely distant point †, and parallel planes as passing through the same straight line at an infinite distance ‡.

The same idea was thrown into full light and made generally known by PONCELET, who, starting from the postulates of the Euclidian Geometry, arrived at the conclusion that the points in space which lie at an infinite distance must be regarded as all lying in the same plane §.

DESARGUES || and NEWTON ¶ considered the asymptotes of the hyperbola as tangents whose points of contact lie at an infinite distance.

The name *homology* is due to PONCELET. Homology, with reference to plane figures, is found in some of the earlier treatises on perspective, for example in LAMBERT ** or perhaps even in DESARGUES ††, who enunciated and proved the theorem concerning triangles and quadrilaterals in perspective or homology. This theorem, for the particular case of two triangles (Art. 17), is however really of much older date, as it

* 'On ne peut se flatter d'avoir le dernier mot d'une théorie, tant qu'on ne peut pas l'expliquer en peu de paroles à un passant dans la rue' (cf. CHASLES, *Aperçu historique*, p. 115).

† *Œuvres de* DESARGUES, *réunies et analysées par* M. POUDRA (Paris, 1864), tome i. *Brouillon-projet d'une atteinte aux événements des rencontres d'un cône avec un plan* (1639), pp. 104, 105, 205.

‡ *Loc. cit.*, pp. 105, 106.

§ *Traité des propriétés projectives des figures* (Paris, 1822), Arts. 96, 580.

|| *Loc. cit.*, p. 210.

¶ *Philosophiae naturalis principia mathematica* (1686), lib. i. prop. 27, scholium.

** *Freie Perspective*, 2nd edition (Zurich, 1774).

†† *Loc. cit.*, pp. 413–416.

is substantially identical with a celebrated porism of EUCLID (Art. 114), which has been handed down to us by PAPPUS *. Homological figures in space were first studied by PONCELET †.

The law of duality, as an independent principle, was enunciated by GERGONNE ‡; as a consequence of the theory of reciprocal polars (under the name *principe de réciprocité polaire*) it is due to PONCELET §.

The geometric forms (range of points, flat pencil) are found, the names excepted, in DESARGUES and the later geometers. STEINER ‖ has defined them in a more explicit manner than any previous writer.

The complete quadrilateral was considered by CARNOT ¶; the idea was extended by STEINER ** to polygons of any number of sides and to figures in space.

Harmonic section was known to geometers of the most remote antiquity; the fundamental properties of it are to be found for example in APOLLONIUS ††. DE LA HIRE ‡‡ gave the construction of the fourth element of a harmonic system by means of the harmonic property of the quadrilateral, *i.e.* by help of the ruler only.

From 1832 the construction of projective forms was taught by STEINER §§.

The complete theory of the anharmonic ratios is due to MÖBIUS ‖‖, but before him EUCLID, PAPPUS ¶¶, DESARGUES ***, and BRIANCHON ††† had demonstrated the fundamental proposition of Art. 63. DESARGUES ‡‡‡ was the author of the theory

* CHASLES, *Les trois livres de porismes d'Euclide, &c.* (Paris, 1860), p. 102.

† *Loc. cit.*, pp. 369 sqq.

‡ *Annales de Mathématiques*, vol. xvi. (Montpellier, 1826), p. 209.

§ Ibid., vol. viii. (Montpellier, 1818), p. 201.

‖ *Systematische Entwickelung*, pp. xiii, xiv. Collected Works, vol. i. p. 237.

¶ *De la corrélation des figures de Géométrie* (Paris, 1801), p. 122.

** *Loc. cit.*, pp. 72, 235; §§ 19, 55.

†† *Conicorum* lib. i. 34, 36, 37, 38.

‡‡ *Sectiones conicae* (Parisiis, 1685), i. 20.

§§ *Loc. cit.*, p. 91.

‖‖ *Der barycentrische Calcul* (Leipzig, 1827), chap. v.

¶¶ *Mathematicae Collectiones*, vii. 129.

*** *Loc. cit.*, p. 425.

††† *Mémoire sur les lignes du second ordre* (Paris, 1817), p. 7.

‡‡‡ *Loc. cit.*, pp. 119, 147, 171, 176.

of *involution*, of which a few particular cases were already known to the Greek geometers *.

The generation of conics by means of two projective forms was set forth, forty years ago, by STEINER and by CHASLES; it is based on two fundamental theorems (Arts. 149, 150) from which the whole theory of these important curves can be deduced. The same method of generation includes the organic description of NEWTON † and various theorems of MACLAURIN.

But the projectivity of the pencils formed by joining two fixed points on a conic to a variable point on the same had already been proved, in other words, by APOLLONIUS ‡.

When only sixteen years old (in 1640) PASCAL discovered his celebrated theorem of the *mystic hexagram* §, and in 1806 BRIANCHON deduced the correlative theorem (Art. 153) by means of the theory of pole and polar.

The properties of the quadrilateral formed by four tangents to a conic and of the quadrangle formed by their points of contact are to be found in the Latin appendix (*De linearum geometricarum proprietatibus generalibus tractatus*) to the *Algebra* of MACLAURIN, a posthumous work (London, 1748). He deduced from these properties methods for the construction of a conic by points or by tangents in several cases where five elements (points or tangents) are given. This problem, in its full generality, was solved at a later date by BRIANCHON.

The idea of considering two projective ranges of points on the same conic was explicitly set forth by BELLAVITIS ||.

To CARNOT ¶ we owe a celebrated theorem (Art. 385) concerning the segments which a conic determines on the sides of

* PAPPUS, *Mathematicae Collectiones*, lib. vii. props. 37–56, 127, 128, 130–133.

† *Loc. cit.*, lib. i. lemma xxi.

‡ *Conicorum* lib. iii. 54, 55, 56. I owe this remark to Prof. ZEUTHEN (1885).

§ *Letter of* LEIBNITZ *to* M. PÉRIER in the *Œuvres de B. Pascal* (BOSSUT'S edition, vol. v. p. 459).

|| *Saggio di geometria derivata* (Nuovi Saggi dell' Accademia di Padova, vol. iv. 1838), p. 270, note.

¶ *Géométrie de position* (Paris, 1803), Art. 379.

a triangle. Of this theorem also certain particular cases were known long before *.

In the *Freie Perspective* of LAMBERT we meet with elegant constructions for the solution of several problems of the first and second degrees by means of the ruler, assuming however that certain elements are given; but the possibility of solving all problems of the second degree by means of the ruler and a fixed circle was made clear by PONCELET; afterwards STEINER, in a most valuable little book, showed the manner of practically carrying this out (Arts. 238 sqq.).

The theory of pole and polar was already contained, under various names, in the works already quoted of DESARGUES † and DE LA HIRE ‡; it was perfected by MONGE §, BRIANCHON ‖, and PONCELET. The last-mentioned geometer derived from it the theory of polar reciprocation, which is essentially the same thing as the law of duality, called by him the 'principe de réciprocité polaire.'

The principal properties of *conjugate diameters* were expounded by APOLLONIUS in books ii and vii of his work on the Conics.

And lastly, the fundamental theorems concerning *foci* are to be found in book iii of APOLLONIUS, in book vii of PAPPUS, and in book viii of DE LA HIRE.

Those who desire to acquire a more extended and detailed knowledge of the progress of Geometry from its beginnings until the year 1830 (which is sufficient for what is contained in this book) have only to read that classical work, the *Aperçu historique* of CHASLES.

* APOLLONIUS, *Conicorum* lib. iii. 16–23. DESARGUES, *loc. cit.*, p. 202. DE LA HIRE, *loc. cit.*, book v. props. 10, 12. NEWTON, *Enumeratio linearum tertii ordinis* (*Opticks*, London, 1704), p. 142.

† *Loc. cit.*, pp. 164, 186, 190 sqq.

‡ *Loc. cit.*, i. 21–28; ii. 23–30.

§ *Géométrie descriptive* (Paris, 1795), Art. 40.

‖ *Journal de l'École Polytechnique*, cahier xiii. (Paris, 1806).

AUTHOR'S PREFACE TO THE ENGLISH EDITION

In April last year, when I was in Edinburgh on the occasion of the celebration of the tercentenary festival of the University there, Professor Sylvester did me the honour of saying that in his opinion a translation of my book on the Elements of Projective Geometry might be useful to students at the English Universities as an introduction to the modern geometrical methods. The same favourable judgement was shown to me by other mathematicians, especially in Oxford, which place I visited in the following month of May at the invitation of Professor Sylvester. There Professor Price proposed to me that I should assist in an English translation of my book, to be carried out by Mr. C. Leudesdorf, Fellow of Pembroke College, and to be published by the Clarendon Press. I accepted the proposal with pleasure, and for this reason. In my opinion the English excel in the art of writing text-books for mathematical teaching; as regards the clear exposition of theories and the abundance of excellent examples, carefully selected, very few books exist in other countries which can compete with those of Salmon and many other distinguished English authors that could be named. I felt it therefore to be a great honour that my book should be considered by such competent judges worthy to be introduced into their colleges.

Unless I am mistaken, the preference given to my *Elements* over the many treatises on modern geometry published on the Continent is to be attributed to the circumstance that in it I have striven, to the best of my ability, to imitate the English models. My intention was not to produce a book of high theories which should be of interest to the advanced mathematician, but to construct an elementary text-book of modest dimensions, intelligible to a student whose knowledge need not extend further than the first books of Euclid. I aimed therefore at simplicity and clearness of exposition; and I was careful to supply an abundance of examples of a kind suitable to encourage the beginner, to make him seize the spirit of the methods, and to render him capable of employing them.

My book has, I think, done some service in Italy by helping to spread a knowledge of projective geometry; and I am encouraged to believe that it has not been unproductive of results even elsewhere, since I have had the honour of seeing it translated into French and into German.

If the present edition be compared with the preceding ones, it will be seen that the book has been considerably enlarged and amended. All the improvements which are to be found in the French and the German editions have been incorporated; a new Chapter, on Foci, has been added; and every Chapter has received modifications, additions, and elucidations, due in part to myself, and in part to the translator.

In conclusion, I beg leave to express my thanks to the eminent mathematician, the Savilian Professor of Geometry, who advised this translation; to the Delegates of the Clarendon Press, who undertook its publication; and to Mr. Leudesdorf, who has executed it with scrupulous fidelity.

L. CREMONA.

Rome, May 1885.

TABLE OF CONTENTS

PAGE

PREFACE v

CHAPTER I.

Definitions (Arts. 1–7) 1

CHAPTER II.

CENTRAL PROJECTION; FIGURES IN PERSPECTIVE.

Figures in perspective (Arts. 8–11) 3

Point at infinity on a straight line (12) 4

Line at infinity in a plane (13). 5

Triangles in perspective; theorems of DESARGUES (14–17) . 6

CHAPTER III.

HOMOLOGY.

Figures in homology (Arts. 18–21) 9

Locus of the centre of perspective of two figures, when one is turned round the axis of perspective (22) 12

Construction of homological figures (23) 13

Homothetic figures (23) 18

CHAPTER IV.

HOMOLOGICAL FIGURES IN SPACE.

Relief-perspective (Art. 24) 20

Plane at infinity (26) 21

CHAPTER V.

GEOMETRIC FORMS.

PAGE
The geometric prime-forms (Arts. 27–31) 22
Their dimensions (32) 24

CHAPTER VI.

THE PRINCIPLE OF DUALITY.

Correlative figures and propositions (Arts. 33–38) 26

CHAPTER VII.

PROJECTIVE GEOMETRIC FORMS.

First notions (Arts. 39–41) 33
Forms in perspective (42, 43) 35
Fundamental theorems (44, 45) 36

CHAPTER VIII.

HARMONIC FORMS.

Fundamental theorem (Arts. 46, 49) 39
Harmonic forms are projective (47, 48, 50, 51) 41
Elementary properties (52–57) 44
Constructions (58–60) 47

CHAPTER IX.

ANHARMONIC RATIOS.

Distinction between metrical and descriptive (Art. 61) . . 50
Rule of signs: elementary segment-relations (62) . . . 50
Theorem of PAPPUS, and converse (63–66) 52
Properties of harmonic forms (68–71) 57
The twenty-four anharmonic ratios of a group of four elements (72) 59
In two projective forms, corresponding groups of elements are equianharmonic (73) 62
Metrical property of two projective ranges (74) 62
Properties of two homological figures (75–77) 63

CHAPTER X.

CONSTRUCTION OF PROJECTIVE FORMS.

PAGE

Two forms are projective if corresponding groups of elements are equianharmonic (Art. 79) 66

Forms in perspective; self-corresponding elements (80) . . . 67

Superposed projective forms (81, 82) 68

A geometric form of four elements, when harmonic (83) . . . 69

Constructions (84–86, 88–90) 70

Hexagon whose vertices lie on two straight lines; theorem of PAPPUS (87) 75

Properties of two projective figures (91–94) 79

Construction of projective plane figures (95, 96) 81

Any two such figures can be placed so as to be in homology (97) 84

CHAPTER XI.

PARTICULAR CASES AND EXERCISES.

Similar ranges and pencils (Arts. 99–103) 86

Equal pencils (104–108) 89

Metrical properties of two collinear projective ranges (109) . 91

Examples (110–112) 93

Porisms of EUCLID and of PAPPUS (113, 114) 95

Problems solved with the ruler only (115–118) 96

Figures in perspective; theorems of CHASLES (119, 120) . . 98

CHAPTER XII.

INVOLUTION.

Definition; elementary properties (Arts. 121–124) 100

Metrical property; double elements (125, 126) 102

Two pairs of conjugate elements determine an involution (127) . 104

The two kinds of involutions (128) 105

PAGE
Another metrical property (130) 106
Property of a quadrangle cut by a transversal (131, 132) . . 107
The middle points of the diagonals of a complete quadrilateral are collinear (133) 108
Constructions (134) 109
Theorems of CEVA and MENELAUS (136–140) 110
Particular cases (142) 113

CHAPTER XIII.

PROJECTIVE FORMS IN RELATION TO THE CIRCLE.

Circle generated by two directly equal pencils (Art. 143) . . 114
Fundamental property of points on a circle (144) . . . 114
Fundamental property of tangents to a circle (146) . . . 115
Harmonic points and tangents (145, 147, 148) . . . 115

CHAPTER XIV.

PROJECTIVE FORMS IN RELATION TO THE CONIC SECTIONS.

Fundamental theorems (Art. 149) 118
Generation of conics by means of two projective forms (150) . 119
Anharmonic ratio of four points or tangents of a conic (151) . 122
Five points or five tangents determine a conic (152) . . . 123
Theorems of PASCAL and BRIANCHON (153, 154) . . . 124
Theorems of MÖBIUS (155) and MACLAURIN (156) . . . 126
Properties of the parabola (157, 158) 127
Properties of the hyperbola; theorems of APOLLONIUS (158–160) 129

CHAPTER XV.

CONSTRUCTIONS AND EXERCISES.

PASCAL'S AND BRIANCHON'S theorems applied to the construction of a conic by points or by tangents (Art. 161) . . . 131
Cases in which one or more of the elements lie at infinity (162, 163) 132

CHAPTER XVI.

DEDUCTIONS FROM THE THEOREMS OF PASCAL AND BRIANCHON.

PAGE

Theorem on the inscribed pentagon (Art. 164) 136

Application to the construction of conics (165) . . . 137

Theorem on the inscribed quadrangle (166, 168) . . . 138

Application to the construction of conics (167, 168) . . . 139

The circumscribed quadrilateral and the quadrangle formed by the points of contact of the sides (169–173) . . . 140

Theorem on the inscribed triangle (174) 143

Application to the construction of conics (175) . . . 143

The circumscribed triangle and the triangle formed by the points of contact of the sides (176–178) 144

Theorem on the circumscribed pentagon (179) 145

Construction of conics subject to certain conditions (180–182) . 146

CHAPTER XVII.

DESARGUES' THEOREM.

DESARGUES' theorem and its correlative (Art. 183) . . . 148

Conics circumscribing the same quadrangle, or inscribed in the same quadrilateral (184) 149

Theorems of PONCELET (186–188) 151

Deductions from DESARGUES' theorem (189–194) . . . 152

Group of four harmonic points or tangents (195, 196) . . 157

Property of the hyperbola (197) 158

Theorem 'ad quatuor lineas' quoted by PAPPUS (198) . . 158

Correlative theorem (199) 159

CHAPTER XVIII.

SELF-CORRESPONDING ELEMENTS AND DOUBLE ELEMENTS.

Projective ranges of points on a conic (Art. 200) . . . 161

Projective series of tangents to a conic (201) 163

Involution of points or tangents of a conic (202, 203) . . 165

PAGE

Harmonic points and tangents (204, 205) 168

Construction of the self-corresponding elements of two superposed projective forms, and of the double elements of an involution (206) 169

Orthogonal pair of rays of a pencil in involution (207) . . 172

Construction for the common pair of two superposed involutions (208, 209) 173

Other constructions (210, 211) 174

CHAPTER XIX.

PROBLEMS OF THE SECOND DEGREE.

Construction of a conic determined by five points or tangents (Arts. 212, 213, 216, 217) 176

Particular cases (214, 215) 178

Construction of a conic determined by four points and a tangent, or by four tangents and a point (218) 180

Case of the parabola (219, 220) 181

Construction of a conic determined by three points and two tangents, or by three tangents and two points (221) . . 182

Construction of a polygon satisfying certain conditions (222–225) 184

Construction of the points of intersection, and of the common tangents, of two conics (226–230) 188

Various problems (231–236) 190

Geometric method of false position (237). 193

Solution of problems of the second degree by means of the ruler and a fixed circle (238) 194

Examples of problems solved by this method (239–249) . . 194

CHAPTER XX.

POLE AND POLAR.

Definitions and elementary properties (Arts. 250–254) . . 201

Conjugate points and lines with respect to a conic (255, 256) . 204

Constructions (257) 205

PAGE

Self-conjugate triangle (258–262) 206

Involution of conjugate points and lines (263, 264) . . . 209

Complete quadrangles and quadrilaterals having the same diagonal triangle (265–267) 210

Conics having a common self-conjugate triangle (268, 269) . 212

Properties of conics inscribed in the same quadrilateral, or circumscribing the same quadrangle (270–273) . . 213

Properties of inscribed and circumscribed triangles (274, 275) . 215

CHAPTER XXI.

THE CENTRE AND DIAMETERS OF A CONIC.

The diameter of a system of parallel chords (Arts. 276–278) . 217

Case of the parabola (279–280) 218

Centre of a conic (281–283, 285) 218

Conjugate diameters (284, 286–288, 290) 219

Case of the circle (289) 222

Theorem of Möbius (291) 224

Involution property of a quadrangle inscribed in a conic (292) . 224

Ideal diameters and chords (290, 294) 226

Involution of conjugate diameters: the axes (296–298) . . 227

Various properties of conjugate diameters; theorems of Apollonius (299–315) 228

Conics inscribed in the same quadrilateral; theorem of Newton (317, 318) 236

Constructions (285, 290, 293, 301, 307, 311, 316, 319) . . 238

CHAPTER XXII.

POLAR RECIPROCAL FIGURES.

Polar reciprocal curves (Arts. 320, 321) 239

The polar reciprocal of a conic with regard to a conic (322, 323) 240

Polar reciprocal figures are correlative figures (324, 325) . . 241

Two triangles which are self-conjugate with regard to the same conic (326) 242

PAGE

Triangles self-conjugate with regard to one conic, and inscribed in, or circumscribed to, another (327–329) 243

Two triangles circumscribed to the same conic (330, 331). . 244

Triangles inscribed in one conic and circumscribed to another (332) 244

HESSE's theorem (334) 245

Reciprocal triangles with regard to a conic are homological (336) 246

Conic with regard to which two given triangles are reciprocal (338) 247

Polar system (339) 248

CHAPTER XXIII.

FOCI.

Foci of a conic defined (Arts. 340–341) 249

The involution determined by pairs of orthogonal conjugate rays on each of the axes (342) 250

The foci are the double points of this involution (343) . . 251

Focal properties of tangents and normals (344–346, 349, 360–362) 252

The circle circumscribing the triangle formed by three tangents to a parabola passes through the focus (347) 253

The directrices (348, 350, 351) 254

The latus rectum (352) 257

The focal radii (353) 258

The eccentricity (354, 355) 259

Locus of the feet of perpendiculars from a focus on the tangents to a conic (356–359) 260

Constructions (363, 364, 366, 367). 264

Confocal conics (365) 266

Locus of intersection of orthogonal tangents to a conic (368, 369) 268

Property of the director circle in relation to the self-conjugate triangles of a conic (370–375). 270

The polar reciprocal of a circle with respect to a circle (376–379) 273

CHAPTER XXIV.

COROLLARIES AND CONSTRUCTIONS.

PAGE

Various properties and constructions connected with the hyperbola and the parabola (Arts. 380–384) 277

CARNOT's theorem, and deductions from it (385–389, 391) . 279

Constructions of conics (390, 392–394) 284

The rectangular hyperbola (395) 285

Method of determining to which kind of conic a given arc belongs (396) 289

Constructions of conics (397–404) 289

Trisection of an arc of a circle (406) 294

Construction of a conic, given three tangents and two points, or three points and two tangents (408) 295

NEWTON's organic description of a conic (409, 410) 296

Various problems and theorems (411–418, 421) 297

Problems of the second degree solved by means of the theory of pole and polar (419) 300

Problems solved by the method of polar reciprocation (420) . 301

Exercises (422) 302

INDEX 303

ELEMENTS OF PROJECTIVE GEOMETRY.

CHAPTER I.

DEFINITIONS.

1. By a *figure* is meant any assemblage of points, straight lines, and planes; the straight lines and planes are all to be considered as extending to infinity, without regard to the limited portions of space which are enclosed by them. By the word *triangle*, for example, is to be understood a system consisting of three points and three straight lines connecting these points two and two; a *tetrahedron* is a system consisting of four planes and the four points in which these planes intersect three and three, &c.

In order to secure uniformity of notation, we shall always denote points by the capital letters A, B, C, ..., straight lines by the small letters a, b, c, ..., planes by the Greek letters α, β, γ, ... Moreover, AB will denote that part of the straight line joining A and B which is comprised between the points A and B; Aa will denote the plane which passes through the point A and the straight line a; $a\alpha$ the point common to the straight line a and the plane α; $\alpha\beta$ the straight line formed by the intersection of the planes α, β; ABC the plane of the three points A, B, C; $\alpha\beta\gamma$ the point common to the three planes α, β, γ; $\alpha.BC$ the point common to the plane α and the straight line BC; $A.\beta\gamma$ the plane passing through the point A and the straight line $\beta\gamma$; $\alpha.Bc$ the straight line common to the plane α and the plane Bc; $A.\beta c$ the straight line joining the point A to the point βc, &c. The notation $\alpha.BC \equiv A'$ we shall use to express that the point common to the plane α and the straight line BC coincides with the point A';

$u \equiv ABC$ will express that the straight line u contains the points A, B, C, &c.

2. *To project from a fixed point S* (the *centre of projection*) a figure ($ABCD\ldots$, $abcd\ldots$) composed of points and straight lines, is to construct the straight lines or *projecting rays* SA, SB, SC, SD, ... and the planes (*projecting planes*) Sa, Sb, Sc, Sd, ... We thus obtain a new figure composed of straight lines and planes which all pass through the centre S.

3. *To cut by a fixed plane σ* (*transversal plane*) a figure ($\alpha\beta\gamma\delta$, ... $abcd\ldots$) made up of planes and straight lines, is to construct the straight lines or *traces* $\sigma\alpha$, $\sigma\beta$, $\sigma\gamma$, ... and the points or *traces* σa, σb, σc, ... By this means we obtain a new figure composed of straight lines and points lying in the plane σ.

4. *To project from a fixed straight line s* (the *axis*) a figure $ABCD\ldots$ composed of points, is to construct the planes sA, sB, sC, ... The figure thus obtained is composed of planes which all pass through the axis s.

5. *To cut by a fixed straight line s* (a *transversal*) a figure $\alpha\beta\gamma\delta\ldots$ composed of planes, is to construct the points $s\alpha$, $s\beta$, $s\gamma$, ... In this way a new figure is obtained, composed of points all lying on the fixed transversal s.

6. If a figure is composed of straight lines $a, b, c, \ldots$ which all pass through a fixed point or *centre* S, it can be *projected* from a straight line or *axis* s passing through S; the result is a figure composed of planes sa, sb, sc, ...

7. If a figure is composed of straight lines a, b, c, ... all lying in a fixed plane, it may be cut by a straight line (transversal) s lying in the same plane; the figure which results is formed by the points sa, sb, sc, ... *.

* The operations of projecting and cutting (*projection* and *section*) are the two fundamental ones of the Projective Geometry.

CHAPTER II.

CENTRAL PROJECTION; FIGURES IN PERSPECTIVE.

8. CONSIDER a plane figure made up of points $A, B, C, \ldots$ and straight lines $AB, AC, \ldots, BC, \ldots$. Project these from a centre S not lying in the plane (σ) of the figure, and cut the rays $SA, SB, SC, \ldots$ and the planes $SAB, SAC, \ldots, SBC, \ldots$ by a transversal plane σ' (Fig. 1). The traces on the plane σ' of the projecting rays and planes will form a second figure, a picture of the first. When we carry out the two operations by which this second figure is derived from the first, we are said *to project from a centre* (or *vertex*) S *a given figure* σ *upon a plane of projection* σ'. The new figure σ' is called the *perspective image* or the *central projection* of the original one. Of course, if the second figure be projected back from the centre S upon the plane σ, the first figure will be formed again; *i.e.* the first figure is the projection of the second from the centre S upon the picture-plane σ. The two figures σ and σ' are said to be *in perspective position*, or simply *in perspective.*

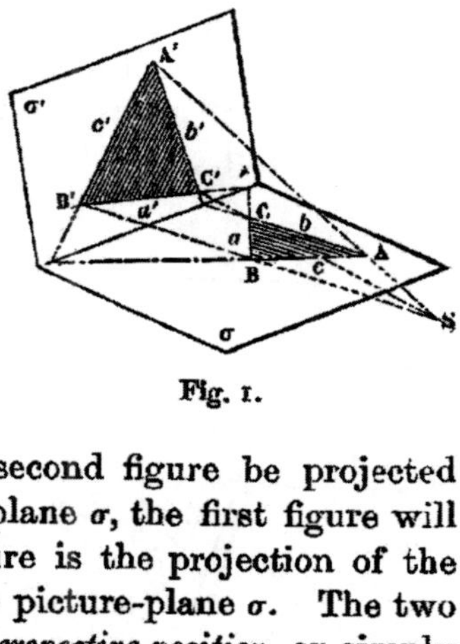

Fig. 1.

9. If $A', B', C', \ldots$ are the traces of the rays $SA, SB, SC, \ldots$ on the plane σ', we may say that to the points $A, B, C, \ldots$ of the first figure *correspond* the points $A', B', C', \ldots$ of the second, with the condition that two corresponding points always lie on a straight line passing through S. If the point A describe a straight line a in the plane σ, the ray SA will describe a plane Sa; and therefore A' will describe a straight line a', the intersection of the planes Sa and σ'. The straight lines a and a',

in which the planes σ and σ' are cut by any the same projecting plane, may thus be called *corresponding lines*. It follows from this that to the straight lines $AB, AC, \ldots, BC, \ldots$ correspond the straight lines $A'B'$, $A'C'$, ..., $B'C'$, ... and that to all straight lines which pass through a given point A of the plane σ correspond straight lines which pass through the corresponding point A' of the plane σ'.

10. If the point A describe a curve in the plane σ, the corresponding point A' will describe another curve in the plane σ', which may be said to *correspond* to the first curve. Tangents to the two curves at corresponding points are clearly corresponding straight lines; and again, the two curves are cut by corresponding straight lines in corresponding points. Two corresponding curves are therefore of the same degree *.

11. The two figures may equally well be generated by the simultaneous motion of a pair of corresponding straight lines a, a'. If a revolve about a fixed point A, then a' will always pass through the corresponding point A'.

Similarly, if a envelop a curve, then a' will envelop the corresponding curve. The lines a and a', in corresponding positions, touch the two curves at corresponding points; and again, to the tangents to the first curve from a point A correspond the tangents to the second from the corresponding point A'. Two corresponding curves are therefore of the same class †.

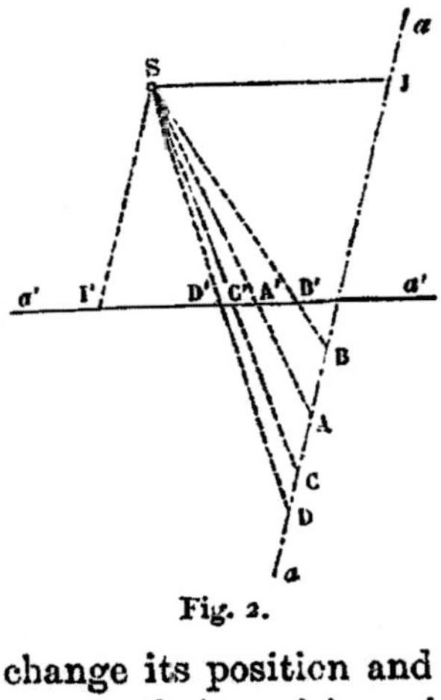

Fig. 2.

12. Consider two straight lines a and a' which correspond to one another in the figures σ, σ' (Fig. 2). Every ray drawn through S in their plane meets them in two points, say A and A', which correspond to one another. If the ray change its position and revolve round S, the points A and A' change their positions simultaneously; when the ray is about to

* The *degree* of a curve is the greatest number of points in which it can be cut by any arbitrary plane. In the case of a plane curve, it is the greatest number of points in which it can be cut by any straight line in the plane.

† The *class* of a plane curve is the greatest number of tangents which can be drawn to it from any arbitrary point in the plane.

become parallel to a, the point A' approaches I' (the point where a' is cut by the straight line drawn through S parallel to a) and the point A moves away indefinitely. In order that the property that to one point of a' corresponds one point of a may hold universally, we say that the line a has a *point at infinity* I, with which the point A coincides when A' coincides with I', viz. when the ray, turning about S, becomes parallel to a. The straight line a has only one point at infinity, it being assumed that we can draw through S only one ray parallel to a*.

The point I', the image of the point at infinity I, is called the *vanishing point* of a'.

Similarly, the straight line a' has a point J' at infinity, which corresponds to the point J where a is cut by the ray drawn through S parallel to a'.

Two parallel straight lines have the same point at infinity. All straight lines which are parallel to a given straight line must be considered as having a common point of intersection at infinity.

Two straight lines lying in the same plane always intersect in a point (finite or infinitely distant).

13. If now the straight line a takes all possible positions in the plane σ, the corresponding straight line a' will always be determined by the intersection of the planes σ' and Sa. As a moves, the ray SI traces out a plane π parallel to σ and the point I' describes the straight line $\pi\sigma'$, which we may denote by i'. This straight line i' is then such that to any point lying on it corresponds a point at infinity in the plane σ, which point belongs also to the plane π.

We assume that the locus of these points at infinity in the plane σ is a straight line i because it may be considered as the intersection of the planes π and σ. But this locus must correspond to the straight line i' in the plane σ'; thus the law that to every straight line in the plane σ' corresponds a straight line in the plane σ holds without exception.

The plane σ has only one straight line at infinity, because through the point S only one plane parallel to σ can be drawn. The straight line i', the image of the straight line at infinity, is called the *vanishing line* of σ'. It is parallel to $\sigma\sigma'$.

* This is one of the fundamental hypotheses of the Euclidian Geometry.

In the same way, the plane σ' has a straight line at infinity which corresponds to the intersection of the plane σ with the plane π' drawn through S parallel to σ'.

Two parallel planes have the same straight line at infinity in common. All planes parallel to a given plane must be considered as passing through a fixed straight line at infinity.

If a straight line is parallel to a plane, the straight line at infinity in the plane passes through the point at infinity on the line. If two straight lines are parallel, they meet in the same point the straight line at infinity in their plane.

Two planes always cut one another in a straight line (finite or infinitely distant).

A straight line and a plane (not containing the line) always intersect in a point (finite or infinitely distant).

Three planes which do not contain the same straight line have always a common point (finite or infinitely distant).

14. THEOREM. *If two plane figures $ABC\ldots$, $A'B'C'\ldots$,* (Fig. 1) *lying in different planes σ and σ', are in perspective, i. e. if the rays AA', BB', CC',... meet in a point O, then the corresponding straight lines AB and $A'B'$, AC and $A'C'$,..., BC and $B'C'$,... will cut one another in points lying on the same straight line, viz. the intersection of the planes of the two figures.*

It is to be shown that if M is a point lying on the straight line $\sigma\sigma'$, and if a straight line a, lying in the plane σ, passes through M, then the corresponding straight line a' will also pass through M. But this is evidently the case, since the two straight lines a and a' are the intersections of the same projecting plane with the two planes σ and σ', and consequently the three straight lines $\sigma\sigma'$, a, and a' meet in a point, viz. that common to the three planes. The straight line $\sigma\sigma'$ is the locus of the points which correspond to themselves in the two figures.

The vanishing line i' in the plane σ' is parallel to the straight line $\sigma\sigma'$, since i' and the corresponding straight line i, which lies entirely at an infinite distance in the plane σ, must intersect one another on $\sigma\sigma'$. Similarly, the vanishing line j of the plane σ is parallel to $\sigma\sigma'$.

If each of the figures is a triangle, the theorem reads as follows:—

If two triangles ABC and $A'B'C'$, lying respectively in the planes σ and σ', are such that the straight lines AA', BB', CC' meet in a point S, then the three pairs of corresponding sides, BC and $B'C'$, CA and $C'A'$, AB and $A'B'$, intersect in points lying on the straight line $\sigma\sigma'$.

15. Conversely, *if to the points A, B, C, .., and to the straight lines AB, AC, .., BC, ... of a plane figure σ correspond severally the points A', B', C', .. and the straight lines $A'B'$, $A'C'$, .., $B'C'$, .. of another plane figure σ'*, in such a way that the corresponding lines AB and $A'B'$, AC and $A'C'$, ..., BC and $B'C'$, ... meet in points lying on the line of intersection ($\sigma\sigma'$), of the planes σ and σ', then the two figures are in perspective.*

For if S be the point which is common to the three planes $AB . A'B'$, $AC . A'C'$, $BC . B'C'$, the three edges AA', BB', CC' of the trihedral angle formed by the same planes will meet in S. Similarly, the three planes $AB . A'B'$, $AD . A'D'$, $BD . B'D'$ meet in a point which is common to the edges AA', BB', DD', and this point is again S, since the two straight lines AA', BB' suffice to determine it. Therefore all the straight lines AA', BB', CC', DD'... pass through the same point S; that is, the two given figures are in perspective, and S is their centre of projection.

If each of the figures is a triangle, we have the theorem: If two triangles ABC and $A'B'C'$, lying respectively in the planes σ and σ', are such that the sides BC and $B'C'$, CA and $C'A'$, AB and $A'B'$ intersect one another two and two in points lying on the straight line $\sigma\sigma'$, then the straight lines AA', BB', CC' meet in a point S.

16. THEOREM. *If two triangles $A_1B_1C_1$ and $A_2B_2C_2$, lying in the same plane, are such that the straight lines A_1A_2, B_1B_2, C_1C_2 meet in the same point O, then the three points of intersection of the sides B_1C_1 and B_2C_2, C_1A_1 and C_2A_2, A_1B_1 and A_2B_2 lie on a straight line.* (Fig. 3.)

Through the point O which is common to the straight lines A_1A_2, B_1B_2, C_1C_2, draw any straight line outside the plane σ, and in this straight line take two points S_1 and S_2. Project the triangle $A_1B_1C_1$ from S_1 and the triangle $A_2B_2C_2$ from S_2. The points A_1, A_2, O, S_2, S_1 lie in the same plane; therefore S_1A_1 and S_2A_2 meet one another (in A suppose); similarly S_1B_1 and S_2B_2 (in B suppose) and S_1C_1 and S_2C_2 (in C suppose).

* The planes σ and σ' are to be regarded as distinct from each other.

Thus the triangle ABC is in perspective both with $A_1B_1C_1$ and with $A_2B_2C_2$. The straight lines BC, B_1C_1, B_2C_2 intersect in pairs and therefore meet in one and the same point A_0*. Similarly CA, C_1A_1, and A_2C_2 meet in a point B_0, and AB, B_1A_1, and A_2B_2 in a point C_0. The three points A_0, B_0, C_0 lie on the straight line which is common to the planes σ and ABC. The theorem is therefore proved.

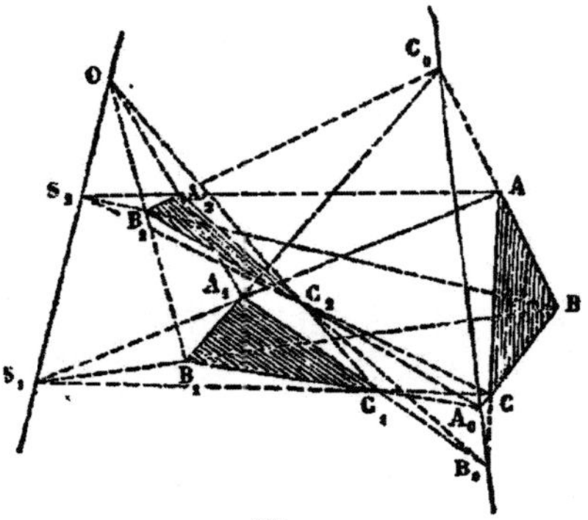

Fig. 3.

17. Conversely, *If two triangles $A_1B_1C_1$ and $A_2B_2C_2$, lying in the same plane, are such that the sides B_1C_1 and B_2C_2, C_1A_1 and C_2A_2, A_1B_1 and A_2B_2 cut one another in pairs in three collinear points A_0, B_0, C_0, then the straight lines A_1A_2, B_1B_2, C_1C_2, which join corresponding angular points, will pass through one and the same point O.* (Fig. 3.)

Through the straight line $A_0B_0C_0$ draw another plane, and project, from an arbitrary centre S_1, the triangle $A_1B_1C_1$ upon this plane. If ABC be the projection, the straight lines BC, B_1C_1 will cut one another in the point A_0, through which B_2C_2 will also pass; similarly AC will pass through B_0 and AB through C_0. The straight lines AA_2, BB_2, CC_2 intersect in pairs, without however all three lying in *the same* plane; they will therefore all meet in one point S_2. The straight lines S_1S_2 and A_1A_2 lie in the same plane, since S_1A_1 and S_2A_2 intersect in A; therefore S_1S_2 meets the three straight lines A_1A_2, B_1B_2, C_1C_2, *i.e.* A_1A_2, B_1B_2, C_1C_2 all meet in one point O, viz. that which is common to the plane σ and the straight line S_1S_2†.

* BC is the intersection of the planes $S_1B_1C_1$ and $S_2B_2C_2$, which do not coincide; so that the straight lines BC, B_1C_1, and B_2C_2 do not all three lie in one plane. The three planes $BC \cdot B_1C_1$, $BC \cdot B_2C_2$, and $B_1C_1 \cdot B_2C_2$ (or σ) intersect in the same point A_0.

† PONCELET, *Propriétés projectives des figures* (Paris, 1822), Art. 168. The theorems of Arts. 11 and 12 are due to DESARGUES (*Œuvres*, ed. Poudra, vol. i. p. 413).

CHAPTER III.

HOMOLOGY.

18. CONSIDER a plane σ and another plane σ', in which latter lies any given figure made up of points and straight lines. Take two points S_1 and S_2 lying outside the given planes, and project from each of them as centre the given figure σ' on to the plane σ. In this way two new figures (σ_1 and σ_2 say) will be formed, which lie in the plane σ, and which are the projections of one and the same figure σ' upon one and the same plane σ, but from different centres of projection. Let two points A_1 and A_2, or two straight lines a_1 and a_2, in the figures σ_1 and σ_2 be said to *correspond* to each other when they are the images of one and the same point A' or of one and the same straight line a' of the figure σ'. We have thus two figures σ_1 and σ_2 lying in the same plane σ, and so related that to the points A_1, B_1, C_1, ... and the lines A_1B_1, $A_1C_1, \ldots, B_1C_1, \ldots$, of the one correspond the points $A_2, B_2, C_2, \ldots$, and the lines $A_2B_2, A_2C_2, \ldots, B_2C_2, \ldots$, of the other. Since any two corresponding straight lines of σ' and σ_1 intersect in a point lying on the straight line $\sigma\sigma'$, and again any two corresponding straight lines of σ' and σ_2 intersect in a point lying on the same straight line $\sigma\sigma'$, it follows that three corresponding straight lines of σ, σ_1, and σ_2 meet in one and the same point, which is determined as the intersection of the straight line of σ' with the straight line $\sigma\sigma'$. That is to say, two corresponding straight lines of the figures σ_1 and σ_2 always intersect on a fixed straight line, the trace of σ' on σ. If moreover A_1 and A_2 are a pair of corresponding points of σ_1 and σ_2, the rays S_1A_1, S_2A_2 have a point A' in common, and therefore lie in the same plane: consequently A_1A_2 and S_1S_2 intersect in a point O. Thus we arrive at the property that every straight line, such as A_1A_2, which connects a pair of

corresponding points of the figures σ_1 and σ_2, passes through a fixed point O, which is the intersection of S_1S_2 and σ. From this we conclude that two figures σ_1 and σ_2 which are the projections of one and the same figure on one and the same plane, but from different centres of projection, possess all the properties of figures in perspective (Art. 8) although they lie in the same plane. To the points and the straight lines of the first correspond, each to each, the points and the straight lines of the second figure; two corresponding points always lie on a ray passing through a fixed point O; and two corresponding straight lines always intersect on a fixed straight line s. Such figures are said to be *homological*, or *in homology*; O is termed the *centre of homology*, and s the *axis of homology**. They may also be said to be *in plane perspective*; O being called the *centre of perspective*, and s the *axis of perspective*.

19. THEOREM. *In the plane σ are given two figures σ_1 and σ_2 which are such that to the points A_1, B_1, C_1, ... and to the straight lines A_1B_1, A_1C_1, ..., B_1C_1, ... of the one correspond, each to each, the points A_2, B_2, C_2, ... and the straight lines A_2B_2, A_2C_2, ..., B_2C_2, ... of the other. If the points of intersection of corresponding straight lines lie on a fixed straight line, then the straight lines which join corresponding points will all pass through a fixed point O.*

Let A_1 and A_2, B_1 and B_2, C_1 and C_2 be three pairs of corresponding points; they form two triangles $A_1B_1C_1$ and $A_2B_2C_2$ whose corresponding sides B_1C_1 and B_2C_2, C_1A_1 and C_2A_2, A_1B_1 and A_2B_2 intersect in three collinear points. By the theorem of Art. 17 the rays A_1A_2, B_1B_2, C_1C_2 will therefore meet in the same point O; but two rays A_1A_2 and B_1B_2 suffice to determine this point; in whatever way then the third pair of points C_1, C_2 may be chosen, the ray C_1C_2 will always pass through O.

The figures σ_1, σ_2 are therefore in homology, O being the centre, and s the axis, of homology.

Corollary.—It follows that if two figures lying either in the same or in different planes are in perspective, and if the plane of one of the figures be made to turn round the axis of perspective, then corresponding straight lines A_1A_2, B_1B_2, &c., will always be

* PONCELET, *Propriétés projectives*, Arts. 297 seqq.

concurrent; *i.e.* the two figures will remain always in perspective. The centre of perspective will of course change its position; it will be seen further on (Art. 22) that it describes a certain circle.

20. THEOREM. *If to the straight lines a, b, c, ... and to the points ab, ac, ..., bc, ..., of a figure correspond severally the straight lines a′, b′, c′,... and the points a′b′, a′c′, ..., b′c′, ... of another coplanar figure, so that the pairs of corresponding points ab and a′b′, ac and a′c′, bc and b′c′, ... are collinear with a fixed point O; then the corresponding straight lines a and a′, b and b′, c and c′, will intersect in points which lie on a straight line.*

Let a and a', b and b', c and c' be three pairs of corresponding straight lines; since by hypothesis the straight lines which join the corresponding vertices of the triangles abc, $a'b'c'$ all meet in a point O, it follows (Art. 16) that the corresponding sides a and a', b and b', c and c' intersect in three points lying on a straight line. But two points aa', bb', suffice to determine this straight line; it remains therefore the same if instead of c and c' any other two corresponding rays are considered. Two corresponding straight lines therefore always intersect on a fixed straight line, which we may call s; thus the given figures are in homology, O being the centre, and s the axis, of homology.

21. Consider two homological figures σ_1 and σ_2 lying in the plane σ; let O be their centre, s their axis of homology. Through the point O and outside the plane σ draw any straight line, and on this take a point S_1, from which as centre project the figure σ_1 upon a new plane σ' drawn in any way through s. In this manner we construct in the plane σ' a figure $A'B'C'$... which is in perspective with the given one $\sigma_1 \equiv A_1B_1C_1 \ldots$. If we consider two points A' and A_2 of the figures σ' and σ_2, which are derived from one and the same point A_1 of σ_1, as corresponding to each other, then to every point or straight line of σ' corresponds a single point or straight line of σ_2, and *vice versa*; and every pair of corresponding straight lines, such as $A'B'$ and A_2B_2, intersect on a fixed straight line $\sigma\sigma'$ or s. Consequently (Art. 15) the figures σ' and σ_2 are in perspective, and the rays $A'A_2$, $B'B_2$, ... all pass through a fixed point S_2. Moreover every ray $A'A_2$ meets the straight line OS_1, since the points A', A_2 lie on the sides S_1A_1, OA_1 of the triangle OA_1S_1. The rays $A'A_2$, $B'B_2$, ... do not all lie in the same plane, because the points A_2, B_2, ... lie arbitrarily in the plane σ; the point S_2 therefore lies on the straight line OS_1.

From this we conclude that two homological figures may be regarded, in an infinite number of ways, as the projections, from two distinct points, of one and the same figure; this figure lying in a plane passing through the axis of homology, and the two points being collinear with the centre of homology.

22. Consider two figures in perspective, lying in the planes σ, σ' respectively (or two figures in plane perspective in the same plane σ); let O (Fig. 4) be the centre and s the axis of perspective, and let j and i' be the vanishing lines of the two figures. If J and I' are points lying on these vanishing lines, the points J' and I which correspond to each of them respectively in the other figure will be at infinity on the rays OJ, OI' respectively. Further, the two corresponding straight lines IJ, $I'J'$ must meet in some point on s; there are consequently an infinite number of parallelograms having one vertex at O, the opposite one on s, and the other two vertices on j and i' respectively.

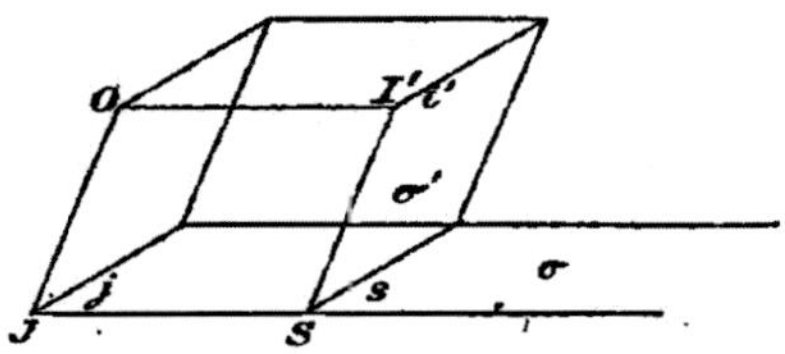

Fig. 4.

Now, supposing the two figures to keep their positions in their planes unaltered, let the plane σ' be made to turn round $\sigma\sigma'$ or s. Every pair of corresponding straight lines must always meet on s; consequently the two figures will always remain in perspective (Arts. 15, 19), and the point O will describe some curve in space.

In order to determine this curve, consider any one of the above-mentioned parallelograms $OJSI'$. It remains always a parallelogram, and the length of $I'S$ is invariable; therefore also OJ is of constant length. The locus of the centre of perspective O is therefore a circle whose centre lies on the vanishing line j and whose plane is perpendicular to this line and therefore to the axis of perspective s*.

* MÖBIUS, *Barycentrische Calcul* (Leipzig, 1827), § 230 (note, p. 326).

23. (1) *Given the centre O and the axis s of homology, and two corresponding points A and A′ (collinear with O); to construct the figure homological with a given figure.*

Take a second point B of the given figure (Fig. 5). To obtain the corresponding point B', we notice that the ray BB' must pass through O and that the straight lines AB, $A'B'$ which correspond to one another must intersect on s; thus B' will be the point where OB meets the straight line joining A' to the intersection of AB with s*. In the same way we can construct any number of pairs of corresponding points; in order to draw the straight line r' which corresponds to a given straight line r, we have only to find the point B' which corresponds to a point B lying on the line r, and to join the points B' and rs.

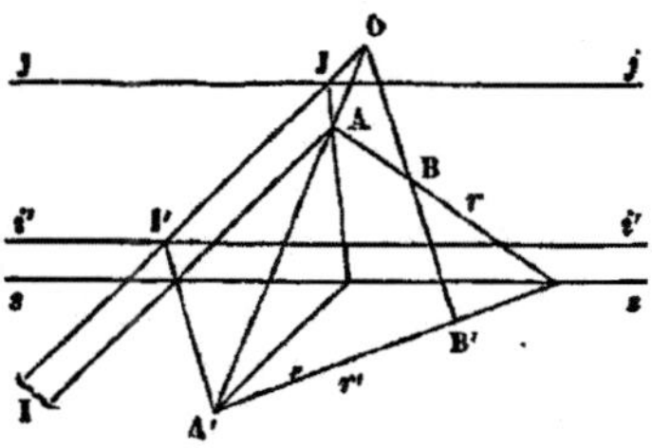

Fig. 5.

In order to find the point I' (the vanishing point) which corresponds to the infinitely distant point I on a given straight line (a ray OI, for example, drawn from O), we repeat the construction just given for the point B'; *i.e.* we join another point A of the first figure to the point at infinity I on OI (that is, we draw AI parallel to OI), and then join A' to the point where AI meets s, and produce the joining line to cut OI in I'. Then I' is the required point.

All points analogous to I' (*i.e.* those which correspond to the points at infinity in the given figure) fall on a straight line i', parallel to s; i' is the vanishing line of the second figure. If, in the preceding construction, we interchange the points A and A'†, we shall obtain a point J (a vanishing point) lying on the vanishing line j of the first figure.

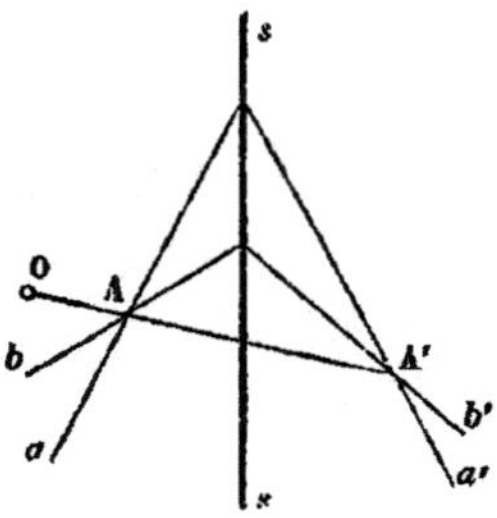

Fig. 6.

(2) Suppose that instead of two corresponding points A, A' there are given (Fig. 6) *two corresponding straight lines* a, a'. These will of course intersect on s; and every ray passing through O will cut them in two corresponding

* This construction shows that if B lies upon s, then B' will coincide with B; *i.e.* that every point of s is its own correspondent.

† Otherwise: Draw through A' any straight line $J'A'$, then through A and the intersection of $J'A'$ with s draw a straight line JA, and through O draw OJ' parallel to $A'J'$. Then the intersection of OJ' and JA is the vanishing point J, and a straight line j drawn through J parallel to s is the vanishing line of the first figure.

points A, A'. In order to obtain the straight line b' which corresponds to any straight line b in the first figure, we have only to join the point bs to the point of intersection of a' with the ray passing through O and ab *.

(3) The data of the problem may also be *the centre O, the axis s, and the vanishing line j of the first figure* (Fig. 7). In this case, if a straight line a of the first figure cuts j in J and s in P, the point J' corresponding to J will be collinear with J and O and at an infinite distance from O. And as the straight line a' corresponding to a must pass both through J' and through P, it is the parallel drawn through P to OJ.

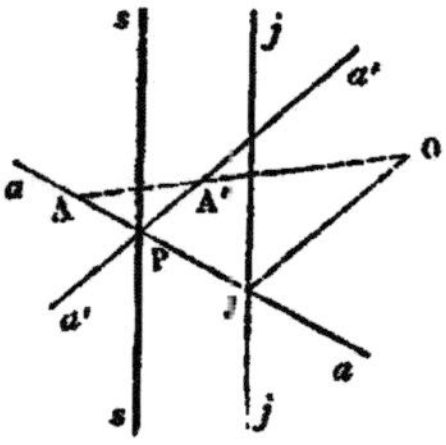

Fig. 7.

To find the point A' corresponding to a given point A, we must draw the straight line a' which corresponds to a straight line a drawn arbitrarily through A; the intersection of a' with OA is the required point A'.

(4) Assuming a knowledge of the constructions just given, let again O be the centre, s the axis, of homology, and j the vanishing line of the first figure.

In the first figure let a circle C be given (Figs. 8, 9, 10); to this circle will correspond in the second figure a curve C' which we can construct by determining, according to the method above, the points and straight lines which correspond to the points and tangents of C.

Two corresponding points will always be collinear with O, and two corresponding chords (*i.e.* straight lines MN, $M'N'$, where M and M', N and N', are two pairs of corresponding points) will always intersect on s; as a particular case two corresponding tangents m and m' (*i.e.* tangents at corresponding points M and M') will meet in a point lying on s.

It follows clearly from this that the curve C' possesses, in common with the circle, the two following properties:

(1) Every straight line in its plane either cuts it in two points, or is a tangent to it, or has no point in common with it.

(2) Through any point in the plane can be drawn either two tangents to the curve, or only one (if the point is on the curve), or none.

Since two homological figures can be considered as arising from the superposition of two figures in perspective lying in different planes (Art. 22), *the curve C' is simply the plane section of an oblique cone on a circular base; i.e. the cone which is formed by the straight lines which run from any point in space to all points of a circle.*

* It follows from this that if a passes through O, then a' will coincide with a; *i.e.* every straight line passing through O corresponds to itself.

For this reason the curve C' is called a *conic section* or simply a *conic*; thus *the curve which is homological with a circle is a conic.*

The points on the straight line j correspond to the points at infinity in the second figure. Now the circle C may cut j in two

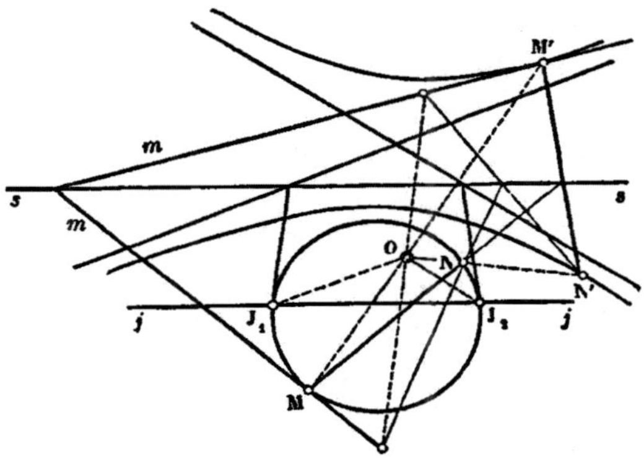

Fig. 8.

points J_1, J_2 (Fig. 8), or it may touch j in a single point J (Fig. 9), or it may have no point in common with j (Fig. 10).

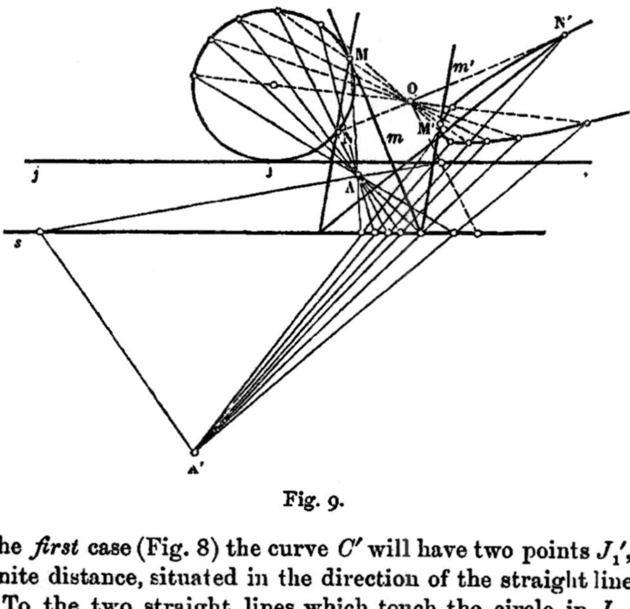

Fig. 9.

In the *first* case (Fig. 8) the curve C' will have two points J_1', J_2', at an infinite distance, situated in the direction of the straight lines OJ_1, OJ_2. To the two straight lines which touch the circle in J_1 and J_2 will correspond two straight lines (parallel respectively to OJ_1 and

OJ_2) which must be considered as tangents to the curve C' at its points at infinity J_1', J_2'. These two tangents, whose points of contact lie at infinity, are called *asymptotes* of the curve C'; the curve itself is called a *hyperbola*.

In the *second* case (Fig. 9) the curve C' has a single point J' at infinity; this must be regarded as the point of contact of the straight line at infinity j', which is the tangent to C' corresponding to the

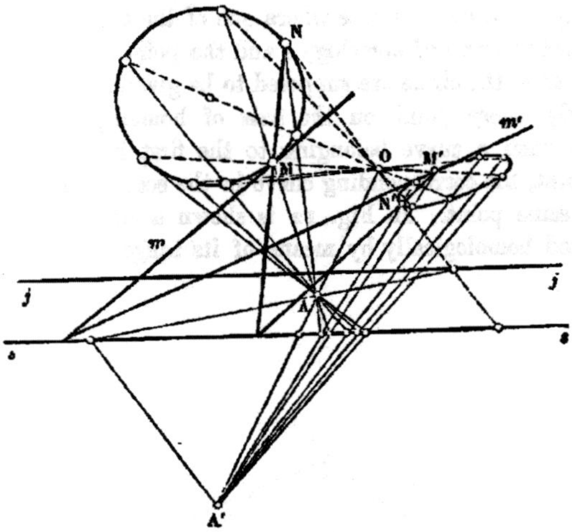

Fig. 10

tangent j at the point J of the circle. This curve C is called a *parabola*.

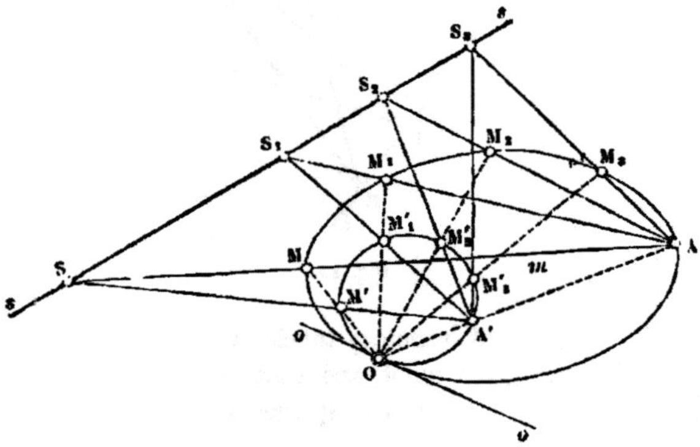

Fig. 11.

In the *third* case (Fig. 10) the curve has no point at infinity; it is called an *ellipse*.

In the same way it may be shown that if in the first figure a conic C is given, the corresponding curve C' in the second figure will be a conic also.

(5). The centre of homology is a point which corresponds to itself, and every ray which passes through it corresponds to itself. If then a curve C pass through O, the corresponding curve C' will also pass through O, and the two curves will have a common tangent at this point. Fig. 11 shows the case where one of the curves is taken to be a circle, and the axis of homology s and the point A corresponding to the point A' of the circle are supposed to be given.

Similarly, every point on the axis of homology corresponds to itself. If then a curve belonging to the first figure touch s at a certain point, the corresponding curve in the second figure will touch s at the same point. In Fig. 12 is shown a circle which is to be transformed homologically by means of its tangents; moreover it is

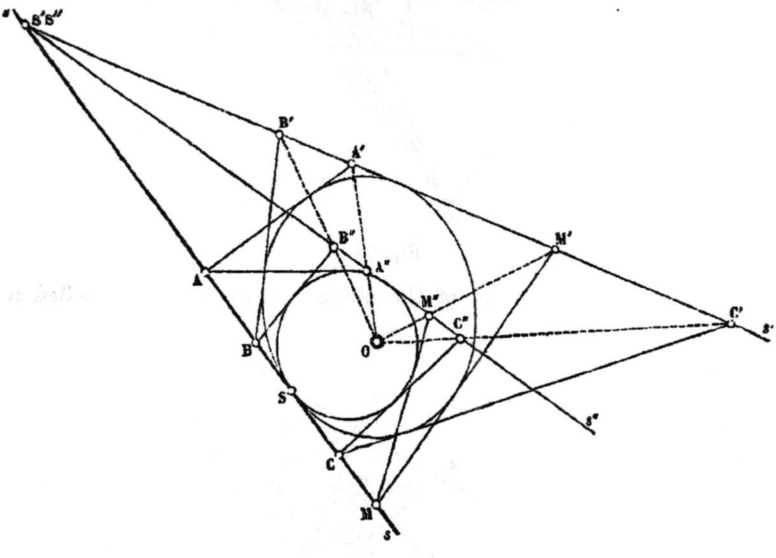

Fig. 12.

supposed that the axis of homology touches the circle, that the centre of homology is any given point, and that the straight line a of the second figure is given which corresponds to the tangent a' of the circle.

(6). Two particular cases may be noticed:

(1) The axis of homology s may lie altogether at infinity; then two corresponding straight lines are always parallel, or, what amounts to the same thing, two corresponding angles are always equal. In this

case the two figures are said to be ***similar and similarly placed***, or ***homothetic*** *, and the point O is called the ***centre of similitude***.

Let M_1, M_1' and M_2, M_2' be two pairs of corresponding points of two homothetic figures, so that M_1M_1', M_2M_2' meet in O, while M_1M_2, $M_1'M_2'$ are parallel. By similar triangles

$$OM_1 : OM_1' = OM_2 : OM_2' = M_1M_2 : M_1'M_2',$$

so that *the ratio $OM : OM'$ is constant for all pairs of corresponding points M and M'*. This constant ratio is called the ***ratio of similitude*** of the two figures.

The tangents at two corresponding points M, M' must meet on the axis of homology s, *i.e.* they are parallel to one another. If then the tangent at M pass through O, it must coincide with the tangent at M'. It follows that if the two figures are such that common tangents can be drawn to them, *every common tangent passes through a centre of similitude*.

Take two points C, C' collinear with O and such that

$$\frac{OC}{OC'} = \frac{OM}{OM'} = \text{ratio of similitude.}$$

Then if CM, $C'M'$ be joined, they will evidently be parallel, and $CM : CM' =$ ratio of similitude. Therefore if M lie on a circle, centre C and radius ρ, M' will lie on another circle whose centre is C' and whose radius ρ' is such that $\rho : \rho' =$ ratio of similitude. In two homothetic figures then to a circle always corresponds a circle. Further, if CC' be again divided at O', so that

$$O'C : O'C' = OC : OC' = \rho : \rho' = \text{ratio of similitude,}$$

it is clear that O' will be a second centre of similitude for the two circles. It can be proved in a similar manner that any two central conics (see Chap. XXI) which are homothetic, and for which a point O is the centre of similitude, have a second centre of similitude O'; and that O, O' are collinear with the centres C, C' of the two conics, and divide the segment CC' internally and externally in the ratio of similitude. If the conics have real common tangents, O and O' will be the points of intersection of these taken in pairs—the two external tangents together, and the two internal tangents together.

(2) The point O, on the other hand, may lie at an infinite distance; then the straight lines which join pairs of corresponding points are parallel to a fixed direction. In this case the figures have been termed *homological by affinity* †, the straight line s being termed the *axis of*

* Homothetic figures may be regarded as sections of a pyramid or a cone made by parallel planes; s, the line of intersection of the two planes, lies at an infinite distance. This is the case in Art. 8 if σ and σ' are parallel planes.

† EULER, *Introductio* ... ii. cap. 18; MÖBIUS, *Baryc. Calcul*, § 144 et seqq.

affinity *. To a point at infinity corresponds in this case a point at infinity, and the straight line at infinity corresponds to itself. It follows from this that to an ellipse corresponds an ellipse, to a hyperbola a hyperbola, to a parabola a parabola, to a parallelogram a parallelogram.

* If two figures are so related, they may be regarded as plane sections of a prism or of a cylinder. This is the case in Art. 8 if the centre S of projection is infinitely distant. The projection is then called *parallel projection*. In the particular case where the parallels SA, SB, SC, ... are perpendicular to the plane of projection it is called *orthogonal projection*.

CHAPTER IV.

HOMOLOGICAL FIGURES IN SPACE.

24. Suppose a figure to be given which is made up of points, planes, and straight lines lying in any manner in space; the *relief-perspective** of this is constructed in the following manner. A point O in space is taken as *centre of perspective* or *homology; a plane of homology* π is taken, every point of which is to be its own image; and in addition to these is taken a point A' which is to be the image of a point A of the given figure, so that AA' passes through O. Let now B be any other point; in order to obtain its image B', the plane OAB is drawn, and we then proceed in this plane as if we had to construct two homological figures, taking O as the centre and the intersection of the planes OAB and π as the axis of homology, and A, A' as two corresponding points. The point B' will be the intersection of OB with the straight line passing through A' and the point where the straight line AB cuts the plane π (Art. 23, Fig. 4). Let C be a third point; its image C' will be the point of intersection of OC with $A'D$ or with $B'E$ (in π), where D and E are the points in which the plane π is met by AC, BC respectively.

This method will yield, for every point of the given figure, the corresponding point of the image, and two corresponding points will always lie on a straight line passing through O. Every plane σ passing through O cuts the two solid figures (the given one and its image) in two homological figures, for which O is the centre, and the straight line $\sigma\pi$ the axis, of homology. It follows from this that to every straight line of the given figure corresponds a straight line in the image, and that two corresponding straight lines lie always in a plane passing through O and meet each other in a point lying on the plane π.

Further: to every plane α, belonging to the given figure, and not passing through O, will correspond a plane α' in the image. For to the straight lines a, b, c,...of the plane α correspond severally the straight

* This problem may present itself in the construction of bas-reliefs and of theatre decorations (Poncelet, *Prop. proj.* 584; Poudra, *Perspective-relief*, Paris, 1860).

lines a', b', c', ...; and to the points ab, ac, ..., bc, ... the points $a'b'$, $a'c'$, ..., $b'c'$, ... In other words, the straight lines a', b', c', ... are such that they intersect in pairs, but do not all meet in the same point; they lie therefore in the same plane α' *. Two corresponding planes α, α' intersect on the plane π; for all the points and all the straight lines of this last plane correspond to themselves, and therefore the straight line $\alpha'\pi$ coincides with the straight line $\alpha\pi$.

The two planes α, α' evidently contain two figures in perspective (like the planes σ, σ' of Arts. 12 and 14).

25. In every plane σ passing through O lies a vanishing line i', which is the image of the point at infinity in the same plane. The vanishing lines of the planes σ_1, σ_2 have a common point, which is the image of the point at infinity on the line $\sigma_1\sigma_2$. The vanishing lines of all the planes σ are therefore such as to cut each other in pairs; and as they do not pass all through the same point (since the planes through O do not pass all through the same straight line), they must lie in one and the same plane ϕ'.

This plane ϕ', which may be called the *vanishing plane*, is parallel to the plane π, since all the vanishing lines of the planes σ are parallel to the same plane π. The vanishing plane ϕ' is thus the locus of the straight lines which correspond to the straight lines at infinity in all the planes of space, and is consequently also the locus of the points which correspond to the points at infinity in all the straight lines of space: for the line at infinity in any plane α is the same thing as the line at infinity in the plane through O parallel to α; so also the point at infinity on any straight line a coincides with the point at infinity on the straight line drawn through O parallel to a.

26. The infinitely distant points of all space are then such that their images are the points of one and the same plane ϕ' (the vanishing plane). It is therefore natural to consider all the infinitely distant points in space as lying in one and the same plane ϕ (the plane at infinity) of which the plane ϕ' is the image †.

The idea of the plane at infinity being granted, the point at infinity on any straight line a is simply the point $a\phi$, and the straight line at infinity in any plane α is the straight line $\alpha\phi$. Two straight lines are parallel if they intersect in a point of the plane ϕ; two planes are parallel if their line of intersection lies in the plane ϕ, &c.

* Since c' cuts both a' and b' without passing through the point $a'b'$, therefore c' has two points in common with the plane $a'b'$, and consequently lies entirely in the plane $a'b'$. And similarly for the other straight lines.

† PONCELET, *Prop. proj.* 580.

CHAPTER V.

GEOMETRIC FORMS.

27. A *range* or *row* of points is a figure $A, B, C, \ldots$ composed of points lying on a straight line (which is called the *base* of the range); such is, for example, the figure resulting from the operations of Art. 5 or Art. 7.

An *axial pencil* is a figure $\alpha, \beta, \gamma, \ldots$ composed of planes all passing through the same straight line (the *axis* of the pencil); such is the figure resulting from the operations of Art. 4 or Art. 6.

A *flat pencil* is a figure $a, b, c, \ldots$ composed of straight lines lying all in the same plane and radiating from a given point (the *centre* or *vertex* of the pencil); such would be the figure obtained by applying the operation of Art. 2 to a range, or that of Art. 3 to an axial pencil.

A *sheaf* (*sheaf of planes, sheaf of lines*) is a figure made up of planes or straight lines, all of which pass through a given point (the *centre* of the sheaf); like that which results from the operation of Art. 2.

A *plane figure* (*plane of points, plane of lines*) is a figure which consists of points or straight lines all of which lie in the same plane; such is the figure resulting from the operation of Art. 3.

28. The first three figures can be derived one from the other by a projection or a section*.

From a range $A, B, C, \ldots$ is derived an axial pencil $s(A, B, C, \ldots)$ by projecting the range from an axis s (Art. 4); and a flat pencil $O(A, B, C, \ldots)$ by projecting it from a centre

* The series of planes $sA, sB, sC, \ldots$; of rays $OA, OB, OC, \ldots$; of points $s\alpha$, $s\beta$, $s\gamma, \ldots$; and of straight lines $\sigma\alpha, \sigma\beta, \sigma\gamma, \ldots$ will be denoted by $s(A, B, C, \ldots)$, $O(A, B, C, \ldots)$, $s(\alpha, \beta, \gamma, \ldots)$, and $\sigma(\alpha, \beta, \gamma, \ldots)$ respectively. To denote the series of points $A, B, C, \ldots$ the symbols $A, B, C, \ldots$ and $ABC \ldots$ will be used indifferently.

O (Art. 2). From an axial pencil $\alpha, \beta, \gamma, \ldots$ is derived a range $s(\alpha, \beta, \gamma, \ldots)$ by cutting the pencil by a transversal line s (Art. 5); and a flat pencil $\sigma(\alpha, \beta, \gamma, \ldots)$ by cutting it by a transversal plane σ (Art. 3). From a flat pencil $a, b, c, \ldots$ is derived a range $\sigma(a, b, c, \ldots)$ by cutting it by a transversal plane σ (Art. 3); and an axial pencil $O(a, b, c, \ldots)$ by projecting it from a centre O (Art. 2).

29. In a similar manner the last two figures of Art. 27 can be derived one from the other by help of one of the operations of Art. 2 or Art. 3; in fact, if we project from a centre O a plane of points or lines we obtain a sheaf of lines or planes; and reciprocally, if we cut a sheaf of lines or planes by a transversal plane we obtain a plane of points or lines. Two plane figures in perspective (Art. 12) are two sections of the same sheaf.

30. The *elements* or *constituents* of the range are the points; those of the axial pencil, the planes; those of the flat pencil, the straight lines or rays.

In the plane figure either the points or the straight lines may be regarded as the elements. If the points are considered as the elements, the straight lines of the figure are so many ranges; if, on the other hand, the straight lines or rays are considered as the elements, the points of the figure are the centres of so many flat pencils.

The plane of points (*i. e.* the plane figure in which the elements are points) contains therefore an infinite number of ranges*, and the plane of lines (*i.e.* the plane figure in which the elements are lines †) contains an infinite number of flat pencils.

In the sheaf either the planes, or the straight lines or rays, may be regarded as the elements. If we take the planes as elements, the rays of the sheaf are the axes of so many axial pencils; if, on the other hand, the rays are considered as the elements, the planes of the sheaf are so many flat pencils.

The sheaf contains therefore an infinite number of axial

* One of these ranges has all its points at an infinite distance; each of the others has only one point at infinity.

† The straight line at infinity belongs to an infinite number of flat pencils, each of which has its centre at infinity, and consequently all its rays parallel.

pencils or an infinite number of flat pencils, according as its planes or its straight lines are regarded as its elements.

31. Space may also be considered as a geometrical figure, whose elements are either points or planes.

Taking the points as elements, the straight lines of space are so many ranges, and the planes of space so many planes of points. If, on the other hand, the planes are considered as elements, the straight lines of space are the axes of so many axial pencils, and points of space are the centres of so many sheaves of planes.

Space contains therefore an infinite number of planes of points* or an infinite number of sheaves of planes †, according as we take the point or the plane as the element in order to construct it.

32. The first three figures, viz. the range, the axial pencil, and the flat pencil, which possess the property that each can be derived from the other by help of one of the operations of Arts. 2, 3,..., are included together under one name, and are termed the *one-dimensional geometric prime-forms.*

The fourth and fifth figures, viz. the sheaf of planes or lines and the plane of points or lines, which may in like manner be derived one from the other by means of one of the operations of Arts. 2, 3,..., and which moreover possess the property of including in themselves an infinite number of one-dimensional prime-forms, are likewise classed together under one title, as the *two-dimensional geometric prime-forms.*

Lastly, space, which includes in itself an infinite number of two-dimensional prime-forms, is considered as constituting the *three-dimensional geometric prime-form.*

There are accordingly six geometric prime-forms; three of one dimension, two of two dimensions, and one of three dimensions ‡.

Note.—With reference to the use of the word *dimension* in the preceding Article, it is clear, from what has been said in Art. 28, that we are justified in considering the range, the flat pencil, and the axial pencil, as of the same dimensions, since to every point in

* One of them lies entirely at infinity.

† Among these, there are an infinite number which have their centre at an infinite distance, and whose rays are consequently parallel.

‡ v. STAUDT, *Geometrie der Lage* (Nürnberg, 1847), Arts. 26, 28.

the first corresponds one ray in the second and one plane in the third. The number of elements in each of these forms is infinite, but it is the same in all three.

Similarly we conclude from Art. 29 that we are justified in considering the plane figure as of the same dimensions with the sheaf.

But the plane of points (lines) contains (Art. 30) an infinite number of ranges (flat pencils); and each of these ranges (flat pencils) itself contains an infinite number of points (rays). Thus the plane figure contains a number of points (lines) which is an infinity of the second order compared with the infinity of points in a range, or of rays in a flat pencil; and must therefore be considered as of two dimensions if the range and flat pencil are taken to be of one dimension.

So too the sheaf of planes (or lines) contains (Art. 30) an infinite number of axial pencils (or of flat pencils), and each of these itself contains an infinite number of planes (or of rays). Therefore also the sheaf of planes or lines must be of double the dimensions of the axial pencil or the flat pencil.

Again, space, considered as made up of points, contains an infinite number of planes of points, and considered as made up of planes, it contains an infinite number of sheaves of planes. Space thus contains an infinite number of forms of two dimensions, which latter, again, contain each an infinite number of forms of one dimension. Space must accordingly be regarded as of three dimensions.

We may put the matter thus:

Forms of one dimension are those which contain a simple infinity (∞) of elements;

Forms of two dimensions are those which contain a double infinity (∞^2) of elements;

Forms of three dimensions are those which contain a triple infinity (∞^3) of elements.

CHAPTER VI.

THE PRINCIPLE OF DUALITY *.

33. GEOMETRY (speaking generally) studies the generation and the properties of figures lying (1) in space of three dimensions, (2) in a plane, (3) in a sheaf. In each case, any figure considered is simply an assemblage of elements; or, what amounts to the same thing, it is the aggregate of the elements with which a moving or variable element coincides in its successive positions. The moving element which generates the figures may be, in the first case, the point or the plane; in the second case the point or the straight line; in the third case the plane or the straight line. There are therefore always two *correlative* or *reciprocal* methods by which figures may be generated and their properties deduced, and it is in this that geometric *Duality* consists. By this duality is meant the co-existence of figures (and consequently of their properties also) in pairs; two such co-existing (*correlative* or *reciprocal*) figures having the same genesis and only differing from one another in the nature of the generating element.

In the Geometry of space the range and the axial pencil, the plane of points and the sheaf of planes, the plane of lines and the sheaf of lines, are correlative forms. The flat pencil is a form which is correlative to itself.

In the Geometry of the plane the range and the flat pencil are correlative forms.

In the Geometry of the sheaf the axial pencil and the flat pencil are correlative forms.

The Geometry of the plane and the Geometry of the sheaf, considered in three-dimensional space, are correlative to each other.

34. The following are examples of correlative propositions

* v. STAUDT, *Geom. der Lage*, Art. 66.

in the Geometry of space. Two correlative propositions are deduced one from the other by interchanging the elements *point* and *plane*.

1. Two points A, B determine a straight line (viz. the straight line AB which passes through the given points) which contains an infinite number of other points.

1. Two planes α, β determine a straight line (viz. the straight line $\alpha\beta$, the intersection of the given planes), through which pass an infinite number of other planes.

2. A straight line a and a point B (not lying on the line) determine a plane, viz. the plane aB which connects the line with the point.

2. A straight line a and a plane β (not passing through the line) determine a point, viz. the point $a\beta$ where the line cuts the plane.

3. Three points A, B, C which are not collinear determine a plane, viz. the plane ABC which passes through the three points.

3. Three planes α, β, γ which do not pass through the same line determine a point, viz. the point $\alpha\beta\gamma$ where the three planes meet each other.

4. Two straight lines which cut one another lie in the same plane.

4. Two straight lines which lie in the same plane intersect in a point.

5. Given four points A, B, C, D; if the straight lines AB, CD meet, the four points will lie in a plane, and consequently the straight lines BC and AD, CA and BD will also meet two and two.

5. Given four planes α, β, γ, δ; if the straight lines $\alpha\beta$, $\gamma\delta$ meet, the four planes will meet in a point, and consequently the straight lines $\beta\gamma$ and $\alpha\delta$, $\gamma\alpha$ and $\beta\delta$, will also meet two and two.

6. Given any number of straight lines; if each meets all the others, while the lines do not all pass through a point, then they must lie all in the same plane (and constitute a plane of lines)*.

6. Given any number of straight lines; if each meets all the others, while the lines do not all lie in the same plane, then they must pass all through the same point (and constitute a sheaf of lines)†.

7. The following problem admits of two correlative solutions: 'Given a plane α and a point A in it, to draw through A a straight line lying in the plane α which shall cut a given straight line r which does not lie in α and does not pass through A.'

* See note to Art. 20.

† For let $a, b, c, \ldots$ be the straight lines; as ab, ac, bc are three planes distinct from each other, the common point must be the intersection of the straight lines $a, b, c, \ldots$.

Join A to the point ra.

Construct the line of intersection of the plane a with the plane rA.

8. *Problem.* Through a given point A to draw a straight line to cut each of two given straight lines b and c (which do not lie in the same plane and do not pass through A).

Solution. Construct the line of intersection of the planes Ab, Ac.

8. *Problem.* In a given plane a, to draw a straight line to cut each of two given straight lines b and c (which do not meet and do not lie in the plane a).

Solution. Join the point ab to the point ac.

35. In the Geometry of Space, the figure correlative to a triangle (system of three points) is a trihedral angle (system of three planes); the vertex, the faces, and the edges of the latter are correlative to the plane, the vertices, and the sides respectively of the triangle; thus the theorem correlative to that of Arts. 15 and 17 will be the following:

If two trihedral angles $\alpha'\beta'\gamma'$, $\alpha''\beta''\gamma''$ are such that the edges $\beta'\gamma'$ and $\beta''\gamma''$, $\gamma'\alpha'$ and $\gamma''\alpha''$, $\alpha'\beta'$ and $\alpha''\beta''$ lie in three planes α_0, β_0, γ_0 which pass through the same straight line, then the straight lines $\alpha'\alpha''$, $\beta'\beta''$, $\gamma'\gamma''$ will lie in the same plane.

The proof is the same as that of Arts. 15 and 17, if the elements point and plane are interchanged. If, for example, the two trihedral angles have different vertices S', S'' (Art. 15), then the points where the pairs of edges intersect are the vertices of a triangle whose sides are $\alpha'\alpha''$, $\beta'\beta''$, $\gamma'\gamma''$; these latter straight lines lie therefore in the same plane (that of the triangle).

So also the proof for the case where the two trihedral angles have the same vertex S will be correlative to that for the analogous case of two triangles $A'B'C'$ and $A''B''C''$ which lie in the same plane (Art. 17). The theorem may also be established by projecting from a point S the figure corresponding to the theorem of Art. 16.

The proof of the theorem correlative to that of Arts. 14 and 16 is left as an exercise for the student. It may be enunciated as follows:

If two trihedral angles $\alpha'\beta'\gamma'$, $\alpha''\beta''\gamma''$ are such that the straight lines $\alpha'\alpha''$, $\beta'\beta''$, $\gamma'\gamma''$ lie in the same plane, then the pairs of edges $\beta'\gamma'$ and $\beta''\gamma''$, $\gamma'\alpha'$ and $\gamma''\alpha''$, $\alpha'\beta'$ and $\alpha''\beta''$ determine three planes which pass all through the same straight line.

36. In the Geometry of the plane, two correlative propositions are deduced one from the other by interchanging the words *point* and *line*, as in the following examples:

1. Two points A, B determine a straight line, viz. the line AB.

1. Two straight lines a, b determine a point, viz. the point ab.

2. Four points A, B, C, D (Fig. 13), no three of which are collinear, form a figure called a *complete quadrangle**. The four

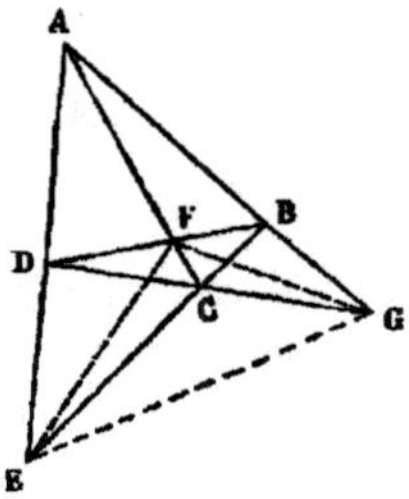

Fig. 13.

points are called the *vertices*, and the six straight lines joining them in pairs are called the *sides* of the quadrangle.

2. Four straight lines a, b, c, d (Fig. 14), no three of which are concurrent, form a figure called *complete quadrilateral**. The four

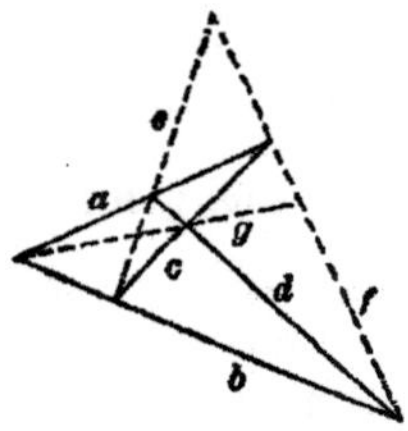

Fig. 14.

straight lines are called the *sides* of the quadrilateral, and the six points in which the sides cut one another two and two are called the *vertices*.

Two sides which do not meet in a vertex are termed *opposite*; there are accordingly three pairs of opposite sides, BC and AD, CA and BD, AB and CD. The

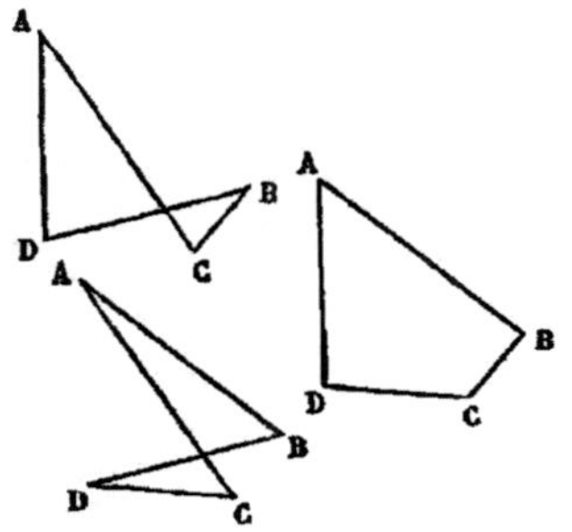

Fig. 15.

points E, F, G in which the opposite sides intersect in pairs are

Two vertices which do not lie on the same side are termed *opposite*; there are accordingly three pairs of opposite vertices, bc and ad, ca and bd, ab and cd.

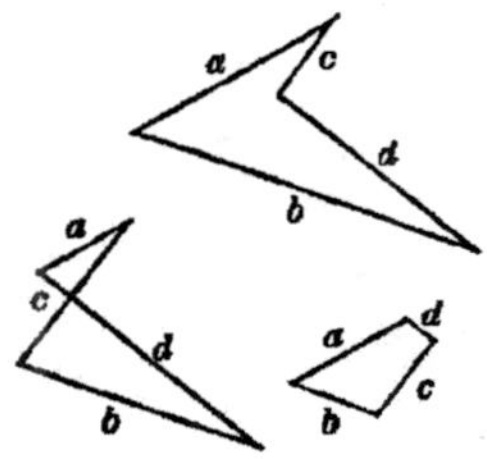

Fig. 16.

The straight lines e, f, g which join pairs of opposite vertices are

* The complete quadrangle has also been called a *tetrastigm*, and the complete quadrilateral a *tetragram*. TOWNSEND, *Modern Geometry*, ch. vii.

termed the *diagonal points*; and the triangle EFG is termed the *diagonal triangle* of the complete quadrangle. The complete quadrangle includes three simple quadrangles, viz. $ACBD$, $ABCD$, and $ABDC$ (Fig. 15).

called the *diagonals*; and the triangle efg is termed the *diagonal triangle* of the complete quadrilateral. The complete quadrilateral includes three simple quadrilaterals, viz. $acbd$, $adcb$, and $acbd$ (Fig. 16).

3. And so, in general:

A *complete polygon* (complete *n-gon*, or *n-point**) is a system of n points or *vertices*, with the $\frac{n(n-1)}{2}$ straight lines or *sides* which join them two and two.

A *complete multilateral* (or *n-side*†) is a system of n straight lines or *sides*, with the $\frac{n(n-1)}{2}$ points or *vertices* in which they intersect one another two and two.

4. The theorems of Arts. 16 and 17 are correlative each to the other.

5. **Theorem.** If two complete quadrangles $ABCD$, $A'B'C'D'$ are such that five pairs of sides AB and $A'B'$, BC and $B'C'$, CA and $C'A'$, AD and $A'D'$, BD and $B'D'$ cut one another in five points lying on a straight line s, then the remaining pair CD and $C'D'$ will also intersect one another on s (Fig. 17).

Theorem. If two complete quadrilaterals $abcd$, $a'b'c'd'$ are such that five pairs of vertices ab and $a'b'$, bc and $b'c'$, ca and $c'a'$, ad and $a'd'$, bd and $b'd'$ lie upon five straight lines which meet in a point S, then the remaining pair cd and $c'd'$ will also lie on a straight line through S (Fig. 18).

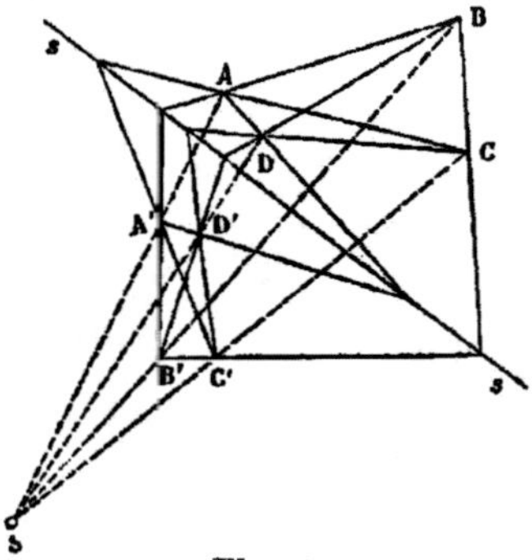

Fig. 17.

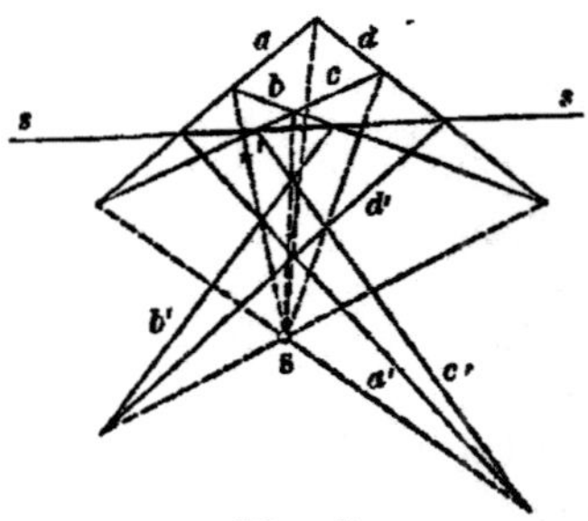

Fig. 18.

Since the triangles ABC, $A'B'C'$ are by hypothesis in

Since the triangles (trilaterals) abc, $a'b'c'$ are by

* Or *polystigm*; TOWNSEND, *loc. cit.*

† Or *polygram*.

perspective (Arts. 17, 18), the straight lines AA', BB', CC' will meet in one point S. So too the triangles ABD, $A'B'D'$ are in perspective; therefore DD' also will pass through S, the point common to AA' and BB'. It follows that the triangles BCD, $B'C'D'$ are also in perspective: therefore CD and $C'D'$ meet in a point on the straight line s, which is determined by the point of intersection of BC and $B'C'$ and by that of BD and $B'D'$*.

hypothesis in perspective (Art. 18), the points aa', bb', cc' will lie on one straight line s. So too the triangles abd, $a'b'd'$ are in perspective; therefore the point dd' lies on the straight line s which passes through the points aa', bb'. It follows that the triangles (trilaterals) bcd, $b'c'd'$ are also in perspective; therefore cd and $c'd'$ lie on a straight line through the point S, which is determined by the straight lines (bc) $(b'c')$ and (bd) $(b'd')$*.

37. In the Geometry of space the following are correlative:

A complete *n-gon* (in a plane).

A complete *n-flat* (in a sheaf); *i.e.* a figure made up of n planes (or *faces*) which all pass through the same point (or *vertex*), together with the $\frac{n(n-1)}{2}$ edges in which these planes intersect two and two.

A complete *multilateral of n sides*, or *n-side* (in a plane).

A complete *n-edge* (in a sheaf); *i.e.* a figure made up of n straight lines radiating from a common point (or *vertex*), together with the $\frac{n(n-1)}{2}$ planes (or *faces*) which pass through these straight lines taken in pairs.

Thus the following theorems are correlative, in the Geometry of space, to the two theorems above (Art. 36, No. 5), which latter are themselves correlative to each other in the Geometry of the plane.

If two complete four-flats in a sheaf (be their vertices coincident or not) $\alpha\beta\gamma\delta$, $\alpha'\beta'\gamma'\delta'$ are such that five pairs of corresponding

If two complete four-edges in a sheaf (be their vertices coincident or not) $abcd$, $a'b'c'd'$ are such that five pairs of corresponding faces

* These two theorems hold good equally when the two quadrangles or quadrilaterals lie in different planes; in fact, the proofs are the same as the above, word for word.

edges lie in five planes which pass all through the same straight line s, then the sixth pair of corresponding edges will lie also in a plane passing through s.

cut one another in five straight lines which lie all in one plane σ, then the line of intersection of the sixth pair of corresponding faces will lie also in the plane σ.

The proofs of these theorems are left as an exercise to the student. They only differ from those of the theorems No. 5, Art. 36 in the substitution for each other of the elements point and plane; and just as theorems 5, Art. 36 follow from those of Arts. 15 and 16, so the theorems enunciated above follow from those of Art. 35. When the two four-flats have the same vertex O, the theorem on the left-hand side may also be established by projecting from the point O (Art. 2) the figure corresponding to the right-hand theorem of No. 5, Art. 36. And in this case we may by the same method deduce the theorem on the right-hand side above from that on the left-hand of No. 5, Art. 36.

38. In the Geometry of the sheaf, two correlative theorems are derived one from the other by interchanging the elements plane and straight line. Just as the Geometry of the sheaf is correlative to that of the plane, with regard to three-dimensional space, so one of the Geometries is derived from the other by the interchange of the elements point and plane. The Geometry of the sheaf may also be derived from that of the plane by the operation of projection from a centre (Art. 2).

From the Geometry of the sheaf may be derived that of spherical figures, by cutting the sheaf by a sphere passing through the centre of the sheaf.

CHAPTER VII.

PROJECTIVE GEOMETRIC FORMS.

39. By means of projection from a centre we obtain from a range a flat pencil, from a flat pencil an axial pencil, from a plane of points or lines a sheaf of lines or planes. Conversely, by the operation of section by a transversal plane we obtain from a flat pencil a range, from an axial pencil a flat pencil, from a sheaf a plane figure. The two operations, projection from a point and section by a transversal plane, may accordingly be regarded as *complementary* to each other; and we may say that if one geometric form has been derived from another by means of one of these operations, we can conversely, by means of the complementary operation, derive the second form from the first. And similarly for the operations: projection from an axis and section by a transversal line.

Suppose now that by means of a series of operations, each of which is either a projection or a section, a form f_2 has been derived from a given form f_1, then another form f_3 from f_2, and so on, until by $n-1$ such operations the form f_n has been arrived at. Conversely, we may return from f_n to f_1 by means of another series of $n-1$ operations which are complementary respectively to the last, last but one, last but two, &c. of tho operations by which we have passed from f_1 to f_n. The series of operations which leads from f_1 to f_n, and the series which leads from f_n to f_1, may be called complementary, and the operations of the one series are complementary respectively to those of the other, taken in the reverse order.

In the above the geometric forms are supposed to lie in space (Art. 31). If we confine ourselves to plane Geometry, the complementary operations reduce to *projection from a centre* and *section by a transversal line.* In the Geometry of the sheaf, section by a plane and projection from an axis are complementary operations.

40. Two geometric prime forms of the same dimensions are said to be *projectively related*, or simply *projective*, when one can be derived from the other by any finite number of projections and sections (Arts. 2, 3, ... 7).

For example, let a range u be given; project it from a centre O, thus obtaining a flat pencil; project this flat pencil from another centre O', by which means an axial pencil with OO' as axis is produced; cut this axial pencil by a straight line u_2, thus obtaining a range of points lying on u_2; project this range from an axis, and cut the resulting axial pencil by a plane, by which means a flat pencil is produced, and so on; then any two of the one-dimensional geometric forms which have been obtained in this manner are projective according to definition.

When we say that a form $A, B, C, D, \ldots$ is projective with another form $A', B', C', D', \ldots$ we mean that, by help of the same series of operations, each of which is either a projection or a section, A' is derived from A, B' from B, C' from C, &c. The elements A and A', B and B', C and C', ... are termed *corresponding elements**.

For example, a plane figure is said to be projective with another plane figure, when from the points $A, B, C, \ldots$ and from the straight lines $AB, AC \ldots, BC, \ldots$ of the one are derived the points $A', B', C', \ldots$ and the straight lines $A'B', A'C', \ldots B'C', \ldots$ of the other, by means of a finite number of projections and sections.

In two projective plane figures, to a range in the one corresponds in the other a range which is projective with the first range; and to a flat pencil in the one figure corresponds in the other a flat pencil which is projective with the first pencil.

41. From what has been said above it is easy to see that two geometric forms which are each projective with

* Two projective forms are termed *homographic* when the elements of which they are constituted are of the same kind; *i.e.* when the elements of *both* are points, or lines, or planes. It will be seen later on (Art. 67) that this definition of homography is equivalent to that given by CHASLES (*Géométrie supérieure*, Art. 99).

a third are projective with one another. For if we first go through the operations which lead from the first form to the third, and then go through those which lead from the third to the second, we shall have passed from the first form to the second.

42. *Geometric forms in perspective.*

The following forms are said to be *in perspective*:

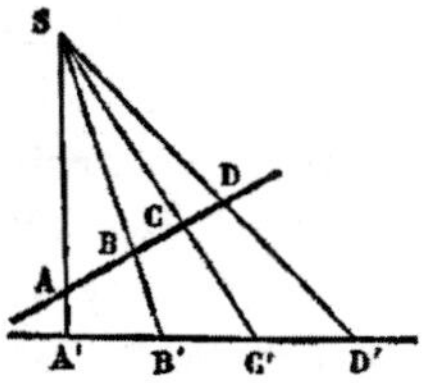

Fig. 19.

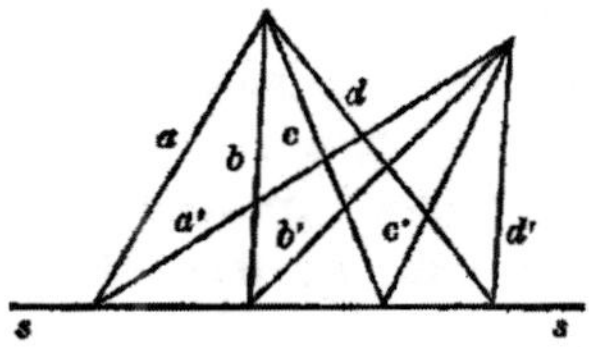

Fig. 20.

Two ranges (Fig. 19), if they are sections of the same flat pencil (Art. 12).

Two flat pencils (Fig. 20), if they project, from different centres, one and the same range; or if they are sections of the same axial pencil.

[*Note.*—If we project a range $u \equiv ABC\ldots$ from two different centres O and O' not lying in the same plane with it, we obtain two flat pencils in perspective. These pencils, again, may be regarded as sections of the same axial pencil made by the transversal planes Ou, Ou'; the axial pencil namely which is composed of the planes $OO'A$, $OO'B$, $OO'C$, ..., and which has for axis the straight line OO'. This is the general case of two flat pencils in perspective; they have not the same centre and they lie in different planes; at the same time, they project the same range and are sections of the same axial pencil. There are two exceptional cases: (1). If we project the row u from two centres O and O' lying in the same plane with u, then the two resulting flat pencils lie in the same plane and are consequently no longer sections of an axial pencil; (2). If an axial pencil is cut by two transversal planes which pass through a common point O on the axis, we obtain two flat pencils which have the same centre O, and which consequently no longer project the same range.]

Two axial pencils, if they project, from two different centres, the same flat pencil.

A range and a flat pencil, a range and an axial pencil, or *a flat pencil and an axial pencil*, if the first is a section of the second.

Two plane figures, if they are plane sections of the same sheaf.

Two sheaves, if they project, from two different centres, the same plane figure.

A plane figure and a sheaf, if the former is a section of the latter.

From the definition of Art. 40 it follows at once that two (one-dimensional) forms which are in perspective are also projectively related; but two projective forms are not in general in perspective position.

43. Two figures in homology are merely two projective plane figures superposed one upon the other, in a particular position; for by Art. 21 two homological figures may always be regarded (and this in an infinite number of ways) as projections of one and the same third figure.

If two projective plane figures are superposed one upon the other in such a manner that the straight line connecting any pair of corresponding points may pass through a fixed point; or, again, in such a manner that any pair of corresponding straight lines may intersect on a fixed straight line; then the two figures are in homology (Arts. 19, 20).

In two homological figures, two corresponding ranges are in perspective (and therefore of course are projectively related); and the same is the case with regard to two corresponding pencils.

44. THEOREM. *Two one-dimensional geometric forms, each consisting of three elements, are always projective.*

To prove this, we notice in the first place that it is enough to consider the case of two ranges ABC, $A'B'C'$; for, if one of the given forms is a pencil, flat or axial, we may substitute for it one of its sections by a transversal.

(1) If the two straight lines ABC, $A'B'C'$ lie in different planes, join AA', BB', CC', and cut these straight lines by a transversal s*. Then the two given forms are seen to be simply two sections of the axial pencil sAA', sBB', sCC'.

(2) If the two straight lines lie in the same plane (Fig. 21), join AA', and take on this straight line any two points, S, S';

* To do this, we have only to draw through any point of AA' a straight line which meets BB' and CC' (Prob. 8, Art. 34).

draw SB, $S'B'$ to cut in B'', and SC, $S'C'$ to cut in C'', and join $B''C''$, cutting SS' in A''. Then $A'B'C'$ may be derived from

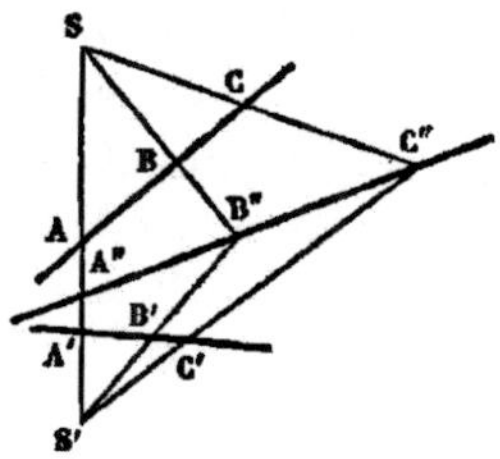

Fig. 21.

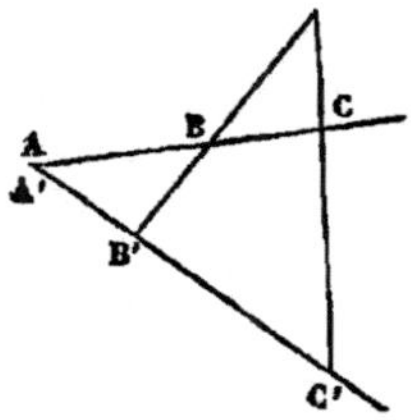

Fig. 22.

ABC by two projections, viz. we first project ABC from S into $A''B''C''$, and then $A''B''C''$ from S' into $A'B'C'$.

(3) In the case where the two points A and A' coincide (Fig. 22), the two given forms are directly in perspective; the centre of perspective is the point where BB' and CC' intersect.

(4) If the two sets of points ABC, $A'B'C'$ lie on the same straight line (Fig. 23), it is only necessary to project one of them $A'B'C'$ on to another straight line $A_1B_1C_1$ (from any centre O); then let any two centres S and S_1 be taken (as in Fig. 21) on AA_1, and let the straight line $A''B''C''$ be constructed in the manner already shown in case (2). Then $A'B'C'$ may be derived from ABC by three projections, viz. we first project ABC from S into $A''B''C''$, then $A''B''C''$ from S_1 into $A_1B_1C_1$, and lastly $A_1B_1C_1$ from O into $A'B'C'$.

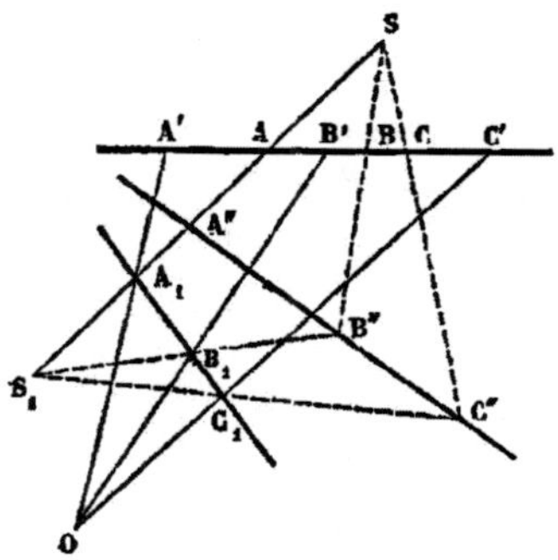

Fig. 23.

(5) If A coincides with A', and B with B', we may make use of a centre S and two transversals s_1, s_2 drawn through A in the plane $SABCC'$. If the triad ABC be projected from S upon s_1 (giving $A_1B_1C_1$), and the triad $A'B'C'$ be projected from S upon s_2 (giving $A_2B_2C_2$); then the triads $A_1B_1C_1$ and $A_2B_2C_2$ will be in perspective, because A_1 coincides with A_2 (in the point AA').

In every case, then, it has been shown that the triads

ABC, $A'B'C'$ can be derived from each other by a finite number of projections and sections; therefore by Art. 40 they are projective.

As a particular case, ABC must be projective with BAC, for example. In order actually to project one of these triads into the other, take (Fig. 24) any two points L and N collinear with C. Join AL, BN, meeting in K, and BL, AN, meeting in M. Then BAC can be derived from ABC by first projecting ABC from K into LNC, and then LNC from M into BAC.

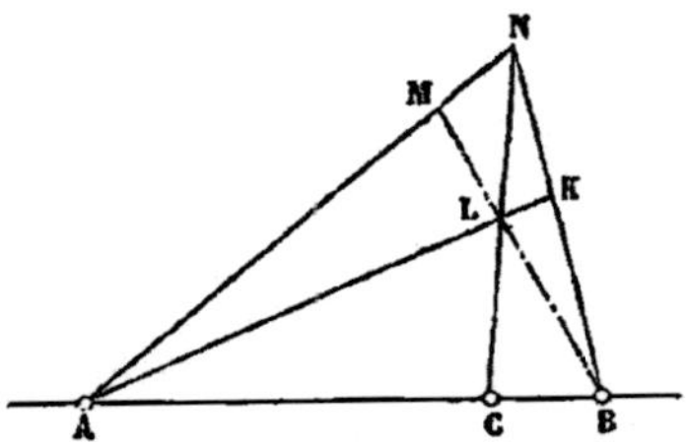

Fig. 24.

In order to project ABC into BCA, we might first project ABC into BAC, and then BAC into BCA.

45. THEOREM. *Any one-dimensional geometric form, consisting of four elements, is projective with any of the forms derived from it by interchanging the elements in pairs. For instance, $ABCD$ is projective with $BADC$.*

Let A, B, C, D be four given points (Fig. 25), and let $EFGD$ be a projection of these points from a centre M on a straight line DF passing through D. If AF, CM meet in N, then $MNGC$ will be a projection of $EFGD$ from centre A, and $BADC$ a projection of $MNGC$ from centre F; therefore (Arts. 40, 41) the form $BADC$ is projective with $ABCD$. In a similar manner it can be shown that $CDAB$ and $DCBA$ are projective with $ABCD$*.

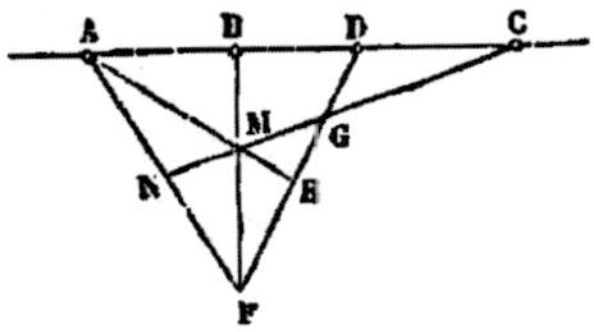

Fig. 25.

From this it follows for example that if a flat pencil $abcd$ is projective with a range $ABCD$, then it is projective also with $BADC$, with $CDAB$, and with $DCBA$; *i.e. if two geometric forms, each consisting of four elements, are projectively related, then the elements of the one can be made to correspond respectively to the elements of the other in four different ways.*

* STAUDT, *Geometrie der Lage*, Art. 59.

CHAPTER VIII.

HARMONIC FORMS.

46. Theorem*.

Given three points A, B, C on a straight line s; if a complete quadrangle ($KLMN$) be constructed (in any plane through s) in such a manner that two opposite sides (KL, MN) meet in A, two other opposite sides (KN, ML) meet in B, and the fifth side (LN) passes through C, then the sixth side (KM) will cut the straight line s in a point D which is determined by the three given points; *i.e.* it does not change its position, in whatever manner the arbitrary elements of the quadrangle are made to vary (Fig. 26).

Given in a plane three straight lines a, b, c which meet in a point S; if a complete quadrilateral ($klmn$) be constructed in such a manner that two opposite vertices (kl, mn) lie on a, two other opposite vertices (kn, ml) lie on b, and the fifth vertex (nl) lies on c, then the sixth vertex (km) will lie on a straight line d which passes through S, and which is determinate; *i.e.* it does not change its position, in whatever manner the arbitrary elements of the quadrilateral are made to vary (Fig. 27).

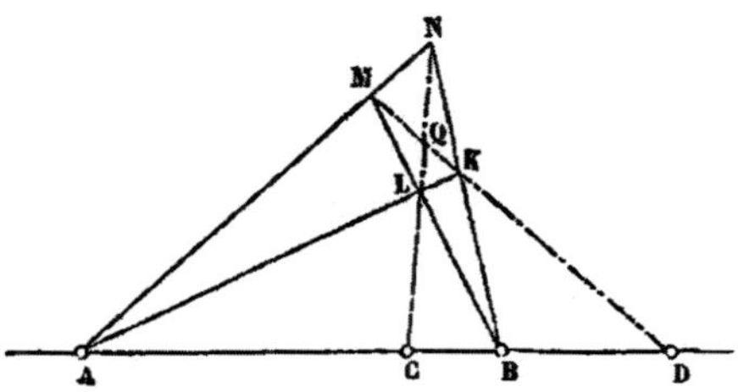

Fig. 26.

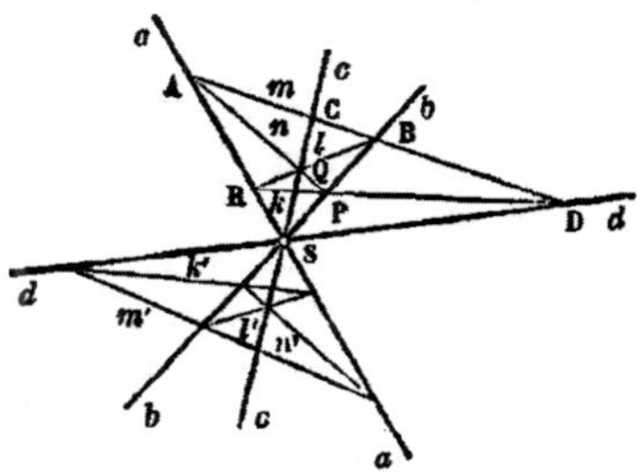

Fig. 27.

For if a second complete quadrangle ($K'L'M'N'$) be con-

For if a second complete quadrilateral ($k'l'm'n'$) be con-

* Staudt, *loc. cit.*, Art. 93.

structed (either in the same plane, or in any other plane through *s*), which satisfies the prescribed conditions, then the two quadrangles will have five pairs of corresponding sides which meet on the given straight line; therefore the sixth pair will also meet on the same line (Art. 36, No. 5, left).

From this it follows that if the first quadrangle be kept fixed while the second is made to vary in every possible way, the point *D* will remain fixed; which proves the theorem.

The four points *ABCD* are called *harmonic*, or we may say that the group or the *geometric form* constituted by these four points is a *harmonic* one, or that *ABCD* form a *harmonic range*. Or again: *Four points ABCD of a straight line, taken in this order, are called harmonic, if it is possible to construct a complete quadrangle such that two opposite sides pass through A, two other opposite sides through B, the fifth side through C, and the sixth through D.* It follows from the preceding theorem that when such a quadrangle exists, *i.e.* when the form *ABCD* is harmonic, it is possible to construct an infinite number of other quadrangles satisfying the same conditions. It further follows that, given three points *ABC* of a range (and also the order in which they are to be taken), the fourth point *D*, which makes with them a harmonic form, is *determinate* and *unique*, and is found by the construction of one of the quadrangles (see below, Art. 58).

structed which satisfies the prescribed conditions, then the two quadrilaterals will have five pairs of corresponding vertices collinear respectively with the given point; therefore the sixth pair will also lie in a straight line passing through the same point (Art. 36, No. 5, right).

From this it follows that if the first quadrilateral be kept fixed while the second is made to vary in every possible way, the straight line *d* will remain fixed; which proves the theorem.

The four straight lines or rays *abcd* are called *harmonic*, or we may say that the group or the *geometric form* constituted by these four lines is a *harmonic* one, or that *abcd* form a *harmonic pencil*. Or again: *Four rays abcd of a pencil, taken in this order, are called harmonic, if it is possible to construct a complete quadrilateral such that two opposite vertices lie on a, two other opposite vertices on b, the fifth vertex on c, and the sixth on d.* It follows from the preceding theorem that when such a quadrilateral exists, *i.e.* when the form *abcd* is harmonic, it is possible to construct an infinite number of other quadrilaterals satisfying the same conditions. It further follows that given three rays *abc* of a pencil (and also the order in which they are to be taken), the fourth ray *d*, which makes with them a harmonic form, is *determinate* and *unique*, and is found by the construction of one of the quadrilaterals (see below, Art. 58).

47. *If from any point S the harmonic range ABCD be projected upon any other straight line, its projection A′B′C′D′ will also be a harmonic range* (Fig. 28).

Imagine two planes drawn one through each of the straight lines AB, $A'B'$, and suppose that in the first of these planes is constructed a complete quadrangle of which two opposite sides meet in A, two other opposite sides meet in B, and a fifth side passes through C; then the sixth side will pass through D (Art. 46), since by hypothesis $ABCD$ is a harmonic range. Now project this quadrangle from the point S on to the second plane; then a new quadrangle is obtained of which two opposite sides meet in A', two other opposite sides meet in B', and whose fifth and sixth sides pass respectively through C' and D'; therefore $A'B'C'D'$ is a harmonic range.

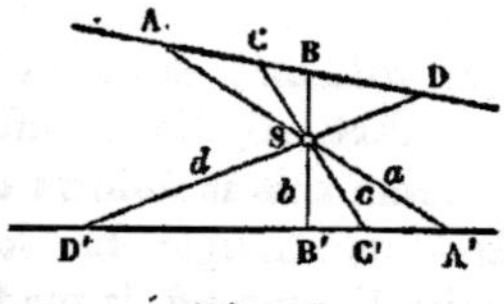

Fig. 28.

48. An examination of Fig. 27 will show that the harmonic pencil $abcd$ is cut by any transversal whatever in a harmonic range. For let S be the centre of the pencil and m be any transversal; in a take any point R; join R to D by the straight line k and to B by the straight line l; and join A to kb or P by the straight line n. As $abcd$ is a harmonic pencil and five vertices of the complete quadrilateral $klmn$ lie on a, b, and d, the sixth vertex ln or Q must lie on the fourth ray c. Then from the complete quadrangle $PQRS$ it is clear that $ABCD$ is a harmonic range.

Conversely, if the harmonic range $ABCD$ (Fig. 27) be given, and any centre whatever of projection S be taken, then the four projecting rays $S\,(A\,,B\,,C\,,D)$ will form a harmonic pencil.

For draw through A any straight line to cut SB in P and SC in Q, and join BQ, cutting AS in R. The quadrangle $PQRS$ is such that two opposite sides meet in A, two other opposite sides in B, and the fifth side passes through C; consequently the sixth side must pass through D (Art. 46, left), since by hypothesis the range $ABCD$ is harmonic. But then we have a complete quadrilateral $klmn$ which has two opposite vertices A and R lying on SA, two other opposite vertices B and P on SB, a fifth vertex Q on SC, and the sixth D on SD; therefore

(Art. 46, right) the four straight lines which project the range $ABCD$ from S are harmonic. We may therefore enunciate the following proposition:

A harmonic pencil is cut by any transversal whatever in a harmonic range; and, conversely, the rays which project a harmonic range from any centre whatever form a harmonic pencil.

Corollary. In two homological figures, to a range of four harmonic points corresponds a range of four harmonic points; and to a pencil of four harmonic rays corresponds a pencil of four harmonic rays.

49. The theorem on the right in Art. 46 is correlative to that on the left in the same Article. In this latter theorem all the quadrangles are supposed to lie in the same plane; but from the preceding considerations it is clear that the theorem is still true and may be proved in the same manner, if the quadrangles are drawn in different planes.

Considering accordingly this latter theorem (Art. 46, left) as a proposition in the Geometry of space, the theorem correlative to it will be the following:

If three planes α, β, γ all pass through one straight line s, and if a complete four-flat (see Art. 37) *$\kappa\lambda\mu\nu$ be constructed, of which two opposite edges $\kappa\lambda$, $\mu\nu$ lie in the plane α, two other opposite edges $\kappa\nu$, $\lambda\mu$ lie in the plane β, and the edge $\lambda\nu$ lies in the plane γ; then the sixth edge $\kappa\mu$ will always lie in a fixed plane δ (passing through s), which does not change, in whatever manner the arbitrary elements of the four-flat be made to vary.*

For if we construct (taking either the same vertex or any other lying on s) another complete four-flat which satisfies the prescribed conditions, the two four-flats will have five pairs of corresponding edges lying in planes which all pass through the same straight line s; therefore (Art. 37, left) the sixth pair also will lie in a plane which passes through s. The four planes, α, β, γ, δ are termed *harmonic planes*; or we may say that the group or the *geometric form* constituted by them is *harmonic*; or again that they form a *harmonic* (*axial*) *pencil.*

50. If a complete four-flat $\kappa\lambda\mu\nu$ be cut by any plane not passing through the vertex of the pencil, a complete quadrilateral is obtained; and the same transversal plane cuts the planes α, β, γ, δ in four rays of a flat pencil of which the first two rays contain each a pair of vertices of the quadrilateral while the other two pass each through one of the remaining vertices. Consequently (Art. 46, right) an axial pencil of four harmonic planes is cut by any transversal plane in a flat pencil of four harmonic rays.

Similarly, if the harmonic axial pencil of four planes $\alpha, \beta, \gamma, \delta$ is cut by any transversal line in four points A, B, C, D, these form a harmonic range. For if through the transversal line a plane be drawn, it will cut the planes $\alpha, \beta, \gamma, \delta$ in four straight lines a, b, c, d. This group of straight lines is harmonic, by what has just been proved; but $ABCD$ is a section of the flat pencil a, b, c, d; consequently (Art. 48) the four points A, B, C, D are harmonic. Conversely, if four points forming a harmonic range be projected from an axis, or if four rays forming a harmonic pencil be projected from a point, the resulting axial pencil is harmonic.

51. If then we include under the title of *harmonic form* the group of four harmonic points (the harmonic range), the group of four harmonic rays (the harmonic flat pencil), and the group of four harmonic planes (the harmonic axial pencil), we may enunciate the theorem:

Every projection or section of a harmonic form is itself a harmonic form: or,

Every form which is projective with a harmonic form is itself harmonic.

Conversely, *two harmonic forms are always projective with one another.*

To prove this proposition, it is enough to consider two groups each of four harmonic points; for if one of the forms were a pencil we should obtain four harmonic points on cutting it by a transversal. Let then $ABCD$, $A'B'C'D'$ be two harmonic ranges, and project ABC into $A'B'C'$ in the manner explained in Art. 44; the same operations (projections and sections) which serve to derive $A'B'C'$ from ABC will give for D a point D_1; from which it follows that the range $A'B'C'D_1$ will be harmonic, since the range $ABCD$ is harmonic. But $A'B'C'D'$ are also four harmonic points, by hypothesis; therefore D_1 must coincide with D', since the three points $A'B'C'$ determine uniquely the fourth point which forms with them a harmonic range (Art. 46, left).

We may add here a consequence of the definitions given in Arts. 49 and 50:

The form which is correlative to a harmonic form is itself harmonic.

52. If a, b, c, d are rays of a pencil (Fig. 28), then a and b are said to be *separated* by c and d, when a straight line passing through the centre of the pencil, and rotating so as to come into coincidence with each of the rays in turn, cannot pass from a to b without coinciding with one and only one of the two other rays c and d*. The same definition applies to the case of four planes of a pencil, and to that of four points of a range (Fig. 26); only it must be granted that we may pass from a point A to a point B in two different ways, either by describing the finite segment AB or the infinite segment which begins at A, passes through the point at infinity, and ends at B.

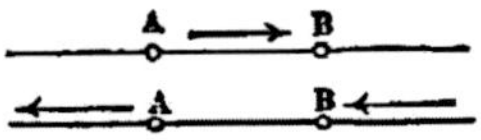

Fig. 29.

This definition premised, the following property may be enunciated as at once evident: Four elements of a one-dimensional geometric form (*i.e.* four points of a range, four rays of a pencil, &c.) can always be so divided into two pairs that one pair is separated by the other, and this can be done in one way only. In Fig. 26, for example, the two pairs which separate one another are AB, CD; and if $A'B'C'D'$ is a form projective with $ABCD$, the pair $A'B'$ will be separated by the pair $C'D'$; for the operations of projection and section do not change the relative position of the elements.

53. Let now $ABCD$ (Fig. 30) be four harmonic points, *i.e.* four points obtained by the construction of Art. 46, left. This allows us to draw in an infinite number of ways a complete quadrangle of which A and B are two diagonal points (Art. 36, No. 2, left), while the other two opposite sides pass through C and D. It is only necessary to state this construction in order to see that the two points A and B are precisely similar in their relation to the system, and that the same is true with regard to C and D. It follows from this that if $ABCD$ is a harmonic range, then $BACD$, $ABDC$, $BADC$, which are obtained by permuting the letters A and B or C and D, or both at the same time, are harmonic ranges also.

* a and b, c and d, may also be termed *alternate* pairs of rays.

Consequently (Art. 51) the harmonic range $ABCD$ for example is projective with $BACD$, *i.e.* we can pass from one range to the other by a finite number of projections and sections. In fact if the range $ABCD$ be projected from K on CQ, we obtain the range $LNCQ$, which when projected from M on AB gives $BACD$.

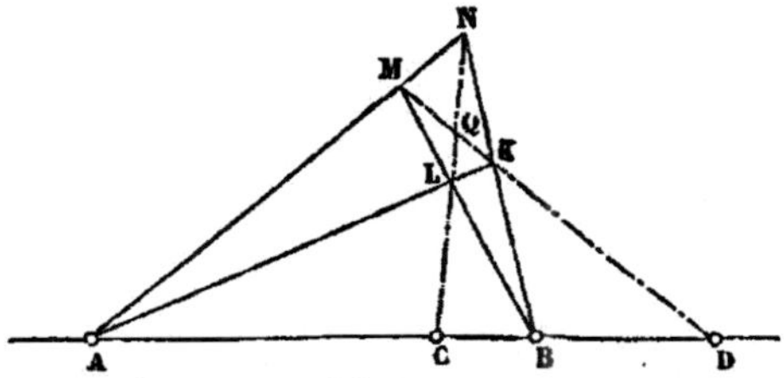

Fig. 30.

54. *If A, B, C, D are four harmonic points, then A and B are necessarily separated by C and D.*

For if (Fig. 30) the group $ABCD$ be projected on the straight line KM, first from the centre L and then from the centre N, the projections are $KMQD$ and $MKQD$ respectively. Now, as already stated in Art. 52, the operations of projection and section do not change the relative position of the elements of the group. If therefore K and Q were separated by M and D, then also M and Q must be separated by K and D; which is impossible. The only possible arrangement is that K and M should be separated by Q and D, and therefore A and B separated by C and D.

55. Let the straight lines AQ, BQ be drawn (Fig. 31), the former meeting MB in U and NB in S, while the latter meets KL in T and MN in V. The complete quadrangle $LTQU$ has two opposite sides meeting in A, two other opposite sides meeting in B, and a fifth side (LQ or LN) passes through C; therefore the sixth side UT will pass through D (Art. 46). In like manner the sixth side VS of the complete quadrangle $NVQS$ must pass through D, and the sixth sides of the complete quadrangles $KSQT$, $MUQV$ through C. We have thus a quadrangle $STUV$ two of whose opposite sides meet in C, two

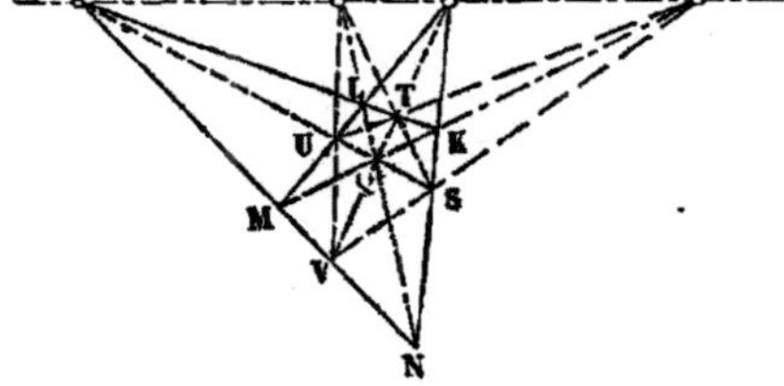

Fig. 31.

other opposite sides in *D*, while the fifth and sixth sides pass respectively through *A* and *B*. This shows that the relation to which the points *C* and *D* are subject (Art. 53) is the same as the relation to which the points *A* and *B* are subject; or, in other words, that the pair *A* , *B* may be interchanged with the pair *C* , *D*. Accordingly, if *ABCD* is a harmonic range, then not only the ranges *BACD* , *ABDC* , *BADC*, but also *CDAB* , *DCAB* , *CDBA* , *DCBA* are harmonic *.

The points *A* and *B* are termed *conjugate* points, as also are *C* and *D*. Or either pair are said to be *harmonic conjugates* with respect to the other. The points *A* and *B* are said to be *harmonically separated* by the points *C* and *D*, or the points *C* and *D* to be harmonically separated by *A* and *B*. We may also say that the segment *AB* is divided harmonically by the segment *CD*, or that the segment *CD* is divided harmonically by *AB*. If two points *A* and *B* (Fig. 30) are separated harmonically by the points *C* and *D* in which the straight line *AB* is cut by two straight lines *QC* and *QD*, we may also say that the segment *AB* is divided harmonically by the straight lines *QC* , *QD*, or by the point *C* and the straight line *QD*, &c.; and that the straight lines *QC* , *QD* are separated harmonically by the points *A* , *B*; &c.

Analogous properties and expressions exist in the case of four harmonic rays or four harmonic planes.

[*Note.*—In future, whenever mention is made of the harmonic system *ABCD*, it is always to be understood that *A* and *B*, *C* and *D*, are conjugate pairs; it being at the same time remembered that (Art. 54) *A* and *B*, *C* and *D*, are necessarily alternate pairs of points.]

56. The following theorem is another consequence of the proposition of Art. 46, left:

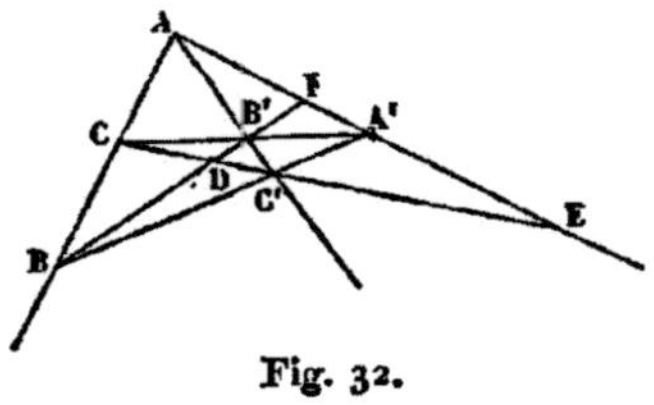

Fig. 32.

In a complete quadrilateral, each diagonal is divided harmonically by the other two †.

Let *A* and *A'*, *B* and *B'*, *C* and *C'* be the pairs of opposite vertices of a complete quadrilateral (Fig. 32), and let the diagonal *AA'* be cut by the other diagonals *BB'* and *CC'* in *F*

* Reye, *Geometrie der Lage* (Hanover, 1866), vol. i. p. 34.

† Carnot, *Géométrie de position* (Paris, 1803), Art. 225.

and E respectively. Consider now the complete quadrangle $BB'CC'$; one pair of its opposite sides meet in A, another such pair in A', a fifth side passes through E, the sixth through F. The points A, A' are therefore harmonically separated by F and E. Similarly a consideration of the two complete quadrangles $CC'AA'$ and $AA'BB'$ will show that B, B' are harmonically separated by F and D; and C, C' by D and E.

57. In the complete quadrangle $BB'CC'$ the diagonal points are A, A', and D; also since the range $BB'FD$ is harmonic, so too is the pencil of four rays which project it from A (Art. 48); therefore:

In a complete quadrangle, any two sides which meet in a diagonal point are divided harmonically by the two other diagonal points.

This theorem is however merely the correlative (in accordance with the principle of Duality in plane Geometry) of that proved in the preceding Article.

58. The theorems of Art. 46 can be at once applied to the solution, by means of the ruler only, of the following problems:

Given three points of a harmonic range, to find the fourth.

Solution. Let A, B, C (Fig. 33) be the given points (lying on a given straight line) and let

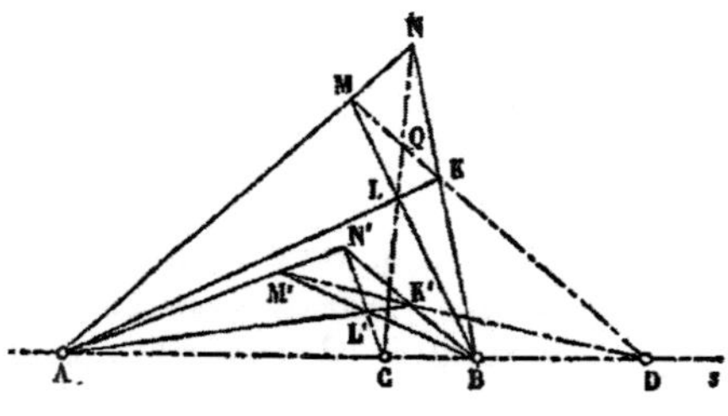

Fig. 33.

A and B be conjugate to each other. Draw any two straight lines through A, and a third through C to cut these in L and

Given three rays of a harmonic pencil, to construct the fourth.

Solution. Let a, b, c (Fig. 34) be the given rays (lying in one plane and passing through a

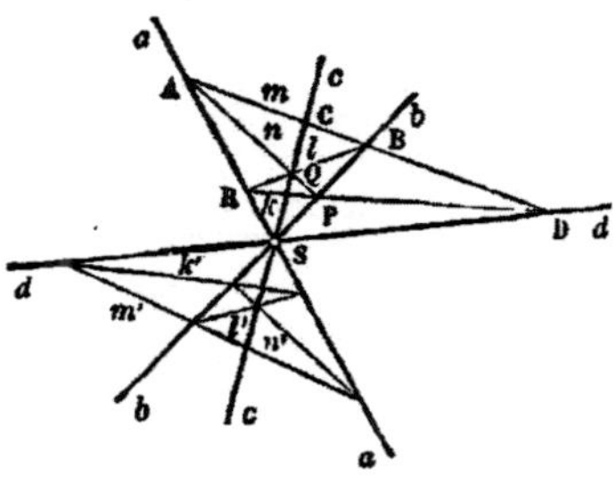

Fig. 34.

given centre S), and let a and b be conjugate to each other. Through any point Q lying on c draw any two straight lines to

N respectively. Join BL cutting AN in M, and BN cutting AL in K; then if KM be joined it will cut the given straight line in the required point D, conjugate to C *.

cut a in A and R, and b in P and B, respectively. Join AB and RP; these will cut in a point D, the line joining which to S is the required ray d, conjugate to c.

59. In the problem of Art. 58, left, let C lie midway between A and B. We can, in the solution, so arrange the arbitrary elements that the points K and M shall move off to infinity; to effect this we must construct (Fig. 35) a parallelogram $ALBN$ on AB as diagonal; then since the other diagonal LN passes through C, the point D will lie at infinity.

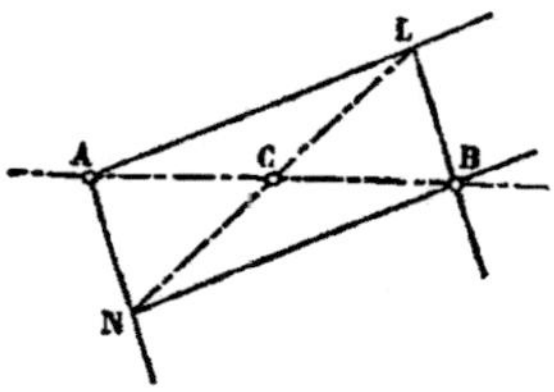

Fig. 35.

If, conversely, the points A, B, D are given, of which the third point D lies at infinity, we may again construct a parallelogram $ALBN$ on AB as diagonal; then the fourth point C, the conjugate of D, must be the point where LN meets the given straight line: that is, it must be the middle point of AB. Therefore:

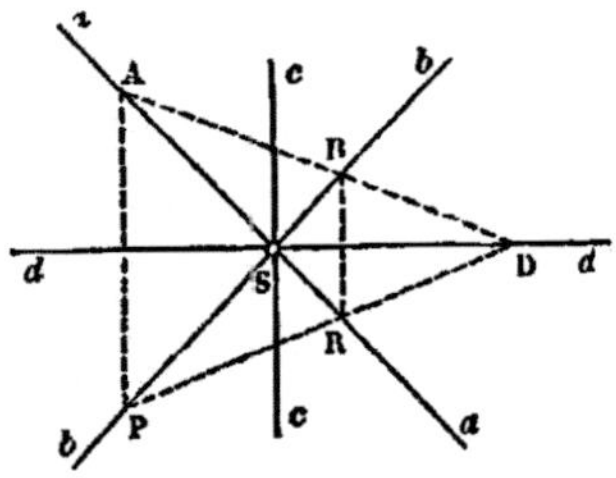

Fig. 36.

If in a harmonic range $ABCD$ the point C lies midway between the two conjugates A and B, then the fourth point D lies at an infinite distance; and conversely, if one of the points D lies at infinity, its conjugate C is the point midway between the two others, A and B.

60. In the problem of Art. 58, right, let c be the bisector of the angle between a and b (Fig. 36). If Q be taken at infinity on c, the segments AB, PR become equal to one another and lie between the parallels AP, BR; consequently the ray d will be perpendicular to c, *i.e. given a harmonic pencil of four rays, abcd; if one of them c bisect the angle between the two conjugates a and b, the fourth ray d will be at right angles to c.*

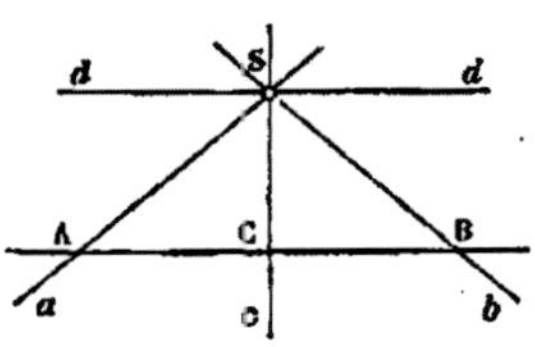

Fig. 37.

Conversely: *if in a harmonic pencil abcd* (Fig. 37) *two conjugate rays c, d are at right angles, then they are the bisectors, internal and external, of the angle between the other two rays a, b.*

* De la Hire, *Sectiones Conicae* (Parisiis, 1685), lib. i, prop. 20.

For if the pencil be cut by a transversal AB drawn parallel to d, the section $ABCD$ will be a harmonic range (Art. 48); and as D lies at infinity, C must lie midway between A and B (Art. 59); consequently, if S be the centre of the pencil, ASB is an isosceles triangle and SC the bisector of its vertical angle.

CHAPTER IX.

ANHARMONIC RATIOS.

61. GEOMETRICAL propositions divide themselves into two classes. Those of the one class are either immediately concerned with the magnitude of figures, as Euc. I. 47, or they involve more or less directly the idea of quantity or measurement, as *e.g.* Euc. I. 12. Such propositions are called *metrical.* The other class of propositions relate merely to the position of the figures with which they deal, and the idea of quantity does not enter into them at all. Such propositions are called *descriptive.* Most of the propositions in Euclid's *Elements* are metrical, and it is not easy to find among them an example of a purely descriptive theorem. Prop. 2, Book XI, may serve as an instance of one. Projective Geometry on the other hand, dealing with projective properties (*i. e.* such as are not altered by projection), is chiefly concerned with descriptive properties of figures. In fact, since the magnitude of a geometric figure is altered by projection, metrical properties are as a rule not projective. But there is one important class of metrical properties (anharmonic properties) which are projective, and the discussion of which therefore finds a place in the Projective Geometry. To these we proceed; but it is necessary first to establish certain fundamental notions.

62. Consider a straight line; a point may move along it in two different directions, one of which is opposite to the other. Let it be agreed to call one of these the positive direction, and the other the negative direction. Let A and B be two points on the straight line; and let it be further agreed to represent by the expression AB the length of the segment comprised between A and B, taken as a positive or as a negative number of units according as the direction is positive or negative in which a point must move in order to describe the segment;

this point starting from A (the first letter of the expression AB) and ending at B.

In consequence of this convention, which is termed the *rule of signs*, the two expressions AB, BA are quantities which are equal in magnitude but opposite in sign, so that $BA = -AB$, or

$$AB + BA = 0. \quad . \quad . \quad . \quad . \quad . \quad . \quad . \quad . \quad (1)$$

Now let A, B, C be three points lying on a straight line. If C lies between A and B (Fig. 38 *a*),

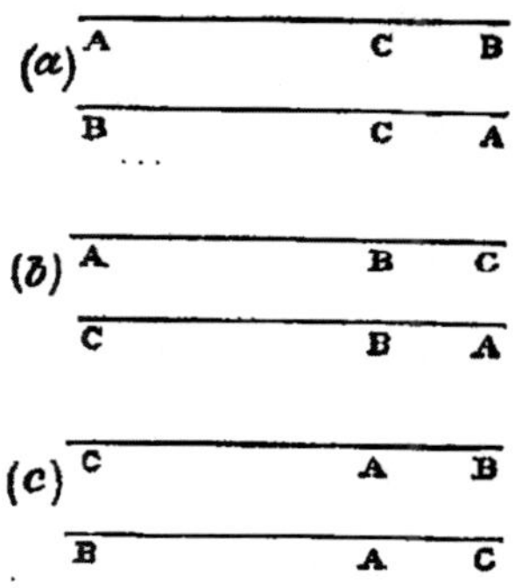

Fig. 38.

we have
$$AB = AC + CB;$$
whence
$$-CB - AC + AB = 0,$$
or
$$BC + CA + AB = 0.$$

Again, if B lies between A and C (Fig. 38 *b*),
$$AC = AB + BC;$$
whence
$$BC - AC + AB = 0,$$
or
$$BC + CA + AB = 0.$$

Lastly, if A lies between B and C (Fig. 38 *c*),
$$CB = CA + AB;$$
whence
$$-CB + CA + AB = 0,$$
or
$$BC + CA + AB = 0.$$

Accordingly:

If A, B, C are three collinear points, then whatever their relative positions may be, the identity

$$BC + CA + AB = 0 \quad . \quad . \quad . \quad . \quad . \quad . \quad . \quad . \quad (2)$$

always holds good.

From this identity may be deduced an expression for the distance between two points A and B in terms of the distances of these points from an origin O chosen arbitrarily on the straight line which joins them.

For since $$OA+AB+BO=0,$$

$$\therefore AB=OB-OA; \quad . \quad . \quad . \quad . \quad . \quad . \quad . \quad . \quad (3)$$

or again, $$AB=AO+OB\text{*}.$$

The results (1) and (2) may be extended; they are in fact particular cases of the following general proposition:

If $A_1, A_2, \ldots A_n$ be n collinear points, then

$$A_1A_2+A_2A_3+\ldots+A_{n-1}A_n+A_nA_1=0,$$

the truth of which follows at once from (3), since the expression on the left hand is equal to

$$(OA_2-OA_1)+(OA_3-OA_2)+\ldots+(OA_1-OA_n),$$

which vanishes.

Another useful result is that if A, B, C, D be four collinear points,

$$BC\,.\,AD+CA\,.\,BD+AB\,.\,CD=0.$$

This again follows from (3), since the left-hand side

$$=(DC-DB)AD+\ldots+\ldots$$
$$=0.$$

Many other relations of a similar kind between segments might be proved, but they are not necessary for our purpose. We will give only one more, viz.

If A, B, C, O be any four collinear points, then

$$OA^2\,.\,BC+OB^2\,.\,CA+OC^2\,.\,AB=-BC\,.\,CA\,.\,AB.$$

For by (3) the left-hand side is equal to

$$(OA^2-OC^2)BC+(OB^2-OC^2)CA$$
$$=CA(OA+OC)BC+CB(OB+OC)CA$$
$$=BC\,.\,CA(OA-OB)$$
$$=-BC\,.\,CA\,.\,AB.$$

It may be noticed that this last theorem is true even if O do not lie on the straight line ABC, but be any point whatever. For if a perpendicular OO' be let fall on ABC,

$$OA^2\,.\,BC+OB^2\,.\,CA+OC^2\,.\,AB$$
$$=(OO'^2+O'A^2)BC+\ldots+\ldots$$
$$=O'A^2\,.\,BC+O'B^2\,.\,CA+O'C^2\,.\,AB$$
$$+OO'^2(BC+CA+AB)$$
$$=-BC\,.\,CA\,.\,AB,$$

by what has just been proved.

63. Consider now Fig. 39, which represents the projection

* Möbius, *Barycentrische Calcul*, § 1.

from a centre S of the points of a straight line a on to another straight line a'; let us examine the relation which exists between the lengths of two corresponding segments AB, $A'B'$.

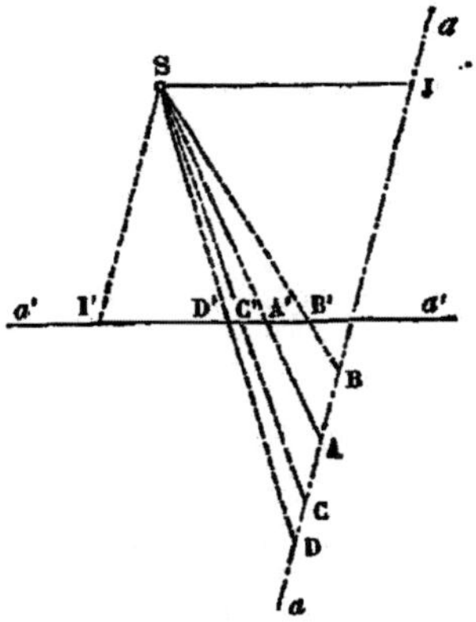

Fig. 39.

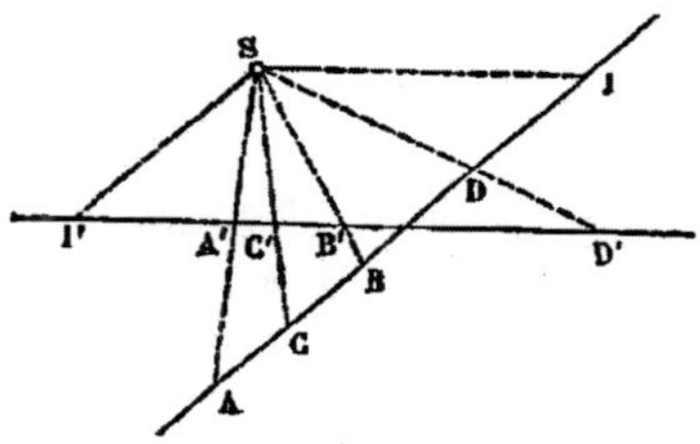

Fig. 40.

From the similar triangles SAJ, $A'SI'$

$$JA : JS :: I'S : I'A';^*$$

so from the similar triangles SBJ, $B'SI'$,

$$JB : JS :: I'S : I'B';$$

$$\therefore\ JA \,.\, I'A' = JB \,.\, I'B' = JS \,.\, I'S;$$

i.e. the rectangle $JA \,.\, I'A'$ has a constant value for all pairs of corresponding points A and A'.

If the constant $JS \,.\, I'S$ be denoted by k, we have

$$I'A' = \frac{k}{JA}, \quad I'B' = \frac{k}{JB};$$

therefore by subtraction,

$$I'B' - I'A' = \frac{k\,(JA - JB)}{JA \,.\, JB}.$$

But $I'B' - I'A' = A'B'$, and $JA - JB = BA = -AB$;

$$\therefore\quad A'B' = \frac{-k}{JA \,.\, JB} \,.\, AB.$$

If we consider four points A, B, C, D (Fig. 40) of the straight line a and their four projections, A', B', C', D', we obtain, in a similar manner,

* We suppose that in all equations involving segments the rule of signs is observed. See MÖBIUS, *Baryc. Calcul*, § 1; TOWNSEND, *Modern Geometry*, chapter v.

$$A'C' = \frac{-k}{JA \cdot JC} \cdot AC,$$

$$B'C' = \frac{-k}{JB \cdot JC} \cdot BC,$$

$$A'D' = \frac{-k}{JA \cdot JD} \cdot AD,$$

$$B'D' = \frac{-k}{JB \cdot JD} \cdot BD,$$

whence by division

$$\frac{A'C'}{B'C'} : \frac{A'D'}{B'D'} = \frac{AC}{BC} : \frac{AD}{BD}.$$

This last equation, which has been proved for the case of projection from a centre S, holds also for the case where $ABCD$ and $A'B'C'D'$ are the intersections of two transversal lines s and s' (not lying in the same plane) with four planes $\alpha, \beta, \gamma, \delta$ which all pass through one straight line u; in other words, when $A'B'C'D'$ is a projection of $ABCD$ made from an axis u (Art. 4). For let the four planes $\alpha, \beta, \gamma, \delta$ be cut in A'', B'', C'', D'' respectively by a straight line s'' which meets s and s'. The straight lines AA'', BB'', CC'', DD'' are the intersections of the planes $\alpha, \beta, \gamma, \delta$ respectively by the plane ss'', and therefore meet in a point S; that namely in which the plane ss'' is cut by the axis u. So also $A'A''$, $B'B''$, $C'C''$, $D'D''$ are four straight lines lying in the plane $s's''$ and meeting in a point S' of the axis u (that namely in which the plane $s's''$ is cut by the axis u). Therefore $A''B''C''D''$ is a projection of $ABCD$ from centre S and a projection $A'B'C'D'$ from centre S'; so that

$$\frac{A''C''}{B''C''} : \frac{A''D''}{B''D''} = \frac{AC}{BC} : \frac{AD}{BD} = \frac{A'C'}{B'C'} : \frac{A'D'}{B'D'}.$$

The number

$$\frac{AC}{BC} : \frac{AD}{BD}$$

is called the *anharmonic ratio* of the four collinear points A, B, C, D. The result obtained above may therefore be expressed as follows:

The anharmonic ratio of four collinear points is unaltered by any projection whatever *.

* PAPPUS, *Mathematicae Collectiones*, book vii. prop. 129 (ed. Hultsch, Berlin, 1877, vol. ii. p. 871).

Or again:

If two ranges, each of four points, are projective, they have the same anharmonic ratio, or, as we may say, *are equianharmonic* *.

64. Dividing one by the other the expressions for $A'C'$ and $B'C'$, we have

$$\frac{A'C'}{B'C'} = \frac{AC}{BC} : \frac{AJ}{BJ}.$$

In this equation the right-hand member is the anharmonic ratio of the four points A, B, C, J; consequently the left-hand member must be the anharmonic ratio of A', B', C', J'; thus *the anharmonic ratio of four points A', B', C', J', of which the last lies at infinity, is merely the simple ratio $A'C' : B'C'$.*

This may also be seen by observing that if A' and B' remain fixed while D' moves off to infinity on the line $A'B'$, then

$$\text{limiting value of } \frac{A'D'}{B'D'} = 1;$$

$$\therefore \text{ limiting value of } \frac{A'C'}{B'C'} : \frac{A'D'}{B'D'} = \frac{A'C'}{B'C'}.$$

Similarly, on the same supposition,

$$\text{limiting value of } \frac{A'D'}{B'D'} : \frac{A'C'}{B'C'} = \frac{B'C'}{A'C'};$$

i.e. the anharmonic ratio of the four points A', B', D', C', of which the third lies at infinity, is equal to the simple ratio $B'C' : A'C'$.

65. From this results the solution of the following

PROBLEM.—*Given three collinear points A, B, C; to find a fourth D so that the anharmonic ratio of the range $ABCD$ may be a number λ given in sign and magnitude* (Fig. 41).

Solution.—Draw any transversal through C, and take on it two points A', B' such that the ratio $CA' : CB'$ is equal to $\lambda : 1$, the given value of the anharmonic ratio; the two points A' and B' lying on the same or on opposite sides of C according as λ is positive or negative. Join AA', BB', meeting in S; the straight line through S parallel to $A'B'$ will cut AB in the point D required †. For if D' be the point at infinity on

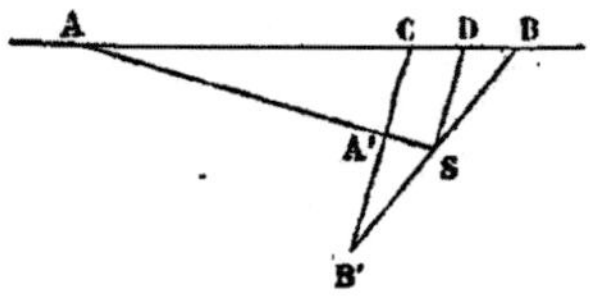

Fig. 41.

* TOWNSEND, *Modern Geometry*, Art. 278.

† CHASLES, *Géométrie supérieure* (Paris, 1852), p. 10.

$A'B'$, and we consider $ABCD$ as a projection of $A'B'C'D'$ (C' coincides with C) from the centre S, then the anharmonic ratio of $ABCD$ is equal to that of $A'B'C'D'$, that is, to the simple ratio $A'C' : B'C'$ or λ.

The above is simply the graphical solution of the equation

$$\frac{AC}{BC} : \frac{AD}{BD} = \lambda,$$

or

$$\frac{AD}{BD} = \frac{AC}{BC} : \lambda = \mu,$$

or in other words of the problem :

Given two points A and B, to find a point D collinear with them such that the ratio of the segments AD, BD to one another may be equal to a number given in sign and magnitude.

As only one such point D can be found, the proposed problem admits of only one solution; this is also clear from the construction given, since only one line can be drawn through S parallel to $A'B'$. Consequently there cannot be two different points D and D_1 such that $ABCD$ and $ABCD_1$ have the same anharmonic ratio. Or :

If the groups $ABCD$, $ABCD_1$ are equianharmonic, the point D_1 must coincide with D.

66. THEOREM. (Converse to that of Art. 63.) *If two ranges $ABCD$, $A'B'C'D'$, each of four points, are equianharmonic, they are projective with one another.*

For (by Art. 44) we can always pass from the triad ABC to the triad $A'B'C'$ by a finite number of projections or sections; let D'' be the point which these operations give as corresponding to D. Then the anharmonic ratio of $A'B'C'D''$ will be equal to that of $ABCD$, and consequently to that of $A'B'C'D'$; whence it follows that D'' coincides with D', and that the ranges $ABCD$, $A'B'C'D'$ are projective with one another.

67. It follows then from Arts. 63 and 66 that the necessary and sufficient condition that two ranges $ABCD$, $A'B'C'D'$, consisting each of four points, should be projective, is the equality (in sign and magnitude) of their anharmonic ratios.

The anharmonic ratio of four points $ABCD$ is denoted by the symbol $(ABCD)$*; accordingly the projectivity of two forms $ABCD$ and $A'B'C'D'$ is expressed by the equation

$$(ABCD) = (A'B'C'D').$$

* MÖBIUS, *Barycentrische Calcul*, § 183.

From what has been proved it is seen that if two pencils each consisting of four rays or four planes are cut by any two transversals in $ABCD$ and $A'B'C'D'$ respectively, the equation $(ABCD) = (A'B'C'D')$ is the necessary and sufficient condition that the two pencils should be projective with one another.

The *anharmonic ratio of a pencil of four rays* a, b, c, d or *four planes* $\alpha, \beta, \gamma, \delta$ may now be defined as the constant anharmonic ratio of the four points in which the four elements of the pencil are cut by any transversal, and may be denoted by $(abcd)$ or $(\alpha\beta\gamma\delta)$.

This done, we can enunciate the general theorem:

If two one-dimensional geometric forms, consisting each of four elements, are projective, they are equianharmonic; and if they are equianharmonic, they are projective.

68. Since two harmonic forms are always projectively related (Art. 51), the preceding theorem leads to the conclusion that the anharmonic ratio of four harmonic elements is a *constant number*. For if $ABCD$ is a harmonic system, $BACD$ is also a harmonic system (Art. 53), and the two systems $ACBD$ and $BCAD$ are projectively related*; thus

$$(ACBD) = (BCAD),$$

i. e.
$$\frac{AB}{CB} : \frac{AD}{CD} = \frac{BA}{CA} : \frac{BD}{CD};$$

whence
$$\frac{AC}{BC} : \frac{AD}{BD} = -1,$$

i. e.
$$(ABCD) = -1;$$

therefore *the anharmonic ratio of four harmonic elements is equal to* -1 †.

69. The equation $(ABCD) = -1$, or

$$\frac{AC}{BC} + \frac{AD}{BD} = 0, \quad \ldots \quad (1)$$

which expresses that the range $ABCD$ is harmonic, may be put into two other remarkable forms.

Since $AD = CD - CA$ (Art. 62) and $BD = CD - CB$, the equation (1) gives

$$CA\,(CD - CB) + CB\,(CD - CA) = 0,$$

or
$$\frac{1}{CD} = \tfrac{1}{2}\left(\frac{1}{CA} + \frac{1}{CB}\right), \quad \ldots \quad (2)$$

* In Fig. 30 $ACBD$ may be projected (from K on NC) into $LCNQ$; and then $LCNQ$ may be projected (from M on AD) into $BCAD$.

† MÖBIUS, *loc. cit.*, p. 269.

i.e. CD is the harmonic mean between CA and CB; a formula which determines the point D when A, B, C are given.

Again, if O is the middle point of the segment CD, so that we have $OD = CO = -OC$, then

$$AC = OC - OA; \quad AD = OD - OA = -(OC + OA);$$
$$BC = OC - OB; \quad BD = -(OC + OB).$$

Substituting these values in (1) or in

$$\frac{AC}{AD} + \frac{BC}{BD} = 0,$$

we have
$$\frac{OC - OA}{OC + OA} = \frac{OB - OC}{OB + OC};$$
$$\therefore \quad \frac{OC}{OA} = \frac{OB}{OC},$$

or
$$OC^2 = OA \, . \, OB, \quad . \quad . \quad . \quad . \quad . \quad . \quad . \quad . \quad (3)$$

i.e. half the segment CD is a mean proportional between the distances of A and B from the middle point of CD.

The equation (3) shows that the segments OA and OB must have the same sign, and that O therefore can never lie between A and B.

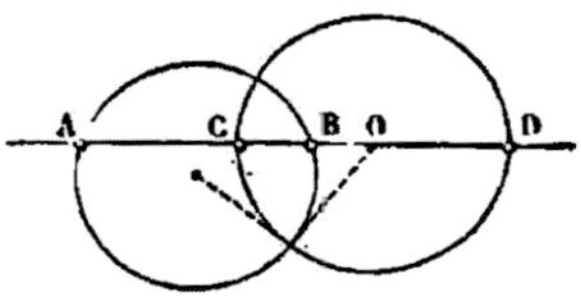

Fig. 42.

If now a circle be drawn to pass through A and B (Fig. 42), O will lie outside the circle, and OC will be the length of the tangent from O to it * (Euc. III. 37). The circle on CD as diameter will therefore cut the first circle (and *all* circles through A and B) orthogonally. Conversely, if two circles cut each other orthogonally, they will cut any diameter of one of them in two pairs of harmonic points †.

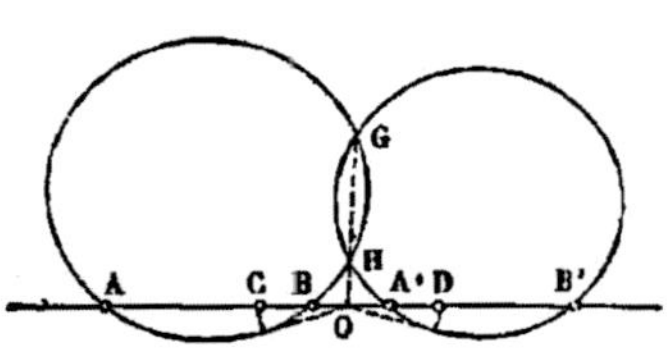

Fig. 43.

70. The same formula (3) gives the solution of the following problem:

Given two collinear segments AB and A′B′; to determine another segment CD which shall divide each of them harmonically (Figs. 43, 44).

Take any point G not lying on the common base AB', and draw the circles GAB, $GA'B'$ meeting

* If through a point O any chord be drawn to cut a circle in P and Q, the rectangle $OP \, . \, OQ$ is called the *power* of the point with regard to the circle. STEINER, *Crelle's Journal*, vol i. (Berlin, 1826); Collected Works, vol. i. p. 22. We may then say that OC^2 is the *power* of the point O with regard to the circle in Fig. 42.

† PONCELET, *Propr. proj.* Art. 79.

again in H. Join GH *, and produce it to cut the axis in O. Then from the first circle

$$OA \cdot OB = OG \cdot OH \text{ (Euc. III. 36)},$$

and from the second

$$OA' \cdot OB' = OG \cdot OH;$$
$$\therefore \quad OA \cdot OB = OA' \cdot OB'.$$

O is therefore the middle point of the segment required; the points C and D will be the intersections with the axis of a circle described from the centre O with radius equal to the length of the tangent from O to either of the circles GAB, $G'A'B'$.

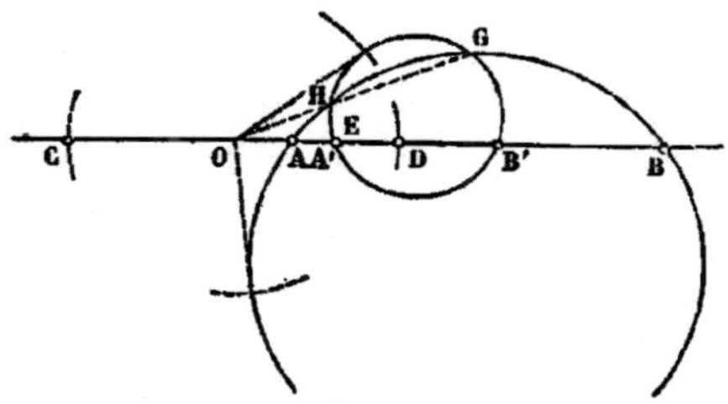

Fig. 44.

The problem admits of a real solution when the point O falls outside both the segments AB, $A'B'$, and consequently outside both the circles GAB, $GA'B'$ (Figs. 43, 44). There is no real solution when the segments AB, $A'B'$ overlap (Fig. 45); in this case O lies within both segments.

71. Let $ABCD$ be a harmonic range, and let A and B (a pair of conjugates) approach indefinitely near to one another and ultimately coincide. If C lie at an infinite distance, then D must coincide with A and B, since it must lie midway between these two points (Art. 59). If C lie at a finite distance, and assume any position not coinciding with that of A or B, then equation (2) of Art. 69 gives $CD = CA = CB$, *i.e.* D coincides with A and B.

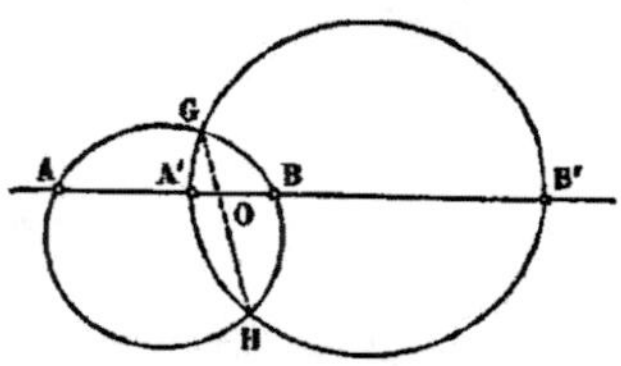

Fig. 45.

Again, let A and C (two non-conjugate points) coincide, and B (the conjugate of A) lie at an infinite distance. In this case A must lie midway between C and D, so that D will coincide with A and C. If B lie at a finite distance, and assume any position not coinciding with that of A or C, then equation (1) of Art. 69 gives $AD = 0$, *i.e.* the point D coincides with A and C. So that:

If, of four points forming a harmonic range, any two coincide, one of the other two points will also coincide with them, and the fourth is indeterminate.

72. The theorem of Art. 45 leads to the following result: given four elements A, B, C, D of a one-dimensional geometric form, the

* GH is the radical axis of the two circles, and all points on it are of equal power with regard to the circles.

anharmonic ratios $(ABCD)$, $(BADC)$, $(CDAB)$, $(DCBA)$ are all equal to one another.

I. Four elements of such a form can be permuted in twenty-four different ways, so as to form the twenty-four different groups

$$\begin{array}{cccc} ABCD, & BADC, & CDAB, & DCBA, \\ ABDC, & BACD, & DCAB, & CDBA, \\ ACBD, & CADB, & BDAC, & DBCA, \\ ACDB, & CABD, & DBAC, & BDCA, \\ ADBC, & DACB, & BCAD, & CBDA, \\ ADCB, & DABC, & CBAD, & BCDA, \end{array}$$

here arranged in six lines of four each. The four groups in each line are projective with one another (Art. 45), and have therefore the same anharmonic ratio. In order to determine the anharmonic ratios of all the twenty-four groups, it is only necessary to consider *one* group in each line; for example, the six groups in the first column. These six groups are so related to each other that when any one of them is known the other five can be at once determined.

II. Consider the two groups $ABCD$ and $ABDC$, which are derived one from the other by interchanging the last two elements. Their anharmonic ratios

$$(ABCD) \quad \text{or} \quad \frac{AC}{BC} : \frac{AD}{BD}$$

and

$$(ABDC) \quad \text{or} \quad \frac{AD}{BD} : \frac{AC}{BC}$$

are one the reciprocal of the other; thus

$$(ABCD)(ABDC) = 1. \qquad (1)$$

Similarly,

$$(ACBD)(ACDB) = 1, \qquad (2)$$

and

$$(ADBC)(ADCB) = 1. \qquad (3)$$

III. Now if A, B, C, D are four collinear points, it has been seen (Art. 62) that the identical relation

$$BC \cdot AD + CA \cdot BD + AB \cdot CD = 0$$

always holds. Dividing by $BC \cdot AD$, we have

$$\frac{AC \cdot BD}{BC \cdot AD} + \frac{AB \cdot CD}{CB \cdot AD} = 1,$$

or

$$\frac{AC}{BC} : \frac{AD}{BD} + \frac{AB}{CB} : \frac{AD}{CD} = 1,$$

that is (Arts. 63, 67),

$$(ABCD) + (ACBD) = 1. \qquad (4)$$

Similarly,

$$(ABDC) + (ADBC) = 1, \qquad (5)$$

and

$$(ACDB) + (ADCB) = 1. \qquad (6)$$

IV. If λ denote the anharmonic ratio of the group $ABCD$, *i.e.* if

$$(ABCD) = \lambda,$$

the formula (1) gives $(ABDC) = \frac{1}{\lambda}$,

and (4) gives $(ACBD) = 1-\lambda$;

then by (2) $(ACDB) = \frac{1}{1-\lambda}$,

and by (6) $(ADCB) = 1 - \frac{1}{1-\lambda} = \frac{\lambda}{\lambda-1}$;

and finally, by (3) or (5)

$$(ADBC) = \frac{\lambda-1}{\lambda}.\text{*}$$

V. The six anharmonic ratios may also be expressed in terms of the angle of intersection θ of the circles described on the segments AB, CD as diameters; it being supposed that A and B are separated by C and D. It will be found that

$$(ABCD) = -\tan^2\frac{\theta}{2}, \qquad (ABDC) = -\cot^2\frac{\theta}{2},$$

$$(ACBD) = \sec^2\frac{\theta}{2}, \qquad (ACDB) = \cos^2\frac{\theta}{2},$$

$$(ADCB) = \sin^2\frac{\theta}{2}, \qquad (ADBC) = \operatorname{cosec}^2\frac{\theta}{2}.\dagger$$

VI. If in the group $ABCD$ two points A and B coincide, then $AC = BC$, $AD = BD$, and therefore

$$(ABCD) = (AACD) = 1.$$

But if $\lambda = 1$, the other anharmonic ratios become

$$(ACAD) = 1-1 = 0, \text{ and } (ACDA) = \infty;$$

thus when of four elements two coincide, the anharmonic ratios have the values 1, 0, ∞.

If $(ABCD) = -1$, *i.e.* if the range $ABCD$ is harmonic, the formulae of (IV) give

$$(ACBD) = 2 \text{ and } (ACDB) = \tfrac{1}{2};$$

so that when the anharmonic ratio of four points has the value 2 or $\frac{1}{2}$, these points, taken in another order, form a harmonic range.

VII. Conversely, the anharmonic ratio of a range $ABCD$, none of whose points lies at infinity, cannot have any of the values 0, 1, ∞, without some two of its points coinciding.

For if in (IV) $\lambda = 0$, $\frac{AC}{BC} : \frac{AD}{BD} = 0$, and either AC or BD must vanish; *i.e.* either A coincides with C, or B with D.

* Möbius, *loc. cit.*, p. 249.

† Casey, *On Cyclides and Sphero-quartics* (Phil. Trans. 1871), p. 704.

If $\lambda = 1$, $(ACBD) = 1 - \lambda = 0$, so that either A coincides with B, or C with D.

And if $\lambda = \infty$, $(ABDC) = \frac{1}{\lambda} = 0$, so that either A coincides with D, or B with C.

VIII. By considering the expressions given for the six anharmonic ratios in (IV) it is clear that whatever be the relative positions of the points A, B, C, D, two of the ratios (and their two reciprocals) are always positive and a third (and its reciprocal) negative; and thus we see that the anharmonic ratios of four points no two of which coincide may have all values positive or negative except $+1$, 0, or ∞.

73. From the theorems of Arts. 63 and 66, which express the necessary and sufficient condition that two ranges, each consisting of four elements, should be projectively related, we conclude that

If two geometric forms of one dimension are projective, then any two corresponding groups of four elements are equianharmonic *.

As a particular case, to any four harmonic elements of the one form correspond four harmonic elements of the other (Art. 51).

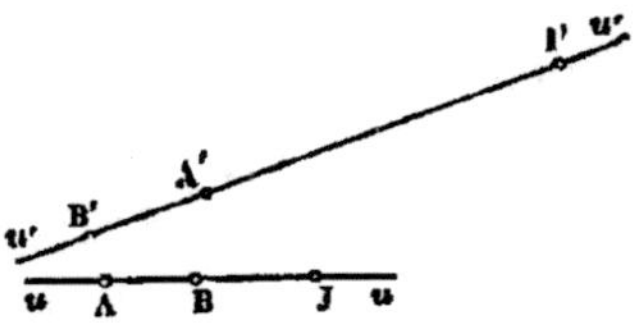

Fig. 46.

74. Let A, A' and B, B' be any two pairs of corresponding points of two projective ranges (Fig. 46); let I be the point at infinity belonging to the first range, and I' the point corresponding to it in the second range; similarly let J' be the point at infinity belonging to the second range, and J its correspondent in the first range. By Art. 73

$$(ABIJ) = (A'B'I'J');$$

$$\therefore \quad (BAJI) = (A'B'I'J') \text{ (Art. 72)};$$

from which, since I and J' lie at infinity,

$$BJ : AJ = A'I' : B'I' \text{ (Art. 64)},$$

and

$$JA \,.\, I'A' = JB \,.\, I'B';$$

i.e. the product $JA \,.\, I'A'$ has a constant value for all pairs of corresponding points †.

[This proposition has already been proved in Art. 63 for the particular case of two ranges in perspective.]

* STEINER, *Systematische Entwickelung*.. (Berlin, 1832), p. 33, § 10; Collected Works, ed. Weierstrass (Berlin, 1881), vol. i. p. 262.

† STEINER, *loc. cit.*, p. 40, § 12; Collected Works, vol. i. p. 267.

75. In two homological figures, four collinear points or four concurrent straight lines of the one figure form a group which is equianharmonic with that consisting of the points or lines corresponding to them in the other figure (Art. 73). Let O be the centre of homology, M and M' any pair of corresponding points in the two figures, N and N' another pair of corresponding points lying on the ray OMM', and X the point in which this ray meets the axis of homology. Since the points $OMNX$, $OM'N'X$ correspond severally to one another,

$$(OXMN) = (OXM'N'),$$

or

$$\frac{OM}{MX} : \frac{ON}{NX} = \frac{OM'}{M'X} : \frac{ON'}{N'X};$$

$$\therefore \quad \frac{OM}{MX} : \frac{OM'}{M'X} = \frac{ON}{NX} : \frac{ON'}{N'X},$$

and consequently the anharmonic ratio $(OXMM')$ is constant for all pairs of corresponding points M and M' taken on a ray OX passing through the centre of homology.

Next let L and L' be another pair of corresponding points, and Y the point in which the ray OLL' cuts the axis of homology. Since the straight lines LM, $L'M'$ must meet in some point Z of the axis XY, it follows that $OYLL'$ is a projection of $OXMM'$ from Z as centre, and therefore

$$(OYLL') = (OXMM');$$

consequently the anharmonic ratio $(OXMM')$ is constant for all pairs of corresponding points in the plane.

Consider now a pair of corresponding straight lines a and a', the axis of homology s, and the ray o joining the centre of homology O to the point aa'. The pencil $osaa'$ is cut by every straight line through O in a range of four points analogous to $OXMM'$; consequently the anharmonic ratio $(osaa')$ is constant for all pairs of corresponding straight lines a and a', and is equal to the anharmonic ratio $(OXMM')$.

This anharmonic ratio is called the *coefficient* or *parameter* of the homology. It is clear that two figures in homology can be constructed when, in addition to the centre and axis, we are given the parameter of the homology.

76. When the parameter of the homology is equal to -1, all ranges and pencils similar to $OXMM'$, $osaa'$, are harmonic.

In this case the homology is called *harmonic* * or *involutorial*, and two corresponding points (or lines) correspond to one another doubly; that is to say, every point (or line) has the same correspondent whether it be regarded as belonging to the first or the second figure. (See below, Arts. 122, 123.)

Harmonic homology presents two cases which deserve special notice: (1) when the centre of homology is at an infinite distance, in the direction perpendicular to the axis of homology; (2) when the axis of homology is at an infinite distance. In the first case we have what is called *symmetry with respect to an axis*; the axis of homology (in this case called also the *axis of symmetry*) bisects orthogonally the straight line joining any pair of corresponding points, and bisects also the angle included by any pair of corresponding straight lines. The second case is called *symmetry with respect to a centre.* The centre of homology (in this case called also the *centre of symmetry*) bisects the distance between any pair of corresponding points, and two corresponding straight lines are always parallel. In each of these two cases the two figures are equal and similar (congruent) †; oppositely equal in the first case, and directly equal in the second.

77. Considering again the general case of two homological figures, let a, b, m, n be four rays of a pencil in the first figure, and a', b', m', n' the straight lines corresponding to them in the second. Then

$$(mnab) = (m'n'a'b').$$

Now let an arbitrary transversal be drawn to cut $mnab$ in $MNAB$, and draw the corresponding (or another) transversal to cut $m'n'a'b'$ in $M'N'A'B'$; then

$$(MNAB) = (M'N'A'B'),$$

or

$$\frac{MA}{MB} : \frac{M'A'}{M'B'} = \frac{NA}{NB} : \frac{N'A'}{N'B'}.$$

Consequently, the ratio $\frac{MA}{MB} : \frac{M'A'}{M'B'}$ depends only on the straight lines ab (and $a'b'$), and not at all on the straight line m (or m').

The ratio $MA : NA$ is equal to that of the distances of the points M, N from the straight line a, which distances we may denote by (M, a), (N, a); thus

* BELLAVITIS, *Saggio di Geometria derivata* (*Nuovi Saggi* of the Academy of Padua, vol. iv. 1838), § 50.

† Two figures are said to be *congruent* when the one may be superposed upon the other so as exactly to coincide with it.

$$\frac{(M, a)}{(M, b)} : \frac{(M', a')}{(M', b')} = \text{constant},$$

that is to say *:

In two homological figures (or, more generally, in two projectively related figures) the ratio of the distances of a variable point M from two fixed straight lines a , b in the first figure bears a constant ratio to the analogous ratio of the distances of the corresponding point M' from the corresponding straight lines a' , b' in the other figure.

Suppose b to pass through the centre of homology O; then M and M' are collinear with O and b' coincides with b, so that

$$(M, b) : (M', b') = OM : OM';$$

and therefore

$$\frac{OM}{OM'} : \frac{(M, a)}{(M', a')} = \text{constant}.$$

If N and N' are another pair of corresponding points, we have then

$$\frac{OM}{OM'} : \frac{(M, a)}{(M', a')} = \frac{ON}{ON'} : \frac{(N, a)}{(N', a')}.$$

Now suppose the straight line a' to move away indefinitely; then a becomes the vanishing line in the first figure; the ratio $\frac{(M', a')}{(N', a')}$ will in the limit become equal to unity, and thus

$$\frac{OM}{OM'} : (M, a) = \frac{ON}{ON'} : (N, a)$$

$$= \text{constant};$$

in other words †:

In two homological figures, the ratio of the distances of any point in the first figure from the centre of homology and from the vanishing line respectively, varies directly as the distance of the corresponding point in the second figure from the centre of homology.

* Chasles, *Géométrie supérieure*, Art. 512.

† Chasles, *Sections coniques*, Art. 267.

CHAPTER X.

CONSTRUCTION OF PROJECTIVE FORMS.

78. LET ABC and $A'B'C'$ be two triads of corresponding elements of two projective forms of one dimension (Fig. 47), and imagine any series of operations (of projection and section) by which we may have passed from ABC to $A'B'C'$. Then whatever this series be*, it will also lead from any other element D of the first form to the element D' which corresponds to it in the second. For if D could give, as the result of these operations, an element D'' different from D', then the anharmonic ratios $(ABCD)$ and $(A'B'C'D'')$ would be equal; but by hypothesis $(ABCD) = (A'B'C'D')$; therefore $(A'B'C'D') = (A'B'C'D'')$, which is impossible unless D'' coincide with D' (Art. 65).

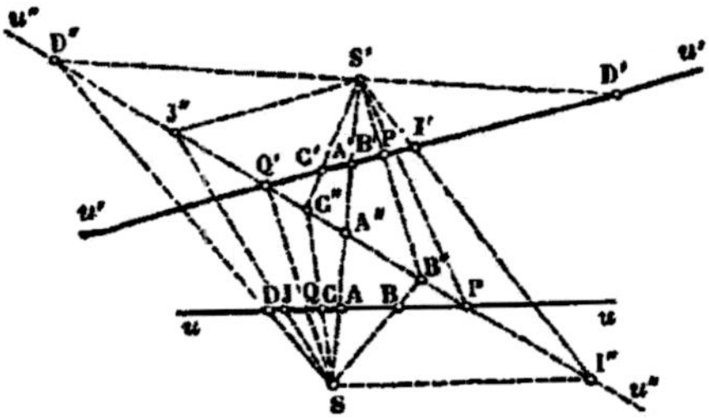

Fig. 47.

79. THEOREM (converse to that of Art. 73):

Given two forms of one dimension; if to the elements $A, B, C, D, \ldots$ of the one correspond respectively the elements $A', B', C', D', \ldots$ of the other in such a manner that any four elements of the first form are equianharmonic with the four corresponding elements of the second, then the two forms are projective.

For every series of operations (of projection or section), which leads from the triad ABC to the triad $A'B'C'$, leads at the same time from the element D to another element D'' such that $(ABCD) = (A'B'C'D'')$. But $(ABCD) = (A'B'C'D')$ by hypothesis; therefore $(A'B'C'D') = (A'B'C'D'')$, and D'' must coincide with D' (Art. 65). And since the same conclusion is

* In Fig. 47 the series of operations is: a projection from S, a section by u'', a projection from S', and a section by u'.

true for any other pair whatever of corresponding elements, it follows that the two forms are projective (Art. 40).

80. From Art. 78 the following may be deduced as a particular case:

If among the elements of two projective forms of one dimension there are two corresponding triads ABC and A′B′C′ which are in perspective, then the two forms themselves are in perspective.

(1). If, for example, the forms are two ranges $ABCD$ ... and $A'B'C'D'$ ...; then if the three straight lines AA', BB', CC' meet in a point S, the other analogous lines DD', ... will all pass through S (Figs. 19, 40).

Suppose, as a particular case, that the points A, A' coincide (Fig. 22), so that the two ranges have a pair of corresponding points A and A' united in the point of intersection of their bases*. The triads ABC, $A'B'C'$ are in perspective, their centre of perspective being the point where BB' and CC' meet; accordingly:

If two projective ranges have a self-corresponding point, they are in perspective.

Conversely it is evident that *two ranges which are in perspective have always a self-corresponding point.*

(2). Again, if the two forms are two flat pencils $abcd$ ... and $a'b'c'd'$ ... lying in the same plane; then if the three points aa', bb', cc' lie on one straight line s, the analogous points dd' ... will all lie on the same straight line (Fig. 20). If the line s lie altogether at infinity, we have the following property:

If, in two projective flat pencils, three pairs of corresponding rays are parallel to one another, then every pair of corresponding rays are parallel to one another.

The hypothesis is satisfied in the particular case where the rays a and a' coincide (Fig. 48),

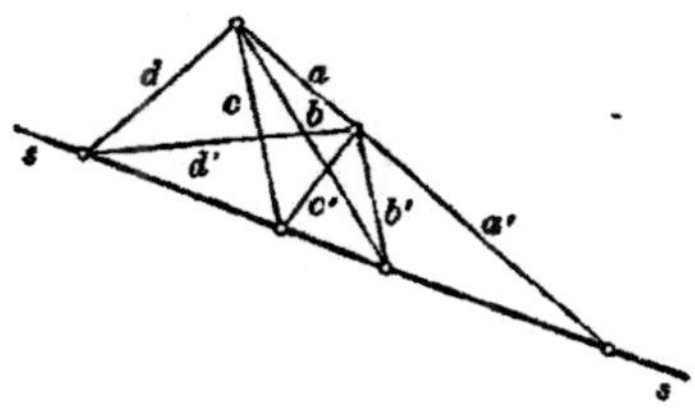

Fig. 48.

so that the two pencils have a self-corresponding ray in the straight line which joins their centres; then s is the straight line joining bb' and cc'. Accordingly:

* In the case of two projective forms we shall in future employ the term *self-corresponding* to denote an element which is such that it coincides with its correspondent; thus A or A' above may be called a *self-corresponding point* of the two ranges.

When two projective flat pencils (lying in the same plane) have a self-corresponding ray, they are in perspective.

Conversely, *two coplanar flat pencils which are in perspective have always a self-corresponding ray.*

(3). If one of the systems is a range $ABCD$ ... and the other a flat pencil $abcd$ ... (Fig. 28), the hypothesis amounts to assuming that the rays a, b, c pass respectively through the points A, B, C; then we conclude that also d, ... will pass through D, ... &c.

81. Two ranges may be *superposed* one upon the other, so as to lie upon the same straight line or base, in which case they may be said to be *collinear.* For example, if two pencils (in the same plane) $S \equiv abc$ ... and $O \equiv a'b'c'$ ... (Fig. 49) are cut by the same transversal, they will determine upon it two ranges ABC ..., $A'B'C'$ ... which will be projectively related if the two pencils are so. The question arises whether there exist in this case any self-corresponding points, *i.e.* whether two corresponding points of the two ranges coincide in any point of the transversal.

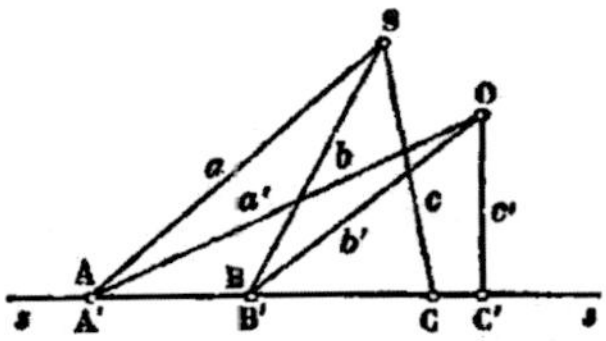

Fig. 49.

If, for instance, the transversal s be drawn so as to pass through the points aa' and bb', then A will coincide with A', and B with B'; in this case consequently there are *two* self-corresponding points.

Again, if a range u be projected (Fig. 50) from two centres S and O (lying in the same plane with u), two flat pencils abc ... and $a'b'c'$ ... will be formed, which have a pair of corresponding rays a, a' united in the line SO. And if a transversal s be drawn through the point in which this line cuts u, we shall obtain two projective ranges ABC ..., $A'B'C'$ ... lying on a common base s, and such that they have *one* self-corresponding point AA'.

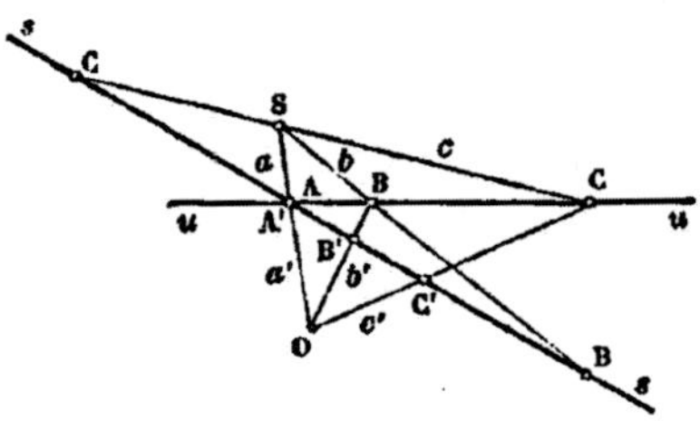

Fig. 50.

And lastly, we shall see hereafter (Art. 109) that it is possible for two collinear projective ranges to be such as to have *no* self-corresponding point.

So also two flat pencils (in the same plane) may have a common centre, in which case they may be termed *concentric*; such pencils are formed when two different ranges are projected from the same centre (Fig. 51). And two axial pencils may have a common axis; such pencils are formed when we project two different ranges from the same axis, or the same flat pencil from different centres. Again, if two sheaves are cut by the same plane, two plane figures are obtained; if, on the other hand, two plane figures are projected from the same centre, two concentric sheaves are formed. In all these cases the forms in question may be said to be *superposed* one upon the other; and the investigation of their *self-corresponding elements*, when the two forms are projectively related, is of great importance. The complete investigation will be given later on, in Chapter XVIII; at present we can only prove the following Theorem.

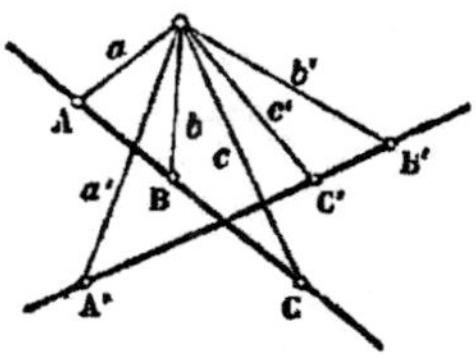

Fig. 51.

82. Theorem. *Two superposed projective (one-dimensional) forms either have* at most two *self-corresponding elements, or else* every *element coincides with its correspondent.*

For if there could be three self-corresponding elements A, B, C suppose; then if D and D' are any other pair of corresponding points, we should have (Art. 73) $(ABCD)=(ABCD')$, and consequently (Art. 65) D would coincide with D'. Unless then the two forms are identical, they cannot have more than two self-corresponding elements.

83. Theorem (converse to that of Art. 53). *If a one-dimensional form consisting of four elements A, B, C, D is projective with a second form deduced from it by interchanging two of the elements (e.g. $BACD$), then the form will be a harmonic one, and the two interchanged elements will be conjugate to each other.*

First Proof. If $(ABCD) = (BACD)$, then (Art. 72. IV) $\lambda = \frac{1}{\lambda}$;

$\therefore \lambda^2 = 1$, and since we cannot take $\lambda = +1$ (Art. 72. VIII) we must have $\lambda = -1$, *i.e.* the form is a harmonic one.

Second Proof. Suppose, for example, that A, B, C, D are four collinear points (Fig. 52). Let K, M, Q, D be a projection of these points on any straight line through D, made from an arbitrary centre L. Since $ABCD$ is projective with $KMQD$ and also (by hyp.) with $BACD$, the forms $KMQD$ and $BACD$ are projective with one another. And they have a self-corresponding point D; consequently they are in perspective (Art. 80), and KB, MA, QC will meet in one point N. But this being the case, we have a complete quadrangle $KLMN$, of which one pair of opposite sides meet in A, another such pair in B, while the fifth and sixth sides pass respectively through C and D. Accordingly (Art. 46) $ABCD$ is a harmonic range.

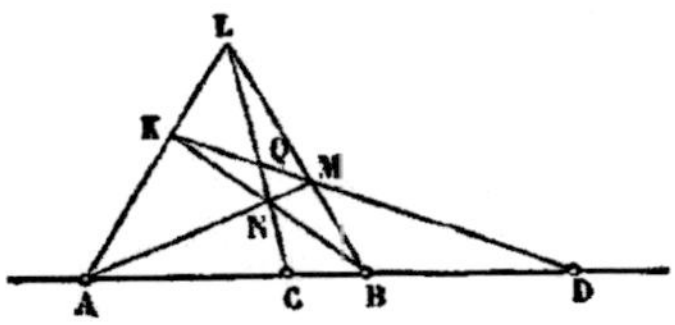

Fig. 52.

84. Let there be given two projectively related geometric forms of one dimension. Any series of operations which suffices to derive three elements of the one from the three corresponding elements of the other will enable us to pass from the one form to the other (Art. 78); and any two given triads of elements are always projective, *i.e.* can be derived one from the other by means of a certain number of projections and sections. Hence we conclude that:

Given three pairs of corresponding elements of two projective forms of one dimension, any number of other pairs of corresponding elements can be constructed.

We proceed to illustrate this by two examples, taking (1) two ranges and (2) two flat pencils; the forms being in each case supposed to lie in one plane.

Given (Fig. 53) *three pairs of corresponding points A and A', B and B', C and C', of the projective ranges u and u'; to construct these ranges.*

We proceed as in Art. 44. On the straight line which joins any two of the corresponding points, say A and A', take two arbitrary points S and S'. Join SB, $S'B'$ cutting one another in B'', and SC, $S'C'$ cutting one another in

Given (Fig. 54) *three pairs of corresponding rays a and a', b and b', c and c', of the projective pencils U and U'; to construct these pencils.*

Through the point of intersection of any two of the corresponding rays, say a and a', draw two arbitrary transversals s and s'. Join the points sb and $s'b'$ by the straight line b'', and the points sc and $s'c'$ by the

C''; join $B''C''$, and let it cut AA' in A''. The operations which enable us to pass from ABC to $A'B'C'$ are: 1. a projection from S; 2. a section by u'' (the line on which lie the points A'', B'', C''); 3. a projection from S'; 4. a section by u'. The same operations lead from any other given point D on u to the corresponding point D' on u', so that the rays SD and $S'D'$ must intersect in a point D'' of the fixed straight line u''.

In this manner a range

$$u'' \equiv A''B''C''D''\ldots$$

is obtained which is in perspective both with u and with u'.

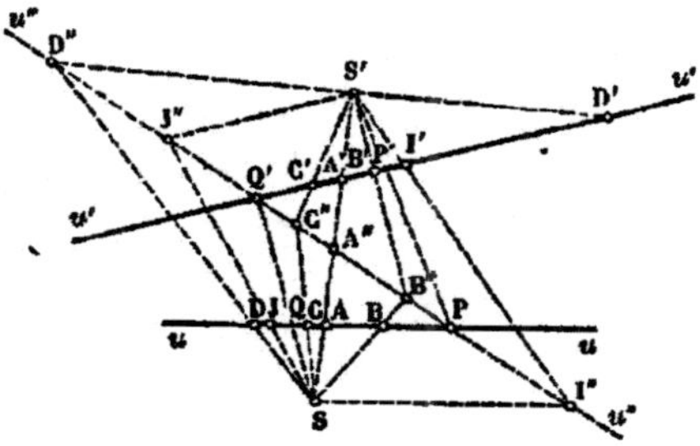

Fig. 53.

straight line c''; and let a'' be the straight line joining the points $b''c''$ and aa'. The operations which enable us to pass from abc to $a'b'c'$ are: 1. a section by s; 2. a projection from the point U'' where a'', b'', c'' meet; 3. a section by s'; 4. a projection from U'. The same operations lead from any other given ray d of the pencil U to the corresponding ray d' of the pencil U'; so that the points sd and $s'd'$ must lie on a straight line d'' which passes through the fixed point U''.

In this manner a pencil

$$U'' \equiv a''b''c''d''\ldots$$

is obtained which is in perspective both with U and with U'.

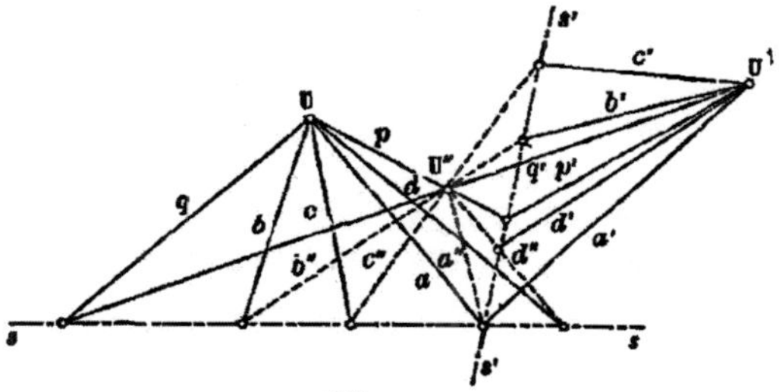

Fig. 54.

In the preceding construction (left), D is any arbitrary point on u. If D be taken to be the point at infinity on u, then (Fig. 53) SD will be parallel to u; in order therefore to find the point on u'

which corresponds to the point at infinity on u, draw SI'' parallel to u to cut u'' in I''; then join $S'I''$, which will cut u' in the required point I'. Similarly, if the ray through S' parallel to u' cuts u'' in J'', and SJ'' be joined, this will cut u in J, the point on u which corresponds to the point at infinity on u'.

If D be taken at P, the point where u and u'' meet, then D'' also coincides with P, and the point P' on u' corresponding to the point P on u is found as the intersection of $S'P$ with u'.

Similarly, if Q' be the point of intersection of u' and u'', the point on u corresponding to Q' on u' is Q, where SQ' cuts u.

85. The only condition to which the centres S and S' are subject is that they are to lie upon the straight line which joins a pair of corresponding points; in other respects their position is arbitrary. We may then for instance take S at A' and S' at A (Fig. 55). Then the ray $S'P$ coincides with u, and P' is accordingly the point of intersection of u and u'. So too the ray SQ' coincides with u, and Q also lies at the point uu'.

If then we take the points A' and A as the centres S and S respectively, the straight line u'' will cut the bases u and u' respectively in P and Q', the points which correspond to the point uu' regarded in the first instance as the point P' of the line u' and in the second instance as the point Q of the line u.

Now in the construction of the preceding Art., the straight line u'' was found at the locus of

In the preceding construction (right), d is any arbitrary ray passing through U. If it be taken to be p, the line joining U to U'', then the corresponding ray p' of the pencil U' is the line joining the point U' to the point $s'p$.

Similarly, if q' be the ray $U'U''$ of the pencil U', the ray q corresponding to it in the pencil U is that which joins the points U and sq'.

The only condition to which the transversals s and s' are subject is that they are to pass through the point of intersection of a pair of corresponding rays; in other respects their position is arbitrary. We may then for instance take a' for s and a for s' (Fig. 56). Then the point $s'p$ coincides with U, and p' is accordingly the straight line UU'. So too the point sq' coincides with U', and q also must be the straight line UU'.

If then we take the rays a' and a as the transversals s and s' respectively, the point U'' will be the intersection of the rays p and q' which correspond to the straight line UU', regarded in the first instance as the ray p' of the pencil U', and in the second instance as the ray q of the pencil U.

Now in the construction of the preceding Art., the point U'' was found as the centre of perspective the points of intersection of pairs of corresponding rays of the pencils in perspective $S(ABCD..)$ and $S'(A'B'C'D'..)$.

The straight line u'' obtained by the construction of the present Art. is in like manner the locus of the points of intersection of pairs of corresponding rays of the pencils $A'(ABCD..)$ and $A(A'B'C'D'..)$, *i.e.* the locus of the points in which the pairs of lines $A'B$ and AB', $A'C$ and AC', $A'D$ and AD', ... intersect.

of the ranges in perspective $s(abcd\ldots)$ and $s'(a'b'c'd'\ldots)$.

The point U'' obtained by the construction of the present Art. is in like manner the centre of perspective of the ranges $a'(abcd\ldots)$ and $a(a'b'c'd'\ldots)$, *i.e.* the point in which the lines joining the pairs of corresponding points $a'b$ and ab', $a'c$ and ac', $a'd$ and ad', ... meet.

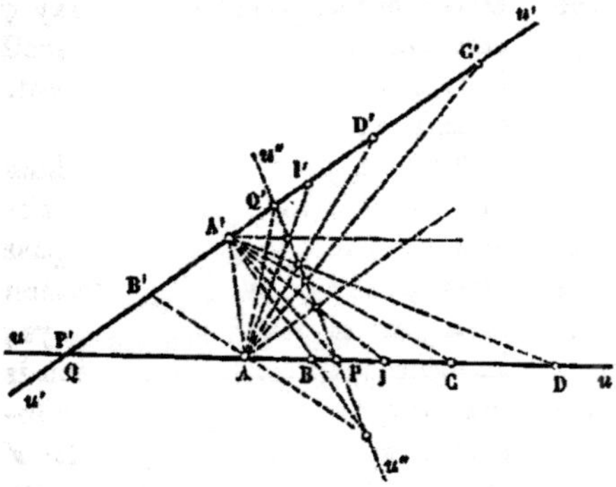

Fig. 55.

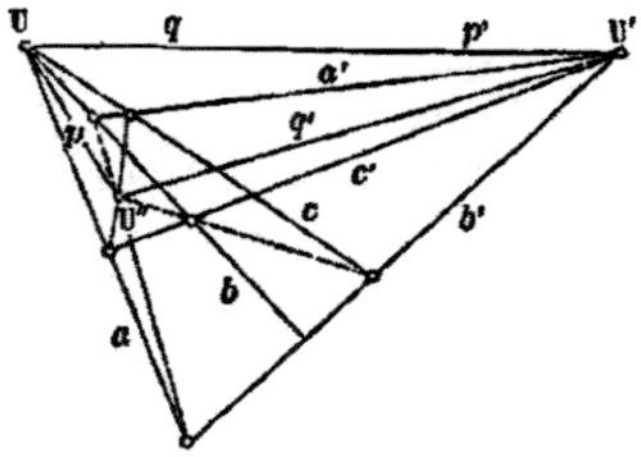

Fig. 56.

If in place of A' and A any other pair of points B' and B, or C' and C, ... be taken as centres of the auxiliary pencils S and S', the straight line u'' must still cut the two bases u and u' in the points P and Q'; *i.e.* the straight line u'' remains the same.

If in place of a' and a any other pair of rays b' and b, or c' and c, ... be taken as transversals, the point U'' must still be the intersection of p and q'; *i.e.* the point U'' remains the same.

If then $ABC \ldots MN \ldots$ and $A'B'C' \ldots M'N' \ldots$ are two projective ranges (in the same plane), every pair of straight lines such as MN' and $M'N$ intersect in points lying on a fixed straight line. This straight line passes through those points which cor-

If then $abc \ldots mn \ldots$ and $a'b'c' \ldots m'n' \ldots$ are two projective pencils (in the same plane) every straight line which joins a pair of points such as mn' and $m'n$ passes through a fixed point. This point is the intersection of those rays which correspond in

respond in each range to the point of intersection of their bases when regarded as a point of the other range.

86. If the two ranges u and u' are in perspective (Fig. 57) the points P and Q' will coincide with the point O in which the bases u and u' meet; and since the straight line which is the locus of the points $(AB', A'B)$, $(AC', A'C)$, $(AD', A'D)$, ... and the straight line which is the locus of the points $(BA', B'A)$, $(BC', B'C)$, $(BD', B'D)$, ... have two points in common, viz. O and $(AB', A'B)$, these straight lines must coincide altogether. This being so, $AA'BB'$ is a complete quadrangle, whose diagonal points are O, S (the point where AA', BB', ... meet), and M (the point of intersection of AB' and $A'B$); consequently (Art. 57) the straight lines u and u' are harmonic conjugates with regard to the straight lines u'' and OS. If therefore two transversals u and u' cut a flat pencil $(a, b, c, \ldots)$ in the points (A, A'), (B, B'), (C, C') ..., then the points of intersection of the pairs of straight lines AB' and $A'B$, AC' and $A'C$, BC' and $B'C$, ... lie on one and the same straight line u'', which passes through the point uu'; and the straight line joining uu' to the centre of the pencil is the harmonic conjugate of u'' with respect to u and u'.

From this follows the solution of the problem:

To draw the straight line connecting a given point M with the

each pencil to the straight line joining the centres of the pencils when regarded as a ray of the other pencil.

If the two pencils U and U' are in perspective (Fig. 59) the rays p and q' will coincide with the straight line UU'; and since through the point of intersection of the rays $(ab', a'b)$, $(ac', a'c)$, $(ad', a'd)$, ... and through the point of intersection of the rays $(ba', b'a)$, $(bc', b'c)$, $(bd', b'd)$, ... pass two different straight lines, viz. UU' and $(ab', a'b)$, these points must coincide. This being so, $aa'bb'$ is a complete quadrilateral, whose diagonals are UU', s (the straight line on which aa', bb', ... intersect), and m (the straight line which joins ab' and $a'b$); consequently (Art. 56) the points U and U' are harmonic conjugates with regard to U'' and the point in which s meets UU'. If therefore a range be projected from two points U and U' by the rays (a, a'), (b, b'), (c, c') ..., then the straight lines which join the pairs of points $(ab', a'b)$, $(ac', a'c)$, $(bc', b'c)$, ... meet in one and the same point U'', which lies on the line UU'; and the point where the straight line UU' cuts the base of the range is the harmonic conjugate of U'' with respect to U and U'.

From this follows the solution of the problem:

To construct the point where a given straight line m would be in-

inaccessible point of intersection of two given straight lines u and u′.

tersected by a straight line (UU′) which cannot be drawn, but which is determined by its passing through two given points U and U′.

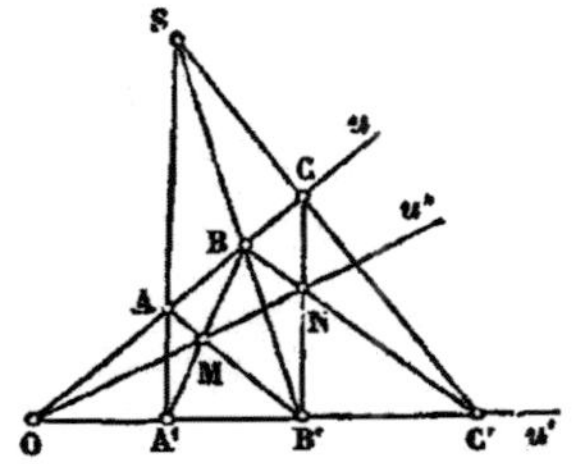

Fig. 57.

Through M (Figs. 57 and 58) draw two straight lines to cut u in A and B, and u' in B' and A' respectively; join AA', BB' meeting in S. Through S draw any straight line to cut u in C and u' in C', and join BC', $B'C$, intersecting in N. The straight line joining M and N will be the line u'' required.

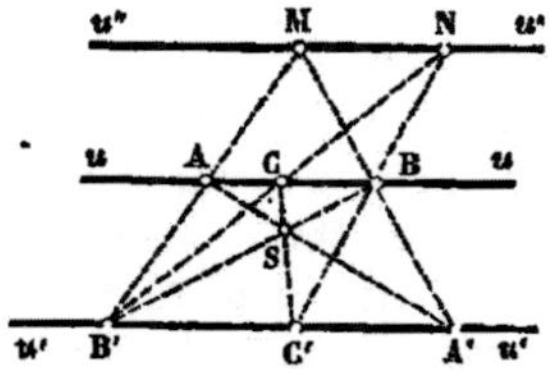

Fig. 58.

On m (Fig. 59) take two points, and join them to U by the straight lines a and b, and to U' by the straight lines b' and a'; let s be the straight line joining the points of intersection of a, a' and b, b'. On s take any other point and join it to U, U' by the straight lines c, c' respectively. The straight line n which joins the points bc' and $b'c$ will cut m in the point U'' required.

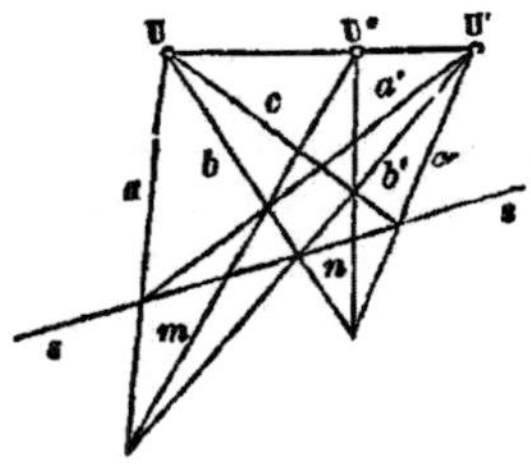

Fig. 59.

If the straight lines u and u' are parallel to one another (Fig. 58), the preceding construction gives the solution of the problem: *given two parallel straight lines, to draw through a given point a straight line parallel to them, making use of the ruler only.*

87. If in the theorem of the preceding article the flat pencil

If in the theorem of the preceding article the range consist

consist of only three rays, the theorem may be enunciated as follows, with reference to the three pairs of points AA', BB', CC':

If a hexagon (six-point) $AB'CA'BC'$ (Fig. 60) has its vertices of odd order (1st, 3rd, and 5th)

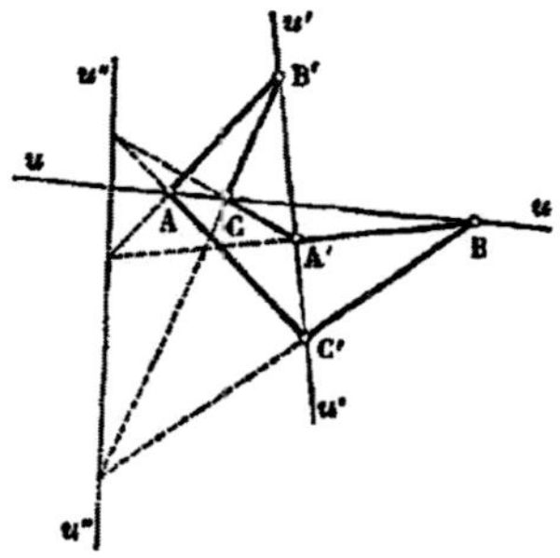

Fig. 60.

on one straight line u, and its vertices of even order (2nd, 4th, and 6th) on another straight line u', then the three pairs of opposite sides (AB' and $A'B$, $B'C$ and BC', CA' and $C'A$) meet in three points lying on one and the same straight line u'' *.

of only three points, the theorem may be enunciated as follows with reference to the three pairs of rays aa', bb', cc':

If a hexagon (six-side) $ab'ca'bc'$ (Fig. 61) be such that its sides of odd order (1st, 3rd, and 5th)

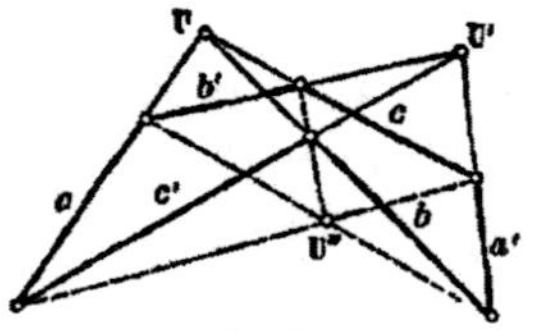

Fig. 61.

meet in one point U, and its sides of even order (2nd, 4th, and 6th) meet in another point U', then the three straight lines which join the pairs of opposite vertices (ab' and $a'b$, $b'c$ and bc', ca' and $c'a$) pass through one and the same point U''.

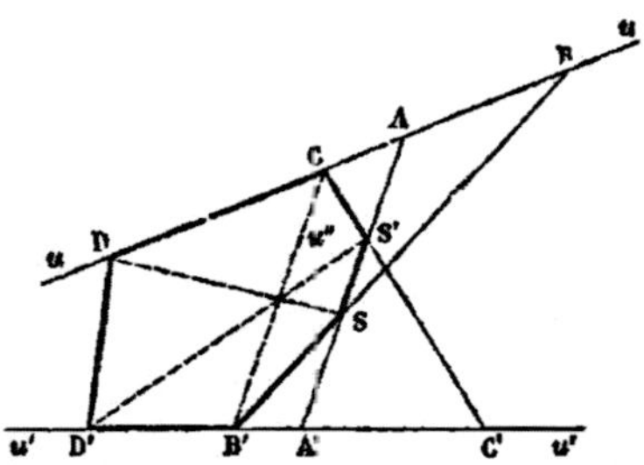

Fig. 62.

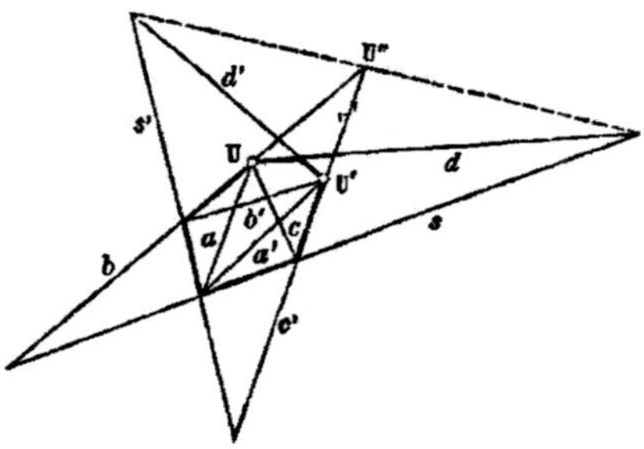

Fig. 63.

88. Returning to the construction of Art. 84 (left), let the

Returning to the construction of Art. 84 (right), let the straight

* PAPPUS, *loc. cit.*, Book vii. prop. 139.

centre S be taken at the point where AA' meets BB', and the centre S' at the point where AA' meets CC' (Fig. 62). Then since SB, $S'B'$ meet in B', and SC, $S'C'$ in C, therefore $B'C$ is the straight line u''. Consequently any other pair of corresponding points D and D' are constructed by observing that the straight lines SD, $S'D'$ must meet on $B'C$.

From a consideration of the figure $SS'CDD'B'$ which is a hexagon, we derive the theorem:

In a hexagon, of which two sides are segments of the bases of two projective ranges, and the four others are the straight lines connecting four pairs of corresponding points, the straight lines which join the three pairs of opposite vertices are concurrent.

89. If in the problem of Art. 84 (left) the three straight lines AA', BB', CC' passed through the same point S (if, for example, A and A' coincided), then the two ranges would be in perspective; we should therefore only have to draw rays through S in order to obtain any number of pairs of corresponding points (Fig. 19).

line joining the points aa', cc' be taken as the transversal s, and that joining the points aa', bb' as the transversal s' (Fig. 63). Then since the line joining the points sb, $s'b'$ is b, and the line joining the points sc, $s'c'$ is c', therefore bc' is the point U''. Consequently any other pair of corresponding rays d and d' are constructed by observing that the points sd, $s'd'$ must be collinear with bc'.

From a consideration of the figure $ss'cdd'b$, which is a hexagon (six-side) we derive the theorem:

In a hexagon, of which two vertices are the centres of two projective pencils, and the four others are the points of intersection of four pairs of corresponding rays, the three points in which the pairs of opposite sides meet one another are collinear.

If the three points aa', bb', cc' in Art. 84 (right) lay on the same straight line s (if, for example, a and a' coincided), then the two pencils would be in perspective; we should therefore only have to connect the two centres of the pencils with every point of s in order to obtain any number of pairs of corresponding rays (Fig. 20).

90. If the two ranges u and u' (Art. 84, left) are superposed one upon the other, *i.e.* if the six given points $AA'BB'CC'$ lie on the same straight line (Fig. 64), we first project u' from an arbitrary centre S' on an arbitrary straight line u_1, and then proceed to make the construction for the case of the ranges $u \equiv (ABC\ldots)$ and $u_1 \equiv (A_1B_1C_1\ldots)$, *i.e.* to construct with regard to the pairs of points (AA_1), (BB_1), (CC_1) in the way shown in Art. 84. A pair of corresponding points D and D_1 of the ranges u and u_1 having been found, the ray $S'D_1$ determines upon u' the point D' which corresponds to D.

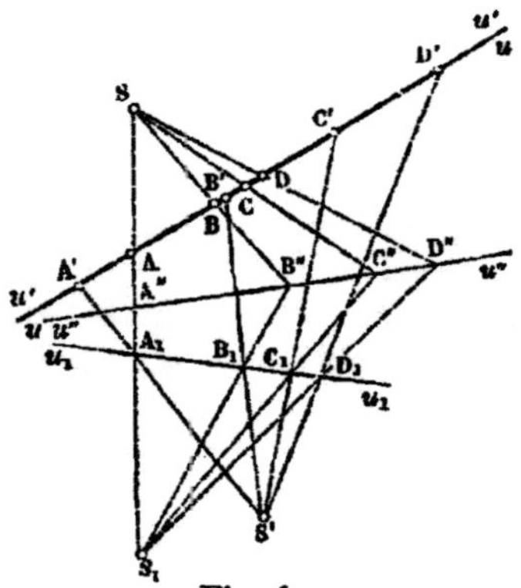

Fig. 64.

The construction is simpler in the case where two corresponding points A and A' coincide (Fig. 65). When this is so, if u_1 be drawn through A, the range u_1 will be in perspective with u; thus, after having projected u' upon u_1 from an arbitrary centre S', if S be the point where BB' and CC_1' meet, it is only necessary further to project u from S upon u_1, and then u_1 from S' upon u'.

The two collinear ranges u and u' have in general two self-corresponding points; one at AA', and a second at the point of intersection of their common base line with the straight line SS'.

If then SS' passes through the point uu_1, the two ranges u and u' have only *one* self-corresponding point. If it were desired to construct upon a given straight line two collinear ranges having A and A' for a pair of corresponding points, and a single self-corresponding point at M (Fig. 66), we should procced as follows. Take

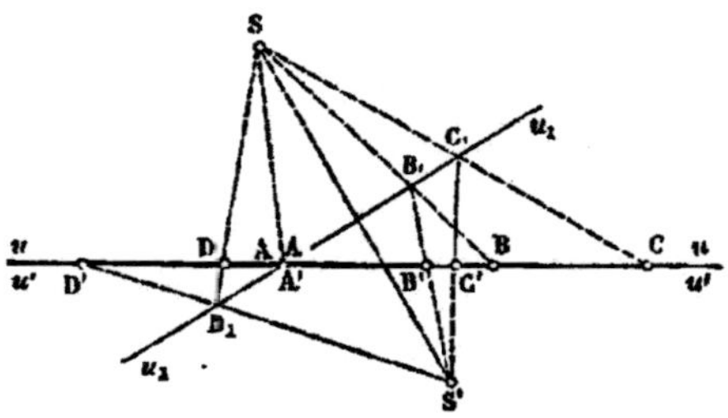

Fig. 65.

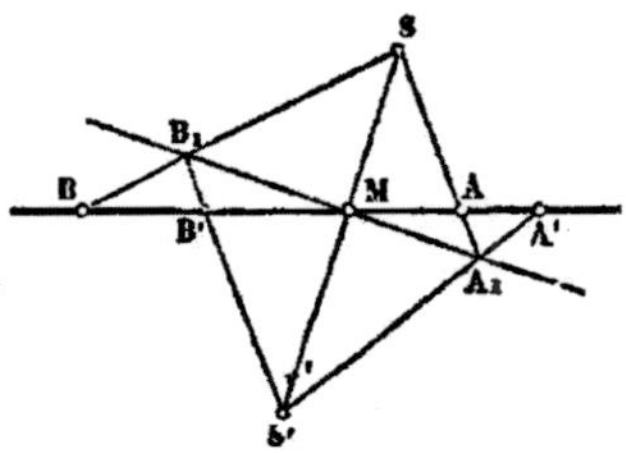

Fig. 66.

any point S', and draw any straight line u_1 through M; project A' from S' on u_1; join the point A_1 so found to A, and let AA_1 meet $S'M$ in S. Then to find the point on u' which corresponds to any point B on u, project B from S into B_1, and then B_1 from S' into B'; this last is the point required.

If the two pencils U, U' (Art. 84, right) are concentric, *i. e.* if the six rays $aa'bb'cc'$ pass all through one point, we first cut $a'b'c'$ by a transversal and then project the points of intersection from an arbitrary centre U_1. If $a_1b_1c_1$ are the projecting rays, we have then only to consider the non-concentric pencils U and $U_1 \equiv (a_1 b_1 c_1)$. Or we may cut abc by a transversal in the points ABC, and $a'b'c'$ by another transversal in $A'B'C'$, and then proceed with the two ranges ABC ..., $A'B'C'$ ... in the manner explained above.

The figures corresponding to these constructions are not given; the student is left to draw them for himself. He will see that in these cases also the constructions admit of considerable simplification if, among the given rays, there be one which is self-corresponding; if, for example, a and a' coalesce and form a single ray, &c.

91. Consider two projective (homographic) plane figures π and π'; as has already been seen (Art. 40), any two corresponding straight lines are the bases of two projective ranges, and any two corresponding points are the centres of two projective pencils.

If the two figures have three self-corresponding points lying in a straight line, this straight line s will correspond to itself; for it will contain two projective ranges which have three self-corresponding points, and every point of the straight line s will therefore (Art. 82) be a self-corresponding point. Consequently every pair of corresponding straight lines of π and π' will meet in some point on s, and therefore the two figures are in perspective (or in homology in the case where they are coplanar).

92. If two projective plane figures which are coplanar have three self-corresponding rays all meeting in a point O, this point will be the centre of two corresponding (and therefore projective) pencils which have three self-corresponding rays; therefore (Art. 82) every ray through O will be a self-corresponding one. Hence it follows that every pair of corresponding points will be collinear with O; therefore the two figures are in homology.

93. *If two projective plane figures which are coplanar have four self-corresponding points A, B, C, D, no three of which are collinear, then will every point coincide with its correspondent.*

For the straight lines AB, AC, AD, BC, BD, CD are all self-corresponding; therefore the points of intersection of AB and CD, AC and BD, BC and AD, *i.e.* the diagonal points of the quadrangle $ABCD$, are all self-corresponding. Since the three points A, B, and (AB) (CD) are self-corresponding, every point on the straight line AB coincides with its correspondent; and the same may be proved true for the other five sides of the quadrangle. If now a straight line be drawn arbitrarily in the plane, there will be six points on it which are self-corresponding, those namely in which it is cut by the six sides of the quadrangle; and therefore every point on the straight line is a self-corresponding one; which proves the proposition.

In a similar manner it may be shown that *if two coplanar projective figures have four self-corresponding straight lines a, b, c, d,* *forming a complete quadrilateral (i.e. such that no three of them are concurrent), then every straight line will coincide with its correspondent.*

94. Theorem. *Two plane quadrangles $ABCD$, $A'B'C'D'$ are always projective.*

(1). Suppose the two quadrangles to lie in different planes π, π'. Join AA', and on it take an arbitrary point S (different from A'), and through A draw an arbitrary plane π'' (distinct from π); then from S as centre project A', B', C', D' upon π'' and let A'', B'', C'', D'' be their respective projections (A'' therefore coinciding with A).

In the plane π join AB, CD, and let them meet in E; so too in the plane π'' join $A''B''$, $C''D''$, and let these meet in E''. The straight lines ABE, $A''B''E''$ lie in one plane since they meet each other in the point $A \equiv A''$; therefore BB'' and EE'' will meet one another in some point S_1.

Now let a new plane π''' (distinct from π) be drawn through the straight line ABE, and let the points A'', B'', C'', D'', E'' be projected from S_1 as centre upon π'''. Let A''', B''', C''', D''', E''' be their respective projections, where A''', B''', E''' are collinear and coincide with A, B, E respectively, and C''', D''', E''' are collinear also, since their correspondents C'', D'', E'' are collinear. The straight lines CDE, $C'''D'''E'''$ lie in one plane since they meet each other in the point $E \equiv E'''$; therefore CC''' and DD''' will meet one another in some point S_2. If now the points A''', B''', C''', D''' be projected from S_2 as centre upon the plane π, their projections will evidently be A, B, C, D.

The quadrangle $ABCD$ may therefore be derived from the quadrangle $A'B'C'D'$ by first projecting the latter from S as centre upon the plane π'', then projecting the new quadrangle so formed in the plane π'' from S_1 upon π''', and lastly projecting the quadrangle so formed in the plane π''' from S_2 upon π; that is to say, by means of three projections and three sections *.

(2). The case of two quadrangles lying in the same plane reduces to the preceding one, if we begin by projecting one of the quadrangles upon another plane.

(3). If the two quadrangles (lying in different planes) have a pair of their vertices coincident, say D and D', then two projections will suffice to enable us to pass from the one to the other; or, what amounts to the same thing, a third quadrangle can be constructed which is in perspective with each of the given ones $ABCD$, $A'B'C'D'$.

For let there be drawn through D two straight lines s and s', one in each of the planes; let s cut the sides of the triangle ABC in

* Grassmann, *Die stereometrischen Gleichungen dritten Grades und die dadurch erzeugten Oberflächen*; Crelle's Journal, vol. 49. § 4 (Berlin, 1855).

L, M, N respectively, and let s' cut the sides of the triangle $A'B'C'$ in L', M', N' respectively. Then in the plane ss' the straight lines LL', MM', NN' will form a triangle which is in perspective at once with ABC and with $A'B'C'$.

(4). If the quadrangles (still supposed to lie in different planes) have two pairs of their vertices $C \equiv C'$, $D \equiv D'$ coincident, then if the straight lines AA', BB' meet one another the quadrangles will be directly in perspective, the point of intersection O of AA' and BB' being the centre of projection; so that we can pass at once from the one quadrangle to the other by one projection from O. If AA', BB' are not in the same plane, so that they do not meet one another, then through CD let an arbitrary plane π'' be drawn, and in it let the straight line be drawn which meets AB and $A'B'$. If in this straight line two arbitrary points A'', B'' be taken, then $A''B''C''D''$ will be a quadrangle which is in perspective at once with $ABCD$ and with $A'B'C'D'$.

95. From the theorem just proved it follows that two projective plane figures π and π' can be constructed when we are given two corresponding quadrangles $ABCD$, $A'B'C'D'$; for the operations (projections and sections) which serve to derive $A'B'C'D'$ from $ABCD$ will lead from any point or straight line whatever of π to the corresponding point or straight line of π'; and *vice versa*.

Or, again, it may be supposed that two corresponding quadrilaterals are given. For if in these two corresponding pairs of opposite vertices be taken, we have thus two corresponding quadrangles; and the operations (projections and sections) which enable us to derive one of these quadrangles from the other will also derive the one quadrilateral from the other.

96. Two plane figures may also be made projective in another manner; leaving out of consideration the relative position of the planes in which they lie, we may operate on each of the figures separately*. Suppose that we are given, as corresponding to one another, two complete quadrilaterals $abcd$, $a'b'c'd'$. We begin by constructing, on each pair of corresponding sides, such as a and a', the projective ranges which are determined by the three pairs of corresponding points ab and $a'b'$, ac and $a'c'$, ad and $a'd'$. This done, to every point of any of the four straight lines a, b, c, d will correspond a determinate point of the corresponding line in the other figure.

(1). Now let in the first figure a transversal m be drawn to cut a, b, c, d in A, B, C, D respectively; then the points A', B', C', D' which correspond to these in the second figure will in like manner lie on a straight line m'.

* Staudt, *Geom. der Lage*, Art. 130.

For, considering the triangle abc, cut by the transversals d and m, the product of the three anharmonic ratios

$$a(bcdm)\,,\; b(cadm)\,,\; c(abdm)$$

is equal to $+1$ (Art. 140); but these anharmonic ratios are equal respectively to the following:

$$a'(b'c'd').A'\,,\; b'(c'a'd').B'\,,\; c'(a'b'd').C',$$

so that the product of these last three is also equal to $+1$. And therefore, since the points $a'd'$, $b'd'$, $c'd'$ are collinear, the points A', B', C' are also collinear (Art. 140).

By considering in the same manner the triangle abd, cut by the transversals c and m, it can be shown that A', B', D' are collinear; it follows then that the four points A', B', C', D' all lie on the same straight line m', the correspondent of m.

This proof holds good also when m passes through one of the vertices of the quadrilateral $abcd$; if for example m pass through cd, the anharmonic ratios $c(abdm)$, $d(abcm)$ will each be equal to $+1$; the reasoning, however, remains unaltered.

Thus every pair of corresponding vertices of the quadrilaterals $abcd$, $a'b'c'd'$ (for example cd and $c'd'$) become the centres of two projective pencils, in which to c, d, $(cd)(ab)$ correspond c', d', $(c'd')(a'b')$ respectively, and to any ray cutting a, b in two points P, Q corresponds a ray cutting a', b' in the two corresponding points P', Q'.

(2). The two ranges $ABCD$, $A'B'C'D'$ in which the sides of the quadrilaterals $abcd$, $a'b'c'd'$ are respectively cut by two corresponding straight lines m, m' are projective.

For, considering the triangle bcm, cut by the transversals a and d, the product of the anharmonic ratios of the three ranges

$$\begin{matrix} bc\,, & B\,, & ba\,, & bd \\ C\,, & cb\,, & ca\,, & cd \\ B\,, & C\,, & A\,, & D \end{matrix}$$

is equal to $+1$. And considering in like manner in the other plane the triangle $b'c'm'$, cut by the transversals a' and d', the product of the anharmonic ratios of the three ranges

$$\begin{matrix} b'c', & B', & b'a', & b'd' \\ C', & c'b', & c'a', & c'd' \\ B', & C', & A', & D' \end{matrix}$$

is also equal to $+1$. But the range in which b is cut by the pencil $cmad$ is equianharmonic with the range in which b' is cut by the pencil $c'm'a'd'$; *i.e.* the ranges

$$\begin{matrix} bc\,, & B\,, & ba\,, & bd \\ b'c', & B', & b'a', & b'd' \end{matrix}$$

are equianharmonic; and for a similar reason the ranges

$$\begin{array}{l} C\ ,\ cb\ ,\ ca\ ,\ cd \\ C',\ c'b',\ c'a',\ c'd' \end{array}$$

are equianharmonic. Therefore the ranges

$$\begin{array}{l} B\ ,\ C\ ,\ A\ ,\ D \\ B',\ C',\ A',\ D' \end{array}$$

will be equianharmonic and therefore projective; whence it follows that the projective ranges m and m' are determined by means of the pairs of corresponding points lying on a and a', b and b', c and c'.

(3). If the straight line m turn round a fixed point M, then m' also will revolve round a fixed point.

For by hypothesis the points A and B, in which m cuts a and b, describe two ranges in perspective whose self-corresponding point is ab. Similarly the points A', B' describe two ranges, which, being respectively projective with the ranges on a, b, are projective with one another; and which are further seen to be in perspective, since they have a self-corresponding point $a'b'$. Consequently the straight line m' will always pass through a fixed point M', the correspondent of M; and will therefore trace out a pencil. The pencils generated by m and m' are projective, since the ranges are projective in which they are cut by a pair of corresponding sides of the quadrilaterals, *e.g.* by a and a'. To the rays of the pencil M which pass respectively through the vertices ab, ac, ad, bc, bd, cd of the quadrilateral $abcd$ correspond the rays of the pencil M' which pass respectively through the vertices $a'b'$, $a'c'$, $a'd'$, $b'c'$, $b'd'$, $c'd'$ of the quadrilateral $a'b'c'd'$.

This reasoning holds good also when the point M, round which m turns, lies upon one of the sides of the quadrilateral, on c for example; because we still obtain two ranges in perspective upon two of the other sides. Since c is now a ray of the pencil M, c' will be the corresponding ray of the pencil M'; that is to say, M' will lie on c'. If M be taken at one of the vertices, as cd, then M' will coincide with $c'd'$, &c.

(4). Now suppose the pencil M to be cut by a transversal n, and the pencil M' to be cut by the corresponding straight line n'. While the point mn describes the range n, the corresponding point $m'n'$ will describe the range n'; and these two ranges will be projective since they are sections of two projective pencils. When the point mn falls on one of the sides of the quadrilateral $abcd$, the point $m'n'$ will fall on the corresponding side of the quadrilateral $a'b'c'd'$; therefore the two projective ranges are the same as those which it has already been shown may be obtained by starting from the pairs of corresponding points on a and a', b and b', c and c'.

In this manner the two planes become related to one another in such a way that there corresponds uniquely to every point in the one a point in the other, to every straight line a straight line, to every range a projective range, to every pencil a projective pencil. The two figures thus obtained are the same as those which can be obtained, as explained above (Art. 95) by means of successive projections and sections, so arranged as to lead from the quadrilateral *abcd* to the quadrilateral $a'b'c'd'$. For the two figures π' derived from π by means of these two processes have four self-corresponding straight lines a', b', c', d' forming a quadrilateral, and therefore (Art. 93) every element (point or straight line) of the one must coincide with the corresponding element in the other; *i.e.* the two figures must be identical.

97. Theorem. *Any two projective plane figures (the straight lines at infinity in which are not corresponding lines) can be superposed one upon the other so as to become homological.*

Let i, j' be the vanishing lines of the two figures—*i.e.* the straight lines in each which correspond respectively to the straight line at infinity in the other. In the first place let one of the figures be superposed upon the other in such a manner that i and j' may be parallel to one another. Since to any point M on i corresponds a point at infinity in the second figure, to the pencil of straight lines in the first figure which meet in M corresponds in the second figure a pencil of parallel rays. Through M draw the straight line m parallel to these rays; then m will be parallel to its correspondent m'. Similarly let a second point N be taken on i and through N let the straight line n be drawn which is parallel to its correspondent n'; let m and n meet in S, and m' and n' in S'. If through S a straight line l be drawn parallel to i, its correspondent l' will pass through S' and will also be parallel to i, since the point at infinity on i corresponds to itself. The corresponding pencils S and S' are therefore such that three rays l, m, n of the one are severally parallel to the three corresponding rays l', m', n' of the other; and consequently (see below, Art. 104) the two pencils are equal. Now let one of the planes be made to slide upon the other, without rotation, until S' comes into coincidence with S; then the two pencils will become concentric; and since they are equal, every ray of the one will coincide with the ray corresponding to it in the other. This being the case, every pair of corresponding points will be collinear with S, and the two figures will be homological, S being the centre of homology.

98. Suppose that in a plane π is given a quadrangle $ABCD$, and in a second plane π' a quadrilateral $a'b'c'd'$. By means of constructions analogous to those explained in Arts. 94–96, the points and straight lines of the one plane can be put into unique correspondence with those of the other, so that to any range in the first plane corresponds in the second plane a pencil projective with the said range, and to any pencil in the first plane corresponds in the second plane a range projective with the said pencil. Two plane figures related to one another in this manner are called *correlative* or *reciprocal.*

CHAPTER XI.

PARTICULAR CASES AND EXERCISES.

99. Two ranges are said to be *similar*, when to the points $A, B, C, D, \ldots$ of the one correspond the points $A', B', C', D', \ldots$ of the other, in such a way that the ratio of any two corresponding segments AB and $A'B'$, AC and $A'C'$, ... is a *constant*.

If this constant is *unity*, the ranges are said to be *equal*.

Two similar ranges are projective, every anharmonic ratio such as $(ABCD)$ being equal to the corresponding ratio $(A'B'C'D')$. For suppose the bases of the two ranges to lie in the same plane (Fig. 67) and let their point of intersection be denoted by P' when considered as a point belonging to u' and by Q when considered as a point belonging to u. Let A, A' be any pair of corresponding points; P that point of u which corresponds to P', and Q' that point of u' which corresponds to Q. Draw AA'' parallel to u', and $A'A''$ parallel to u.

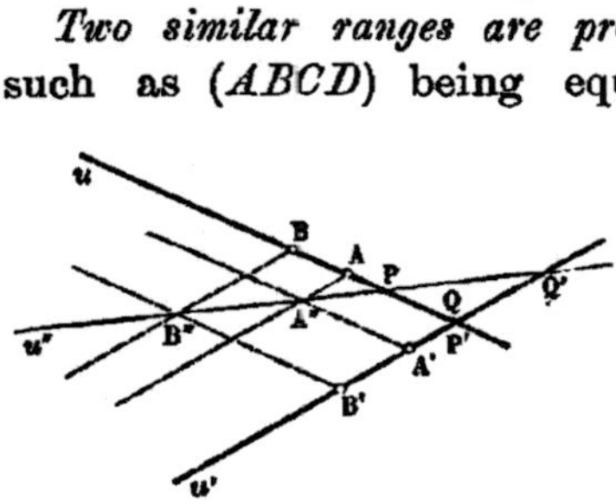

Fig. 67.

The triangles PQQ', PAA'' have the angles at Q and A equal and the sides about these equal angles proportionals, since by hypothesis

$$\frac{PQ}{P'Q'} = \frac{PA}{P'A'} = \frac{PA}{AA''}.$$

Therefore the triangles are similar, and the angles QPQ' and APA'' are equal; and consequently the points P, Q', A'' are collinear. If then the range $ABC\ldots$ be projected upon PQ', by straight lines drawn parallel to u', we shall obtain the range $A''\ B''\ C''\ldots$; and from this last, by projecting it upon

u' by straight lines drawn parallel to u, the range $A'B'C'\ldots$ may be derived.

If $PQ=P'Q'$, *i.e.* if the straight line PQ' makes equal angles with the bases of the given ranges, the ranges are equal.

To the point at infinity of u corresponds the point at infinity of u'.

100. Conversely, *if the points at infinity I and I' of two projective ranges u and u' correspond to each other, the ranges will be similar.* For if (Fig. 67) u be projected from I', and u' from I (as in Art. 85, left), two pencils of parallel rays will be formed, corresponding pairs of which intersect upon a fixed straight line u''. The segments $A''B''$ of u'' will be proportional to the segments AB of u and also to the segments $A'B'$ of u'; consequently the segments AB of u will be proportional to the segments $A'B'$ of u'.

Otherwise: if AA', BB', CC' are three pairs of corresponding points, and I, I' the points at infinity, we have (by Art. 73)

$$(ABCI) = (A'B'C'I');$$

or (by Art. 64), since I and I' are infinitely distant,

$$\frac{AC}{BC} = \frac{A'C'}{B'C'},$$

an equation which shows that corresponding segments are proportional to one another.

Examples. If a flat pencil whose centre lies at a finite distance be cut by two parallel straight lines, two similar ranges of points will be obtained.

Any two sections of a flat pencil composed of parallel rays are similar ranges.

In these two examples the ranges are not only projective, but also in perspective: in the first case the self-corresponding point lies at infinity; in the second case it lies (in general) at a finite distance.

101. Two flat pencils, whose centres lie at infinity, are projective and are called *similar*, when a section of the one is similar to a section of the other. When this is the case any other two sections of the pencils will also be similar to one another.

102. From the equality of the anharmonic ratios we conclude that two equal ranges are projective (Art. 79), and that conversely two projective ranges are equal (Art. 73), when the corresponding segments which are bounded by the points of two corresponding triads ABC and $A'B'C'$ are equal; *i.e.* when $A'B'=AB$, $A'C'=AC$, (and consequently $B'C'=BC$).

Examples. If a flat pencil consisting of parallel rays be cut by two transversals which are equally inclined to the direction of the rays, two directly equal ranges of points will be obtained *.

If a flat pencil of non-parallel rays be cut by two transversals which are parallel to one another, and equidistant from the centre of the pencil, two oppositely equal ranges will be obtained *.

103. Two similar ranges lying on the same base, and which have one self-corresponding point N at infinity, have also a second such point M, which is in general at a finite distance. If AA', BB' are two pairs of corresponding points,

$$MA : MA' = AB : A'B' = \text{a constant.}$$

To find M therefore it is only necessary to divide the segment AA' into two parts MA, MA' which bear to one another a given ratio.

This ratio $MA : MA'$ is equal (Art. 64) to the anharmonic ratio $(AA'MN)$. If its value is -1, the points $AA'MN$ are harmonic (Art. 68), *i.e.* M is the middle point of AA', and similarly also that of every other corresponding segment BB', ...; in other words, the two ranges, which in this case are oppositely equal, are composed of pairs of points which lie on opposite sides of a fixed point M, and at equal distances from it.

But if the constant ratio is equal to $+1$, *i.e.* if MA and MA' are equal in sign and magnitude, the point M will lie at infinity. For since $(AA'MN)=1$, $\therefore$ $(NMA'A)=1$ (Art. 45); consequently the points M and N coincide.

It follows also from the construction of Art. 90 (Fig. 66) that *two ranges on the same base, which have a single self-corresponding point lying at infinity, are directly equal.*

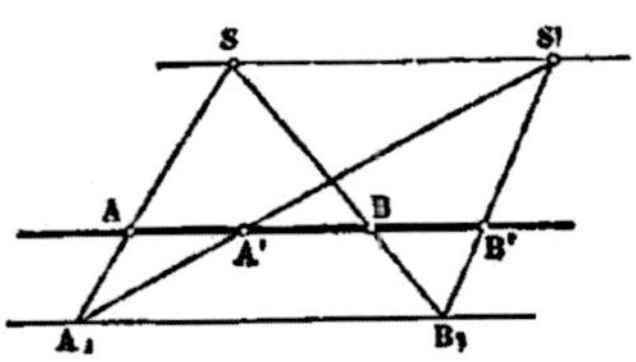

Fig. 68.

For if in Fig. 66 the point M move off to infinity, the straight lines SS' and A_1B_1 become parallel to the given straight line u or u' on which the ranges lie (Fig. 68), and as the triangles SA_1B_1 and $S'A_1B_1$ lie upon the same base and between the same parallels, the segments

* Imagine a moving point P to trace out a range ABC... and its correspondent P' to trace out simultaneously the equal range $A'B'C'$.... Then if P and P' move in the same direction, the two ranges are said to be *directly* equal; if P and P' move in opposite directions, the ranges are said to be *oppositely* equal.

which they intercept upon any parallel to the base are equal; thus $AB = A'B'$, or *two corresponding segments are equal*; consequently $AA' = BB'$, *i.e. the segment bounded by a pair of corresponding points is of constant length.* We may therefore suppose the two ranges to have been generated by a segment given in sign and magnitude, which moves along a given straight line; the one extremity A of the segment describes the one range, and the other extremity A' describes the other range.

Conversely it is evident that if a segment AA', given in sign and magnitude, slide along a given straight line, its extremities A and A' will describe two directly equal (and consequently projective) ranges, which have a single self-corresponding point, lying at an infinite distance.

104. Two flat pencils are said to be *equal* when to the elements of the one correspond the elements of the other in such a way that the angle included between any two rays of the first pencil is equal in sign and magnitude to the angle included between the two corresponding rays of the second.

It is evident that two such pencils can always be cut by two transversals in such a way that the resulting ranges are equal; but two equal ranges are always projective; therefore also *two equal flat pencils are always projective.*

Conversely, *two projective flat pencils $abcd$ ... and $a'b'c'd'$ ... will be equal if three rays abc of the one make with each other angles which are equal respectively to those which the three corresponding rays make with each other.*

This theorem may be proved by cutting the two pencils by two transversals in such a way that the sections ABC and $A'B'C'$ of the groups of rays abc and $a'b'c'$ may be equal. The projective ranges so formed will be equal (Art. 102); consequently also the other corresponding angles ad and $a'd'$,... of the given pencils must be equal to one another.

105. Since two equal forms (ranges or flat pencils) are always projective with one another, it follows that if a range or a flat pencil be placed in a different position in space, without altering the relative position of its elements, the form in its new position will be projective with regard to the same form in its original position.

106. Consider two equal pencils $abcd$... and $a'b'c'd'$... in the same plane or in parallel planes; and suppose a ray of the one pencil to revolve about the centre and to describe the pencil; then the corresponding ray of the other pencil will describe that other pencil, by revolving about its centre. This revolution may take place in the *same* direction as that of the first ray, or it may be in the *opposite* direction; in the first case the pencils are said to be *directly equal*, and in the second case to be *oppositely equal* to one another.

In the first case the angles aa', bb', cc', ... are evidently all equal, in sign as well as in magnitude; consequently a pair of corresponding rays are either always parallel or never parallel.

In the second case two corresponding angles are equal in magnitude, but of opposite signs. If then one of the pencils be shifted parallel to itself until its centre coincides with that of the other pencil, the two pencils, now concentric, will still be projective (Art. 105) and will evidently have a pair of corresponding rays united in each of the bisectors (internal and external) of the angle included between two corresponding rays a and a'. It follows that these rays are also the bisectors of the angle included between any other pair of corresponding rays. If the first pencil be now replaced in its original position, so that the two pencils are no longer concentric, we see that *there are in each pencil two rays, each of which is parallel to its correspondent in the other pencil; and these two rays are at right angles to each other,* since they are parallel to the bisectors of the angle between any pair of corresponding rays.

107. *If two flat pencils $abcd$... and $a'b'c'd'$... are projective, and if the angles aa', bb', cc' included by three pairs of corresponding rays are equal in magnitude and of the same sign, then the angle dd' included by any other pair of corresponding rays will have the same sign and magnitude.*

For if we shift the first pencil parallel to itself until it becomes concentric with the second, and then turn it about the common centre through the angle aa', the rays a, b, c will coincide with the rays a', b', c' respectively. The two pencils, which are still projective (Art. 105), have then three self-corresponding rays; consequently (Art. 82) every other ray will coincide with its correspondent. If now the first pencil be moved back into its original position, the angle dd' will be equal to aa'.

108. As the angles aa', bb', cc', ... of two directly equal

pencils are equal to one another, such pencils, when concentric and lying in the same plane, may be generated by the rotation of a constant angle aa' round its vertex O, supposed fixed; the one arm a traces out the one pencil, while the other arm a' traces out the other pencil.

Conversely, if an angle of constant magnitude turn round its vertex, its arms will trace out two (directly) equal and therefore projective pencils. Evidently these pencils have no self-corresponding rays.

A transversal cutting these two pencils determines on itself two collinear ranges having no self-corresponding points.

What has been said in Arts. 104–108 with respect to two pencils in a plane might be repeated without any alteration for the case of two axial pencils in space.

109. (1). Let $ABC\ldots$, $A'B'C'\ldots$ be two projective ranges lying upon the same base, and let them, by means of the pencils $abc\ldots$, $a'b'c'\ldots$, be projected from different points U, U'. Let i, j' be those rays passing through U, U' respectively, which are parallel to the given base, and let i', j be the rays corresponding to them. The points I', J in which these last two rays cut the given base will then be those points which correspond to the point at infinity (I or J') of the base, according as that point is regarded as belonging to the range $ABC\ldots$ or to the range $A'B'C'\ldots$.

The fact that the two corresponding groups of points are projectively related gives an equation between the anharmonic ratios, from which we deduce (as in Art. 74)

$$JA\,.\,I'A' = JB\,.\,I'B' = \text{a constant}; \quad \ldots \quad (1)$$

i.e. the product $JA\,.\,I'A'$ is constant for every pair of points A, A'.

Let O be the middle point of the segment JI', and O' the point corresponding to O regarded as a point belonging to the first range.

Since the equation (1) holds for every pair of corresponding points, and therefore also for O and O', we have

$$JA\,.\,I'A' = JO\,.\,I'O', \quad \ldots \quad (2)$$

or
$$(OA-OJ)(OA'-OI')+OJ(OO'-OI') = 0;$$

or since
$$OI' = -OJ,$$

$$OA\,.\,OA'-OI'(OA-OA'+OO') = 0. \quad \ldots \quad (3)$$

Let us now enquire whether there are in this case any self-corresponding points. If such a point exist, let it be denoted by E; then replacing both A and A' in (3) by E, we have

$$OE^2 = OI'\,.\,OO'. \quad \ldots \quad (4)$$

We conclude that when $OI'\,.\,OO'$ is positive, *i.e.* when O does not lie between I' and O', there are two self-corresponding points E and F, lying at equal distances on opposite sides of O, and dividing the segment $I'O'$ harmonically (Art. 69).

When O lies between I' and O', there are no such points.

When O' coincides with O, there is only one such point, viz. the point O itself.

(2). Imagine each of the given ranges to be generated by a point moving always in one direction*. If the one range is described in the order ABC, the other range will be described in the order $A'B'C'$; this order may be the same as the first, or may be opposite to it.

If the order of ABC is opposite to that of $A'B'C'$, the same will be the case with regard to the order of IJA and that of $I'J'A'$, and again with regard to the finite segment JA and the infinite segment $J'A'$; *i. e.* the finite segments JA and $I'A'$ have the same sign. In consequence therefore of equation (2), JO and $I'O'$ have the same sign; so that O does not fall between I' and O' (Fig. 69 *a*); there are therefore two self-corresponding points. And these will lie outside the finite segment JI', since OE is a mean proportional between OI' and OO'.

If the order of ABC is the same as that of $A'B'C'$, we arrive in a similar manner at the conclusion that JA and $I'A'$, and again JO and $I'O'$, have opposite signs. In this case then, self-corresponding points exist if O does not lie between I' and O'; that is, if O' lies between O and I' (Fig. 69 *b*). And these will lie within the segment JI', since OE is a mean proportional between OI' and OO'.

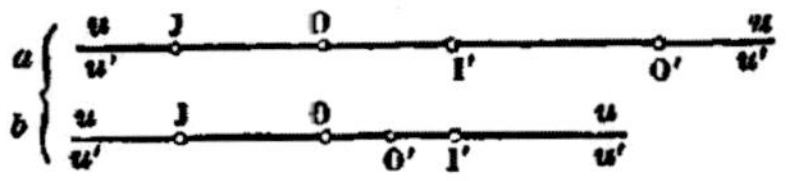

Fig. 69.

(3). Suppose that there are two self-corresponding points E and F (Fig. 70); draw through E any straight line, on which take two points S, S'; and project one of the ranges from S and the other from S'. The two pencils which result are in perspective, since they have a self-corresponding ray SES'; accordingly the corresponding rays SA and $S'A'$, SB and $S'B'$, ... SF and $S'F'$ will intersect in points lying on a straight line u'' which passes through F.

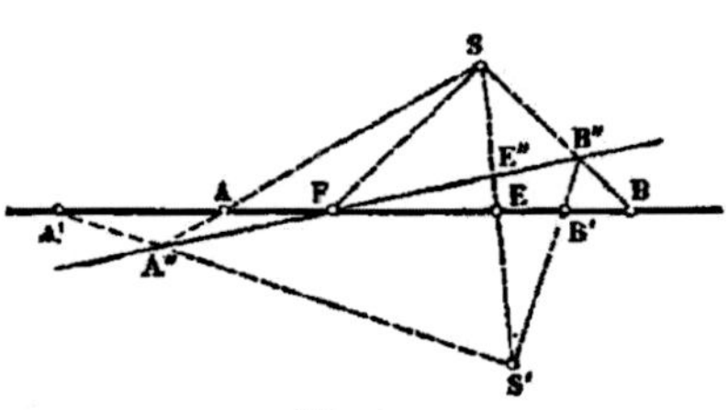

Fig. 70.

Let E'' be the point where this straight line u'' meets SS'. Then

* STEINER, *loc. cit.*, p. 61. § 16, II. Collected Works, vol. i. p. 280.

$EFAA'$ and $EFBB'$ are the projections of $EE''SS'$ from the centres A'' and B'' respectively; therefore $EFAA'$ and $EFBB'$ are projective with one another; thus the anharmonic ratio of the system consisting of any two corresponding points together with the two self-corresponding points is constant.

In other words: *two projective forms which are superposed one upon the other, and which have two self-corresponding elements, are composed of pairs of elements which give with two fixed ones a constant anharmonic ratio* *.

(4). Next suppose that there are no self-corresponding points; so that O lies between O' and I' (Fig. 71). Draw from O a straight line OU at right angles to the given base and make OU the geometric mean between $I'O$ and OO'; thus $I'UO'$ will be a right angle.

Again, draw through U the straight line IUJ' parallel to the given base; then the angle IUI' will be equal to JUJ', and the angle OUO' will be equal to $OI'U$ and therefore to IUI'. Thus in the two projective pencils which project the two given ranges from U, the angles IUI', JUJ', OUO' included by three pairs of corresponding rays are all equal; consequently (Art. 107) the angles AUA', BUB', ... are also all equal to them and to one another, and are all measured in the same direction †.

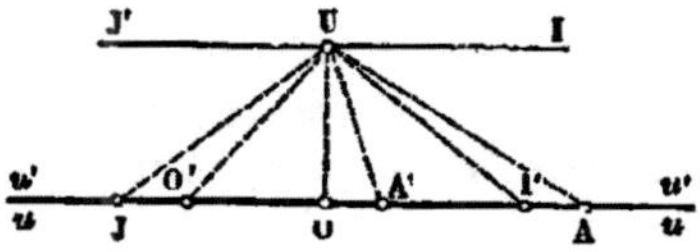

Fig. 71.

Thus: *two collinear ranges which have no self-corresponding points can always be regarded as generated by the intersection of their base line with the arms of an angle of constant magnitude which revolves, always in the same direction, about its vertex.*

110. We have seen (Art. 84) the general solution of the problem: Given three pairs of corresponding elements of two projective one-dimensional forms, to construct any desired number of pairs; or, in other words, to construct the element of the one form which corresponds to a given element of the other. The solution of the following particular cases is left as an exercise to the student:

1. Suppose the two forms to be two ranges u and u' which lie on different bases; and let the given pairs of elements be

(a) P and P', Q and Q' ‡, A and A';

* The above construction gives the solution of the problem: Given two pairs A, A and B, B' of corresponding points, and one of the self-corresponding points E, to find the other self-corresponding point.

† CHASLES, *loc. cit.*, p. 119.

‡ P, P', Q, Q', I, I', J, J' have the same meaning as in Art. 84; A, B, ... are any given points.

(b) P and P', A and A', B and B';
(c) I and I', J and J', P and P';
(d) I and I', J and J', A and A';
(e) I and I', P and P', Q and Q';
(f) I and I', P and P', A and A';
(g) I and I', A and A', B and B'.

2. Solve problems (d) and (g), supposing the ranges to be collinear.

3. Solve the problems correlative to (a) and (b) when the two given forms are two non-concentric pencils.

4. Suppose one of the pencils to have its centre at infinity.

5. Suppose both the pencils to have their centres at infinity.

111. He may also prove for himself the following proposition:

If the three vertices A, A', A'' of a variable triangle slide respectively on three fixed straight lines u, u', u'' which meet in a point, while two of its sides $A'A''$, $A''A$ turn respectively round two fixed points O and O', then will also the third side AA' always pass through a fixed point O'', collinear with O and O'.

It is only necessary to show that the points A, A', A'' in moving describe three ranges which are two and two in perspective. Or the theorem of Art. 16 may be applied to two positions of the variable triangle.

This proposition proved, the following corollary may be at once deduced:

If the four vertices A, A', A'', A''' of a variable quadrangle slide respectively upon four fixed straight lines which all pass through the same point O, while three of its sides AA', $A'A''$, $A''A'''$ turn respectively round three fixed points C', B''', B', then will the fourth side $A'''A$ and the diagonals AA'', $A'A'''$ pass respectively through three other fixed points C''', C'', B'', which are determined by the three former ones. The six fixed points are the vertices of a complete quadrilateral, i.e. they lie three by three on four straight lines (Fig. 72).

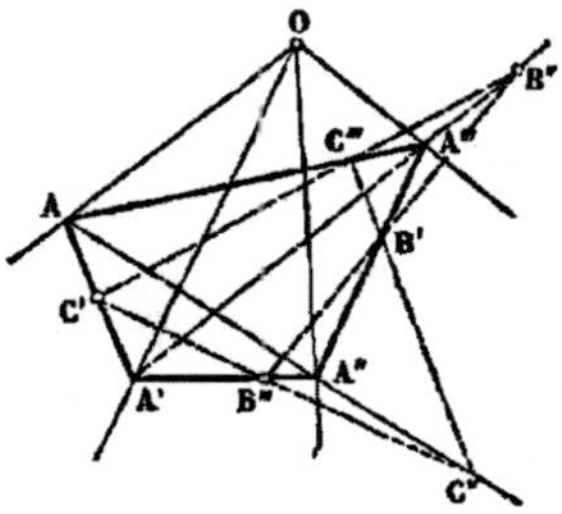

Fig. 72.

In a similar manner may be deduced the analogous corollary relating to a polygon of n vertices.

112. Theorem. *If a triangle $O_1O_2O_3$ circumscribes another triangle $U_1U_2U_3$, there exist an infinite number of triangles each of which is circumscribed about the former and inscribed in the latter* (Fig. 73).

The two pencils

$$O_2(U_1, U_2, U_3 \ldots) \text{ and } O_3(U_1, U_2, U_3 \ldots)$$

obtained by projecting the range $U_2U_3 \ldots$ from O_2 and from O_3, are evidently in perspective. Similarly the pencils

$$O_1(U_1, U_2, U_3 \ldots) \text{ and } O_3(U_1, U_2, U_3 \ldots)$$

obtained by projecting the range $U_1U_3 \ldots$ from O_1 and from O_3, are in perspective. Therefore the pencils

$$O_1(U_1, U_2, U_3 \ldots) \text{ and } O_2(U_1, U_2, U_3 \ldots)$$

are projective (Art. 41); but the rays O_1U_3 and O_2U_3 coincide; therefore (Art. 62) the pencils are in perspective, and their corresponding rays intersect in pairs on U_1U_2.

There are then three pencils O_1, O_2, O_3, which are two and two in perspective; corresponding rays of the first and second, second and third, third and first, intersecting in pairs on the straight lines U_1U_2, U_2U_3, U_3U_1 respectively. This shows that every triad of corresponding rays will form a triangle which is circumscribed about the triangle $O_1O_2O_3$, and inscribed in the triangle $U_1U_2U_3$*.

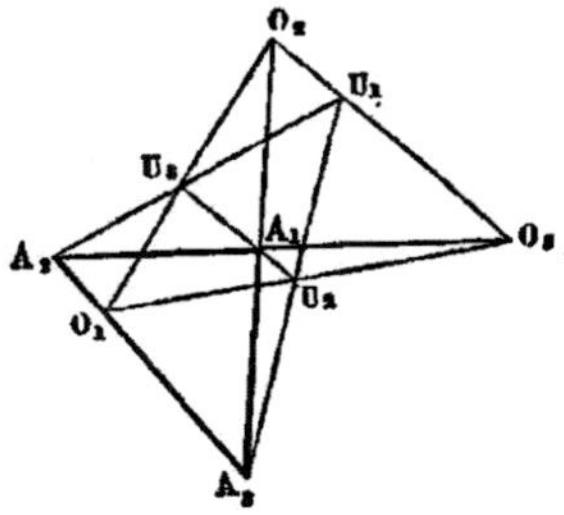

Fig. 73.

113. THEOREM. *A variable straight line turning about a fixed point U cuts two fixed straight lines u and u' in A and A' respectively; if S, S' are two fixed points collinear with uu', and SA, $S'A'$ be joined, the locus of their point of intersection M will be a straight line* †.

To prove this, we observe that the points A and A' trace out two ranges in perspective with one another, and that consequently the pencils generated by the moving rays SA, $S'A'$ are in perspective (Arts. 41, 80).

The demonstration of the correlative theorem is proposed as an exercise to the student.

114. THEOREM. *U, S, S' are three collinear points; a transversal turning about U cuts two fixed straight lines u and u' in A and A' respectively; if SA, $S'A'$ be joined, their point of intersection M will describe a straight line passing through the point uu'* ‡.

The proof is analogous to that of the preceding theorem.

The proposition just stated may also be enunciated as follows:

If the three sides of a variable triangle $AA'M$ turn respectively about three fixed collinear points U, S, S', while two of its vertices A, A'

* STEINER, *loc. cit.*, p. 85. § 23, II. Collected Works, vol. i. p. 297.

† PAPPUS, *loc. cit.*, book VII. props. 123, 139, 141, 143; CHASLES, *loc. cit.*, pp. 241, 242.

‡ CHASLES, *loc. cit.*, p. 242.

slide respectively upon two fixed straight lines u , u′, then will the third vertex M also describe a straight line *.

In a like manner may be demonstrated the more general theorem:

If a polygon of n sides displaces itself in such a manner that each of its sides passes through one of n fixed collinear points, while n − 1 of its vertices slide each on one of n − 1 fixed straight lines, then will also the remaining vertex, and the point of intersection of any two non-consecutive sides, describe straight lines †.

The correlative proposition is indicated in Art. 85.

115. PROBLEM. *Given a parallelogram ABCD and a point P in its plane, to draw through P a parallel to a given straight line EF also lying in the plane, making use of the ruler only.*

First Solution.—Let *E* and *F* (Fig. 74) be the points where the given straight line is cut by *AB* and *AD* respectively. On *AC* take any point *K*; join *EK*, meeting *CD* in *G*, and *FK*, meeting *BC* in *H*.

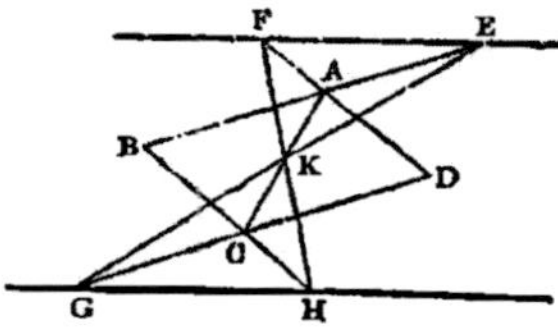

Fig. 74.

The triangles *AEF* , *CGH* are homological (Art. 18), since *AC* , *EG* , *FH* meet in the same point *K*; and the axis of homology is the straight line at infinity, since the sides *AE* , *AF* of the first triangle are parallel respectively to the corresponding sides *CG*, *CH* of the second. Therefore also the remaining sides *EF* and *GH* are parallel to one another ‡.

The problem is thus reduced to one already solved (Art. 86), viz. given two parallel straight lines *EF* and *GH*, to draw through a given point *P* a parallel to them.

Second Solution §.—Produce (Fig. 75) the sides *AB*, *BC*, *CD*, *DA* and a diagonal *AC* of the given parallelogram to meet the given straight line *EF* in *E*, *F*, *G*, *H*, *I* respectively, and join *EP*, *GP*. Through *I* draw any straight line cutting *EP* in *A′* and *GP* in *C′*, and join *HA′*, *FC′*; if these meet in *Q*, then will *PQ* be the required straight line.

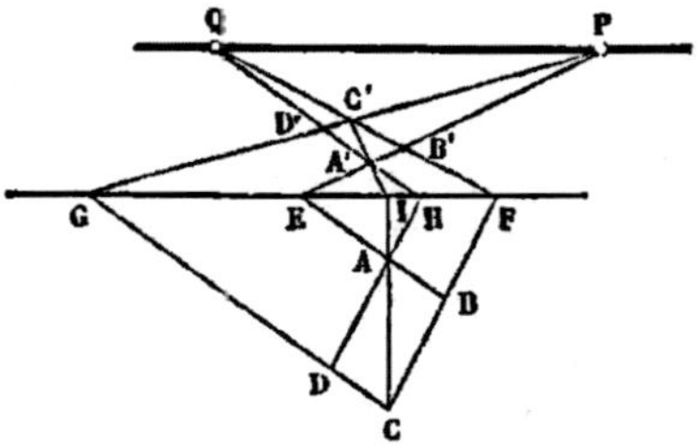

Fig. 75.

For if *B′* denote the point where *EP* cuts *FQ*, and *D′* the point

* This is one of Euclid's porisms. See PAPPUS, *loc. cit.*, preface to book VII.

† This is one of the porisms of PAPPUS; *loc. cit.*, preface to book VII.

‡ PONCELET, *Propriétés projectives*, Art. 198.

§ LAMBERT, *Freie Perspective* (Zürich, 1774), vol. ii. p. 169.

where GP cuts HQ, the parallelograms $ABCD$ and $A'B'C'D'$ are homological, EF being the axis of homology. The point P corresponds to the point of intersection of AB and CD, and the point Q to that of BC and AD; therefore PQ corresponds to the line at infinity in the first figure; accordingly it is the vanishing line of the second figure, and consequently PQ is parallel to EF (Art. 18).

116. Problem. *Given a circle and its centre; to draw a perpendicular to a given straight line, making use of the ruler only.*

Draw two diameters AC, BD of the circle (Fig. 76); the figure $ABCD$ is then a rectangle. Accordingly, if any point K be taken on the circumference, then by means of the last proposition (Art. 115) a parallel KL can be drawn to the given straight line EF. If the point L where this parallel again meets the circumference be joined to the other extremity M of the diameter through K, then evidently LM will be perpendicular to KL, and therefore also to the given straight line.

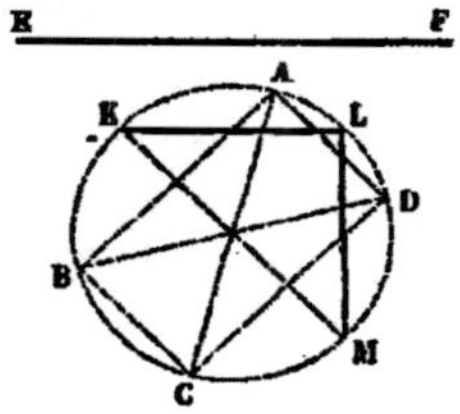

Fig. 76.

117. Problem. *Given a segment AC and its point of bisection B, to divide BC into n equal parts, making use of the ruler only.*

Construct a quadrilateral $ULDN$ (Fig. 77) of which one pair of opposite sides DL, NU meet in A, the other pair LU, DN in C, and of which one diagonal DU passes through B; the other diagonal LN will be parallel to AC (Art. 59), and will be bisected in M by DU.

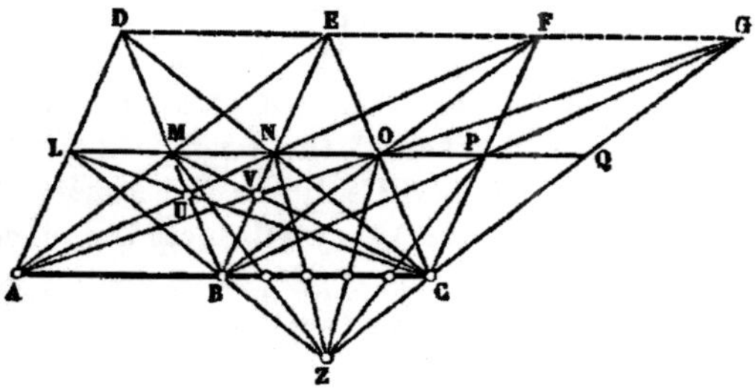

Fig. 77.

Now construct a second quadrilateral $VMEO$ which satisfies the same conditions as the first, and which moreover has M for an extremity and N for middle point of that diagonal which is parallel to AC. To do this it is only necessary to join AM and BN, meeting in E, and to join CE; this last will cut LN produced in a point O

such that $NO=MN=LM$. Now construct a third quadrilateral analogous to the first two, and which has N for an extremity and O for middle point of that diagonal which is parallel to AC. If P is the other extremity of this diagonal, then $OP=NO=MN=LM$. Proceed in a similar manner, until the number of the equal segments LM, MN, NO, OP, ... is equal to n.

If PQ is the segment last obtained, join LB, meeting QC in Z; the straight lines which join Z to the points M, N, O, P, ... will divide BC into n equal parts *.

118. The following problems, to be solved by aid of the ruler only, are left as exercises to the student:

Given two parallel straight lines AB and u; to bisect the segment AB (Art. 59).

Given a segment AB and its point of bisection C; to draw through a given point a parallel to AB (Art. 59).

Given a circle and its centre; to bisect a given angle (Art. 60).

Given two adjacent equal angles AOC, COB; to draw a straight line through O at right angles to OC (Art. 60).

119. THEOREM. *If two triangles ABC, $A'B'C'$, lying in different planes σ, σ', are in perspective, and if the plane of one of them be made to turn round $\sigma\sigma'$, then the point O in which the rays AA', BB', CC' meet will change its position, and will describe a circle lying in a plane perpendicular to the line $\sigma\sigma'$* †.

Let D, E, F (Fig. 78) be the points of the straight line $\sigma\sigma'$ in which the pairs of corresponding sides BC and $B'C'$, CA and $C'A'$, AB and $A'B'$ meet respectively (Art. 18). First consider the planes of the triangles to have any given definite position, and let O be the centre of projection for that position. Through O draw OG, OH, OK parallel respectively to the sides of the triangle $A'B'C'$; as these parallels lie in the same plane (parallel to σ') they will meet the plane σ in three points G, H, K of the line $\pi\sigma$.

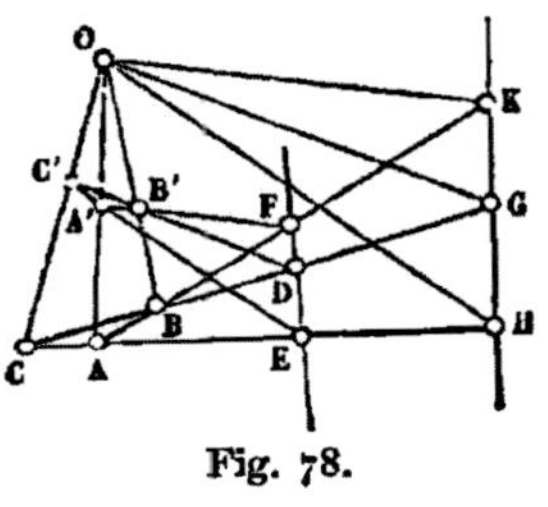

Fig. 78.

Now suppose the plane σ' together with the triangle $A'B'C'$ to turn round the line $\sigma\sigma'$. The range $BCDG$ is in perspective with the range $B'C'DG'$ (where G' denotes the point at infinity on $B'C'$); therefore the anharmonic ratio $(BCDG)$ is equal to the anharmonic ratio $(B'C'DG')$, *i.e.* to the simple ratio $B'D : C'D$ (Art. 64), which is

* These and other problems, to be solved by aid of the ruler only, will be found in the work of LAMBERT quoted above.

† CHASLES, *loc. cit.*, Arts. 368, 369. This proposition has already been proved by a different method in Art. 22.

constant. Since then B, C, D are fixed points, G must also be a fixed and invariable point (Art. 65). From the similar triangles OBG, $B'BD$

$$OG : B'D :: BG : BD,$$

$$\therefore \quad OG = \frac{B'D \cdot BG}{BD},$$

i. e. OG is constant. The point O therefore moves on a sphere whose centre is G and whose radius is the constant value just found for OG.

In a similar manner it may be shown that O moves upon each of two other spheres having their centres at H and K respectively.

Since then the point O must lie simultaneously on several spheres, its locus must be a circle, whose plane is perpendicular to the line of centres of the spheres, and whose centre lies upon this same line.

This line GHK is the line of intersection of the planes π and σ and is consequently parallel to $\sigma\sigma'$ (since π and σ' are parallel planes); it is the vanishing line of the figure σ, regarded as the perspective image of the figure σ' (Art. 13).

120. Theorem. *Two concentric projective pencils lying in the same plane, which have no self-corresponding rays, may be regarded as the perspective image of two directly equal pencils* *.

Let O be the common centre of the two pencils. Cut them by a transversal s, thus forming two collinear projective ranges ABC ... and $A'B'C'$ ... which have no self-corresponding points. Draw through s any plane σ'; we can determine in this plane (Art. 109) a point U such that the segments AA', BB', CC', ... subtend at it a constant angle; thus if the two ranges be projected from U as centre, two directly equal pencils will be obtained. Now let the eye be placed at any point of the straight line OU, and let the given pencils be projected from this point as centre on to the plane σ'. In this way two new pencils will be formed; and these are precisely the two directly equal pencils mentioned in the enunciation.

* Chasles, *loc. cit.*, Art. 180.

CHAPTER XII.

INVOLUTION.

121. CONSIDER two projective flat pencils (Fig. 79) having a common centre O; let them be cut in corresponding points by the transversals u and u', thus giving two projective ranges ABC ... and $A'B'C'$ ...; and let u'' be the straight line on which the pairs of lines AB' and $A'B$, ... (Art. 85, left) intersect. Through O draw any ray (not a self-corresponding ray); it will cut u and u' in two non-corresponding points A and B' and will meet u'' in a point of the line $A'B$. To the ray OA of the first pencil corresponds accordingly the ray OA' of the second, and to the ray OB' of the second pencil corresponds the ray OB of the first. In other words, to the ray OA or OB' correspond two different rays OA', OB according as the first ray is regarded as belonging to the first pencil or to the second. For the line $A'B$ must cut AB' on u'', and cannot pass through O so long as this point does not lie on u''. We see then that

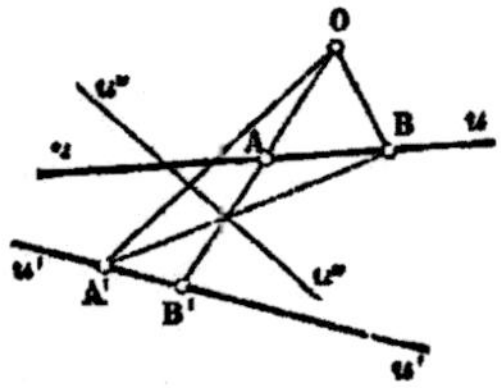

Fig. 79.

In two superposed projective forms (of one dimension) there correspond, in general, to any given element two different elements, according as the given element is regarded as one belonging to the first or to the second form.*

We say *in general*, because in what precedes it has been assumed that O does not lie upon u''.

* We say two forms, because the reasoning which we have made use of in the case of two concentric flat pencils may equally well be applied in the case of two collinear ranges, and of two axial pencils having a common axis. The same result may be arrived at by cutting the two flat pencils by a transversal, and by projecting them from a point lying outside their plane.

122. But in the case where O lies upon u'' (Fig. 80), if a ray be drawn through O to cut u and u' in A and B' respectively, then will also $A'B$ pass through O; in other words, to the ray OA or OB' corresponds the same ray OA' or OB. This property may be expressed by saying that *the two rays correspond doubly to one another*; or we may say that *the two rays are conjugate to one another*.

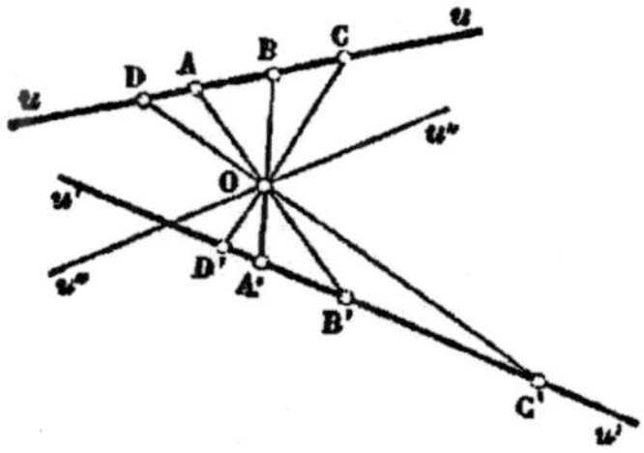

Fig. 80.

Now suppose, reciprocally, that two concentric projective flat pencils have a pair of rays which correspond doubly to one another. Cut the pencils by two transversals u and u', and let A and B' denote the points where these transversals intersect one of the given rays; then A' and B will denote the points where they intersect the other given ray. The straight line u'', the locus of the points of intersection of the pairs of lines such as MN', $M'N$, formed by joining crosswise any two pairs of corresponding points of the ranges u, u' (Art. 85), will pass through O, since the lines AB', $A'B$ meet in that point. If now there be drawn through O any other ray, cutting the transversals say in C and D', then will $C'D$ also pass through O, *i.e.* the rays OCD' and ODC' also correspond doubly to each other. We conclude that

When two superposed projective forms of one dimension are such that any one element has the same correspondent, to whichever form it be regarded as belonging, then every element possesses this property.

123. This particular case of two superposed projective forms of one dimension is called *Involution* *. We speak of an involution of points, of rays, or of planes, according as the elements are points of a range, rays of a flat pencil, or planes of an axial pencil.

In an involution, then, *the elements are conjugate to one another in pairs;* *i.e.* each element has its conjugate. To whichever of the two forms a given element be considered to

* DESARGUES, *Brouillon projet d'une atteinte aux événements des rencontres d'un cône avec un plan* (Paris, 1639): edition POUDRA (Paris, 1864), vol. i. p. 119.

belong, the element which corresponds to it is the same, viz. its conjugate. It follows from this that it is not necessary to regard the two forms as distinct, but that *an involution may be considered as a set of elements which are conjugate to one another in pairs.*

When $AA', BB', CC', \ldots$ are said to form an involution, it is to be understood that A and A', B and B', C and C', ... are pairs of conjugate elements; moreover, any element and its conjugate may be interchanged, so that $AA'\, BB'\, CC' \ldots$ and $A'A\, B'B\, C'C \ldots$ are projective forms.

124. Since an involution is only a particular case of two superposed projective forms, *every section and every projection of an involution gives another involution* *.

Two conjugate elements of the given involution give rise to two conjugate elements of the new involution. It follows (Art. 18) that the figure homological with an involution is also an involution.

125. When two collinear projective ranges form an involution, there corresponds to each point (and consequently also to the point at infinity I or J') a single point (I' or J); *i.e. the two vanishing points coincide in a single point.* Let this point, the conjugate of the point at infinity, be denoted by O. The equation (1) of Art. 109 then becomes

$$OA \,.\, OA' = \text{constant}.$$

In other words, an involution of points consists of pairs of points A, A' which possess the property that the rectangle contained by their distances from a fixed point O, lying on the base, is constant †. This point O is called the *centre* of the involution.

The self-corresponding elements of two forms in involution are called the *double elements* of the involution. In the case of the involution of points $AA', BB', \ldots$ we have

$$OA \,.\, OA' = OB \,.\, OB' = \ldots = \text{constant}.$$

If this constant is positive, *i.e.* if O does not lie between two conjugate points, there are two double points E and F, such that

$$OE^2 = OF^2 = OA \,.\, OA' = OB \,.\, OB' = \ldots ;$$

* DESARGUES, *loc. cit.*, p. 147.

† Ibid., pp. 112, 119.

O therefore lies midway between E and F, and the segment EF divides harmonically each of the segments AA', BB', ... (Art. 69. [3]). Accordingly:

If an involution has two double elements, these separate harmonically any pair of conjugate elements; or: *An involution is made up of pairs of elements which are harmonically conjugate with regard to two fixed elements.*

If, on the other hand, the constant is negative, *i.e.* if O falls between two conjugate points, there are no double points. In this case there are two conjugate points situated at equal distances from O and on opposite sides of it, such that $OE = -OE'$, and

$$OE^2 = OE'^2 = -OE \cdot OE' = -OA \cdot OA'.$$

If the constant is zero, there is only one double point O; but in this case there is no involution properly so called. For since the rectangle $OA \cdot OA'$ vanishes, one out of every pair of conjugate points must coincide with O.

126. The proposition that if an involution has two double elements, these separate harmonically any pair of conjugate elements, may also be proved thus:

Let E and F be the double elements, A and A' any pair of conjugate elements; since the systems $EFAA'$, $EFA'A$ are projective, therefore (Art. 83) each of them is harmonic.

The following is a third proof.

Consider EAA'... and $EA'A$ ... as two projective ranges, and project them respectively from two points S and S' collinear with E (Fig. 81). The projecting pencils $S(EAA'...)$ and $S'(EA'A...)$ are in perspective (since they have a self-corresponding ray in $SS'E$); therefore the straight line which joins the point of intersection of SA and $S'A'$ to that of SA' and $S'A$ will contain the points of intersection of all pairs of corresponding rays, and will consequently meet the common base of the two ranges at the second double point F. But from the figure we see that we have now a complete quadrilateral, one diagonal of which, AA', is cut by the other two in E and F; consequently (Art. 56) $EFAA'$ is a harmonic range.

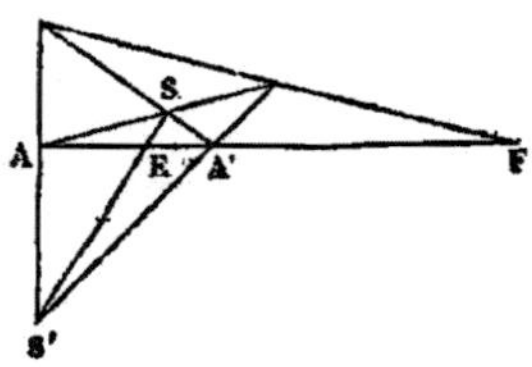

Fig. 81.

The proposition itself is a particular case of that proved in Art. 109 (3). From this we conclude that the pairs of elements (points of a range, rays or planes of a pencil) which, with two fixed elements, give a constant anharmonic ratio, form two superposed projective forms, which become an involution in the case where the anharmonic ratio has the value -1 (Art. 68).

127. *An involution is determined by two pairs of conjugate elements.*

For let A, A' and B, B' be the given pairs. If any element C be taken, its conjugate is determinate, and can be found as in Art. 84, by constructing so that the form $A'A\ B'C'$ shall be projective with $AA'BC$. We then say that *the six elements* AA', BB', CC' *are in involution;* *i.e.* they are three pairs of an involution.

Suppose that the involution with which we have to deal is an involution of points. Take any point G (Fig. 82) outside the base, and describe circles round GAA' and GBB'; if H is the second point in which these circles meet, join GH, and let it cut the base in O. Since $GHAA'$ lie on a circle,

$$OG\,.\,OH = OA\,.\,OA';$$

and since $GHBB'$ lie on a circle,

$$OG\,.\,OH = OB\,.\,OB';$$

$$\therefore\ OA\,.\,OA' = OB\,.\,OB'.$$

O is therefore the centre of the involution determined by the

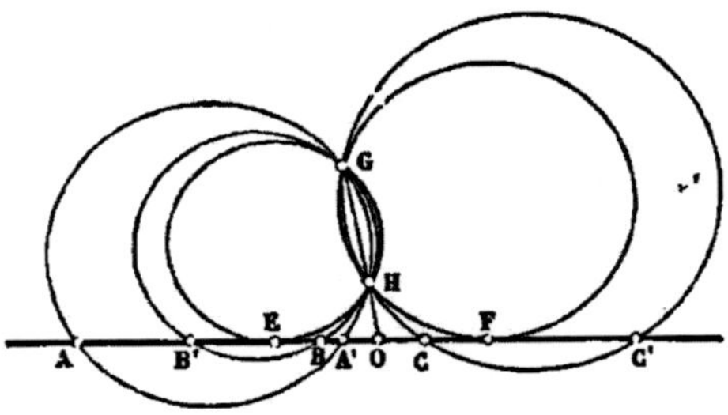

Fig. 82.

pairs of points A, A' and B, B'. If any other circle be drawn through G and H, and cut the base in C and C', we have

$$OG\,.\,OH = OC\,.\,OC';$$

$$\therefore\ OC\,.\,OC' = OA\,.\,OA' = OB\,.\,OB',$$

and C, C' are therefore a pair of conjugate points of the involution. In other words, the circle which passes through two conjugate points C, C' or D, D' and through one of the points G, H always passes through the other. Accordingly:

The pairs of conjugate points of the involution are the points of intersection of the base with a series of circles passing through the points G and H.

128. From what precedes it is evident that if the involution has double points, these will be the points of contact of the base with the two circles which can be drawn to pass through G and H and to touch the base. It has already been seen (Art. 125) that these points are harmonically conjugate with regard to A and A', and also with regard to B and B'. Consequently (Art. 70) *the involution has double points when one of the pairs AA', BB' lies entirely within or entirely without the other, i.e. when the segments AA' and BB' do not overlap* (Fig. 82); *and the involution has no double points when one pair is alternate to the other, i.e. when the segments AA' and BB' overlap* (Fig. 83)*.

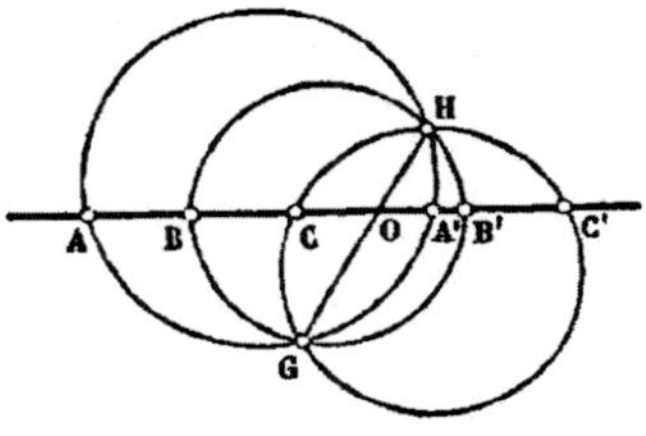

Fig. 83.

In the first case, the involution (as already seen) consists of an infinite number of pairs of points which are harmonically conjugate with regard to a pair of fixed points.

In the second case, on the other hand, the involution is traced out on the base by the arms of a right angle which revolves about its vertex. For since (Fig. 84) the segments AA' and BB' overlap, the circles described on AA' and BB' respectively as diameters will intersect in two points G and H which lie symmetrically with regard to the base; GH being perpendicular to the base, which bisects it at O, the centre of the involution. It follows that

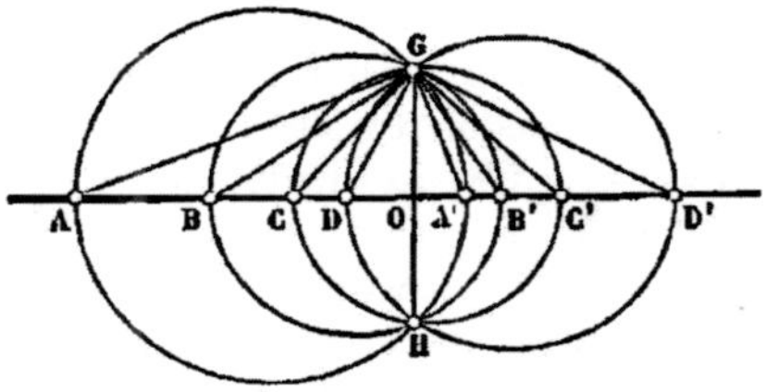

Fig. 84.

* An involution of the kind which has double points is often called a *hyperbolic* involution; one of the kind which has no double points being called an *elliptic* involution.

$$OG^2 = OH^2 = AO \,.\, OA' = BO \,.\, OB',$$

and that all other circles passing through G and H and cutting the base in the other pairs $CC', DD', \ldots$ of the involution will have their centres also on the base, and will have $CC', DD', \ldots$ as diameters. If then we project any of the segments $AA', BB', CC', \ldots$ from G (or H) as centre, we shall obtain in each case a right angle $AGA', BGB', CGC', \ldots$ (or $AHA', BHB', CHC', \ldots$).

We conclude that when an involution of points $AA', BB', \ldots$ has no double points, *i.e.* when the rectangle $OA \,.\, OA'$ is equal to a negative constant $-k^2$, each of the segments $AA', BB', \ldots$ subtends a right angle at every point on the circumference of a circle of radius k, whose centre is at O and whose plane is perpendicular to the base of the involution.

This last proposition is a particular case of that of Art. 109 (4). If then an angle of constant magnitude revolve in its plane about its vertex, its arms will determine on a fixed transversal two projective ranges, which are in involution in the case where the angle is a right angle.

129. Consider an involution of parallel rays; these meet in a point at infinity, and the straight line at infinity is a ray of the involution. The ray conjugate to it contains the centre of the involution of points which would be obtained by cutting the pencil by any transversal; it may therefore be called the *central ray* of the given involution. If, reciprocally, we project an involution of points by means of parallel rays, these rays will form a new involution, whose central ray passes through the centre of the given involution.

When one involution is derived from another involution by means of projections or sections (Art. 124), the double elements of the first always give rise to the double elements of the second.

130. Since in an involution any group of elements is projective with the group of conjugate elements, it follows that if any four points of the involution be taken, their anharmonic ratio will be equal to that of their four conjugates. In the involution $AA', BB', CC', \ldots$ the groups of points $ABA'C'$ and $A'B'AC$, for example, will be projective; therefore

$$\frac{AA'}{BA'} : \frac{AC'}{BC'} = \frac{A'A}{B'A} : \frac{A'C}{B'C},$$

whence

$$AB' \,.\, BC' \,.\, CA' + A'B \,.\, B'C \,.\, C'A = 0.$$

Conversely, if this relation hold among the segments determined by six collinear points $AA'BB'CC'$, these will be three conjugate pairs of an involution. For the given relation shows that the anharmonic ratios $(ABA'C')$ and $(A'B'AC)$ are equal to one another; the groups $ABA'C'$ and $A'B'AC$ are therefore projective. But A and A' correspond doubly to each other; therefore (Art. 122) AA', BB', CC' are three conjugate pairs of an involution.

131. THEOREM. *The three pairs of opposite sides of a complete quadrangle are cut by any transversal in three pairs of conjugate points of an involution* *.

Let $QRST$ (Fig. 85) be a complete quadrangle, of which the pairs of opposite sides RT and QS, ST and QR, QT and RS are cut by any transversal in A and A', B and B', C and C' respectively. If P is the point of intersection of QS and RT, then $ATPR$ is a projection of $ACA'B'$ from Q as centre, and $ATPR$ is also a projection of $ABA'C'$ from S as centre; therefore the group $ACA'B'$ is projective with $ABA'C'$, and therefore (Art. 45) with $A'C'AB$. And since A and A' correspond doubly to one another in the projective groups $ACA'B'$

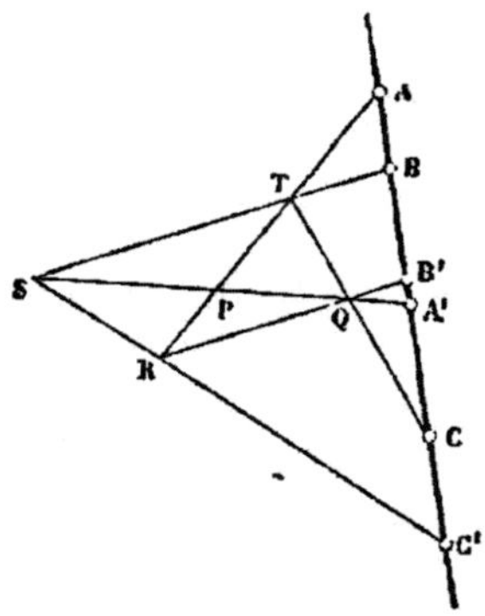

Fig. 85.

CORRELATIVE THEOREM. *The straight lines which connect any point with the three pairs of opposite vertices of a complete quadrilateral are three pairs of conjugate rays of an involution.*

Let $qrst$ (Fig. 86) be a complete quadrilateral, of which the pairs of opposite vertices rt and qs, st and qr, qt and rs are projected from any centre by the rays a and a', b and b', c and c' respectively.

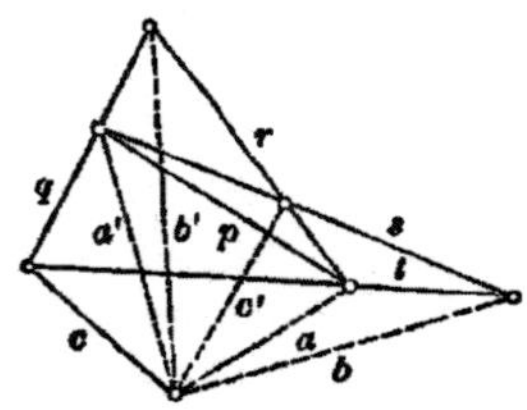

Fig. 86.

Let p be the straight line which joins the points qs and rt. The pencils $atpr$ and $aca'b'$ are in perspective (their corresponding rays intersect in pairs on q); similarly $atpr$ and $aba'c'$ are in perspective (their corresponding rays intersect in pairs on s). The pencil $atpr$ is therefore of course projective with each of the pencils $aca'b'$ and $aba'c'$, and

* DESARGUES, *loc. cit.*, p. 171.

and $A'C'AB$, it follows (Art. 122) that AA', BB', CC' are three conjugate pairs of an involution.

therefore $aca'b'$ is projective with $aba'c'$ or (Art. 45) with $a'c'ab$. And since a and a' correspond doubly to one another in the pencils $aca'b'$ and $a'c'ab$, it follows (Art. 122) that aa', bb', cc' are three pairs of conjugate rays of an involution.

The theorem just proved may also be stated in the following form:

If a complete quadrangle move in such a way that five of its sides pass each through one of five fixed collinear points, then its sixth side will also pass through a fixed point collinear with the other five, and forming an involution with them.

The theorem just proved may also be stated in the following form:

If a complete quadrilateral move in such a way that five of its vertices slide each on one of five fixed concurrent straight lines, then its sixth vertex will also move on a fixed straight line, concurrent with the other five, and forming an involution with them.

132. By combining the preceding theorem (left) with that of Art. 130, we see that

If a transversal be cut by the three pairs of opposite sides of a complete quadrangle in A and A', B and B', C and C' respectively, these determine upon it segments which are connected by the relation

$$AB'.BC'.CA' + A'B.B'C.C'A = 0^*.$$

133. In the theorem of Art. 131 (right) let U and U', V and V', W and W' denote respectively the opposite vertices rt and qs, st and qr, qt and rs of the quadrilateral $qrst$, and let AA', BB', CC' denote respectively the points of intersection of the rays aa', bb', cc' with an arbitrary transversal. With the help of Art. 124 the following proposition may be enunciated:

If the three pairs UU', VV', WW' of opposite vertices of a complete quadrilateral be projected from any centre upon any straight line, the six points AA', BB', CC' so obtained will form an involution.

Suppose now, as a particular case of this, that the centre of projection G is taken at one of the two points of intersection of the circles described on UU', VV' respectively as diameters. Then AGA' and BGB' are right angles, and therefore also (Art. 128) CGC' is a right angle; therefore the circle on WW' as diameter will also pass through G. Hence the three circles which have for diameters the three diagonals of a complete quadrilateral pass all through the same two

* PAPPUS, *loc. cit.*, book VII. prop. 130.

points; that is, they have the same radical axis. The centres of these circles lie in a straight line; hence

The middle points of the three diagonals of a complete quadrilateral are collinear *.

134. The proposition of Art. 131 (left) leads immediately to the

Construction for the sixth point C' of an involution of which five points A, A', B, B', C are given.

For draw through C (Fig 85) an arbitrary straight line, on which take any two points Q and T, and join AT, BT, $A'Q$, $B'Q$; if AT, $B'Q$ meet in R, and BT, $A'Q$ in S, the straight line RS will cut the base of the involution in the required point C'.

The proposition of Art. 131 (right) leads immediately to the

Construction for the sixth ray c' of an involution of which five rays a, a', b, b', c are given.

For take on c (Fig. 86) an arbitrary point, through which draw any two straight lines q and t, and join the point ta to qb', and the point tb to qa'; if the joining lines be called r, s respectively, then the straight line connecting the centre of the pencil with the point rs is the required ray c'.

If, in the preceding problem (left), the point C lies at infinity, its conjugate is the centre O of the involution. In order then *to find the centre of an involution of which two pairs AA', BB' of conjugate points are given*, we construct (Fig. 87) a complete quadrangle $QSTR$ of which one pair of opposite sides pass respectively through A and A', another such pair through B and B', and which has a fifth side parallel to the base; the sixth side will then pass through the centre O.

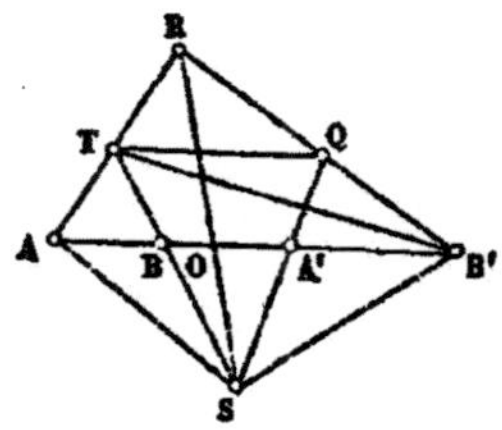

Fig. 87.

The sixth point C' which, together with five given points $AA'BB'C$, forms an involution, is completely determined by the construction; there is only one point C' which possesses the property on which the construction depends (Art. 127). This may be otherwise seen by regarding C' as given by the equation $(AA'BC)=(A'AB'C')$ between anharmonic ratios; for it is known (Art. 65) that there is only one point C' which satisfies this equation.

135. The theorem converse to that of Art. 131 is the following:

If a transversal cut the sides of a triangle RSQ (Fig. 85) *in three points A', B', C' which, when taken together with three other points A, B, C lying on the same transversal, form three conjugate*

* CHASLES, *loc. cit.*, Arts. 344, 345; GAUSS, Collected Works, vol. iv. p. 391.

pairs of an involution, then the three straight lines RA, SB, QC meet in the same point.

To prove it, let RA, SB meet in T, and let TQ meet the transversal in C_1. Applying the theorem of Art. 131 (left) to the quadrangle $QRST$, we have

$$(AA'BC_1) = (A'AB'C').$$

But by hypothesis

$$(AA'BC) = (A'AB'C');$$

$$\therefore \quad (AA'BC_1) = (AA'BC);$$

consequently (Art. 54) C_1 coincides with C, *i.e.* QC passes through T.

The correlative theorem is:

If a point S be joined to the vertices of a triangle rsq (Fig. 86) *by three rays a', b', c' which, when taken together with three other rays a, b, c passing also through S, form three conjugate pairs of an involution, then the points ra, qb, sc lie on the same straight line t.*

136. Take again the figure of the complete quadrangle $QRST$ whose three pairs of opposite sides are cut by a transversal in A and A', B and B', C and C'. Let (Fig. 88) SQ and RT meet in R', QR and ST in S', RS and QT in Q'.

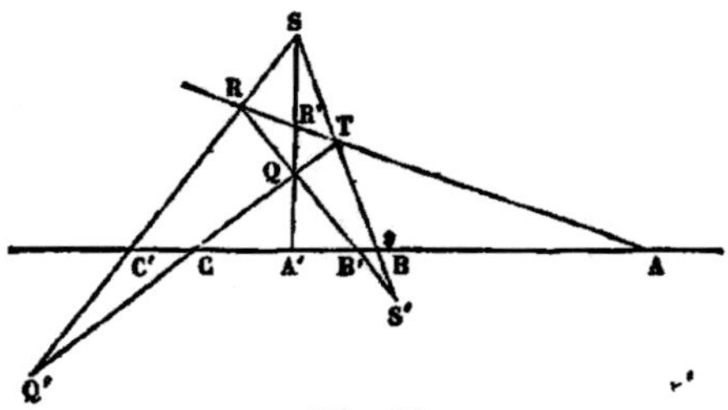

Fig. 88.

Consider the triangle RSQ; on each of its sides we have a group of four points, viz.

$$SQR'A', \quad QRS'B', \quad RSQ'C'.$$

The projections of these from T on the transversal are

$$BCAA', \quad CABB', \quad ABCC'.$$

The product of the anharmonic ratios of these last three groups is

$$\left(\frac{BA}{CA}:\frac{BA'}{CA'}\right)\left(\frac{CB}{AB}:\frac{CB'}{AB'}\right)\left(\frac{AC}{BC}:\frac{AC'}{BC'}\right),$$

or
$$-\frac{CA'.AB'.BC'}{BA'.CB'.AC'},$$
which (Art. 130) is equal to -1. Therefore:

If any transversal meet the sides of a triangle, and if moreover from any point as centre each vertex be projected upon the side opposite to it, the groups of four points thus obtained on each of the sides of the triangle will be such that the product of their anharmonic ratios is equal to -1.

Conversely, *if three pairs of points $R'A'$, $S'B'$, $Q'C'$ be taken, one on each of the sides of a triangle RSQ, such that the product of the anharmonic ratios $(SQR'A')$, $(QRS'B')$, $(RSQ'C')$ is equal to* -1; *then, if the straight lines RR', SS', QQ' are concurrent, the points A', B', C' will be collinear; and conversely, if the points A', B', C' are collinear, the straight lines RR', SS', QQ' will be concurrent.*

137. Suppose now the transversal to lie altogether at infinity; then the anharmonic ratios $(SQR'A')$, $(QRS'B')$, and $(RSQ'C')$ become (Art. 64) respectively equal to $SR' : QR'$, $QS' : RS'$, and $RQ' : SQ'$; so that the preceding proposition reduces to the following*:

If the straight lines connecting the three vertices of a triangle RSQ with any given point T meet the respectively opposite sides in R', S', Q', the segments which they determine on the sides will be connected by the relation
$$\frac{SR'.QS'.RQ'}{QR'.RS'.SQ'} = -1;$$
and conversely:

If on the sides SQ, QR, RS respectively of a triangle RSQ points R', S', Q' be taken such that the above relation holds, then will the straight lines RR', SS', QQ' meet in one point T.

138. Repeating this last theorem for two points T' and T'', we obtain the following:

If the two sets of three straight lines which connect the vertices of a triangle RSQ with any two given points T' and T'' meet the respectively opposite sides in R', S', Q' and R'', S'', Q'', then will the product of the anharmonic ratios $(SQR'R'')$, $(QRS'S'')$, and $(RSQ'Q'')$ be equal to $+1$.

[For each of the expressions
$$\frac{SR'.QS'.RQ'}{QR'.RS'.SQ'},\quad \frac{SR''.QS''.RQ''}{QR''.RS''.SQ''}$$

* CEVA's theorem. See his book, *De lineis rectis se invicem secantibus statica constructio* (Mediolani, 1678), i. 2. Cf. MÖBIUS, *Baryc. Calc.* § 198.

is equal to -1; and the required result follows on dividing one of them by the other.]

139. Considering again the triangle QRS (Fig. 88), and taking the transversal to be entirely arbitrary, let ST, QT be taken so as to be parallel to QR, RS respectively. Then the figure $QRST$ becomes a parallelogram; the points S' and Q' pass to infinity, and R' (being the point of intersection of the diagonals QS, RT) becomes the middle point of SQ. Consequently (Art. 64) the anharmonic ratios $(SQR'A')$, $(QRS'B')$, $(RSQ'C')$ become equal respectively to $-(QA' : SA')$, $(RB' : QB')$, and $(SC' : RC')$. Thus*:

If a transversal cut the sides of a triangle RSQ in A', B', C' respectively, it determines upon them segments which are connected by the relation

$$\frac{QA' . RB' . SC'}{SA' . QB' . RC'} = 1;$$

and conversely:

If on the sides SQ, QR, RS respectively of a triangle points A', B', C' be taken such that the above relation holds, then will these three points be collinear.

140. Repeating the last theorem of the preceding Article for two transversals, we obtain the following:

If the sides of a triangle RSQ are cut by two transversals in A', B', C' and in A'', B'', C'' respectively, the product of the anharmonic ratios $(SQA'A'')$, $(QRB'B'')$, and $(RSC'C'')$ will be equal to $+1$.

[For each of the expressions

$$\frac{QA' . RB' . SC'}{SA' . QB' . RC'}, \quad \frac{QA'' . RB'' . SC''}{SA'' . QB'' . RC''}$$

is equal to 1; dividing one by the other, the required result follows.]

Reciprocally, *if on the sides of a triangle RSQ three pairs of points $A'A''$, $B'B''$, $C'C''$ be taken such that the product of the anharmonic ratios $(SQA'A'')$, $(QRB'B'')$, $(RSC'C'')$ may be equal to $+1$; then, if the points A', B', C' are collinear, the points A'', B'', C'' will also be collinear, and if the lines RA', SB', QC' are concurrent, the lines RA'', SB'', QC'' will also be concurrent.*

141. It has been shown (Art. 122) that if two projective ranges

* Theorem of MENELAUS; *Sphaerica*, iii. 1. Cf. MÖBIUS, *loc. cit.*

$(ABC\ldots)$ and $(A'B'C'\ldots)$, lying in the same plane, are projected from the point of intersection of a pair of lines such as AB' and $A'B$, AC' and $A'C$, ... or BC' and $B'C$..., the projecting rays form an involution. The theorems correlative to this are as follows:

Given two projective, but not concentric, flat pencils $(abc\ldots)$ and $(a'b'c'\ldots)$ lying in the same plane; if they be cut by the straight line which joins a pair of points such as ab' and $a'b$, ac' and $a'c$,... or bc' and $b'c$..., the points so obtained form an involution.

Given two projective axial pencils $(\alpha\beta\gamma\ldots)$ and $(\alpha'\beta'\gamma'\ldots)$ whose axes meet one another; if they be cut by the plane which is determined by passing through a pair of lines such as $\alpha\beta'$ and $\alpha'\beta$, $\alpha\gamma'$ and $\alpha'\gamma$,... or $\beta\gamma'$ and $\beta'\gamma$..., the rays so obtained form an involution.

Given two projective flat pencils $(abc\ldots)$ and $(a'b'c'\ldots)$ which are concentric, but lie in different planes; if they be projected from the point of intersection of a pair of planes such as ab' and $a'b$, ac' and $a'c$..., or bc' and $b'c$..., the projecting planes form an involution.

142. Particular Cases. All points of a straight line which lie in pairs at equal distances on opposite sides of a fixed point on the line, form an involution, since every pair is divided harmonically by the fixed point and the point at infinity.

Conversely, if the point at infinity is one of the double points of an involution of points, then the other double point bisects the distance between any point and its conjugate. If in such an involution the segments AA', BB' formed by any two pairs of conjugate points have a common middle point, then will this point bisect also the segment CC' formed by any other pair of conjugates.

All rectilineal angles which have a common vertex, lie in the same plane, and have the same fixed straight line as a bisector, form an involution, since the arms of every angle are harmonically conjugate with regard to the common bisector and the ray perpendicular to it through the common vertex.

Conversely, if the double rays of a pencil in involution include a right angle, then any ray and its conjugate make equal angles with either of the double rays. If in such an involution the angles included by two pairs of conjugate rays aa' and bb' have common bisectors, these will be the bisectors also of the angle included by any other pair of conjugate rays cc'.

All dihedral angles which have a common edge and which have the same fixed plane as a bisector, form an involution; for the faces of every angle are harmonically conjugate with regard to the fixed plane and the plane drawn perpendicular to it through the common edge.

Conversely, if the double planes of an axial pencil in involution are at right angles to one another, then any plane and its conjugate make equal angles with either of the double planes.

CHAPTER XIII.

PROJECTIVE FORMS IN RELATION TO THE CIRCLE.

143. CONSIDER (Fig. 89) two directly equal pencils $abcd$... and $a'b'c'd'$... in a plane, having their centres at O and O' respectively. The angle contained by a pair of corresponding rays aa', bb', cc',... is constant (Art. 106); the locus of the intersection of pairs of corresponding rays is therefore (Euc. III. 21) a circle passing through O and O'. The tangent to this circle at O makes with OO' an angle equal to any of the angles OAO', OBO', OCO', &c.; but this is just the angle which $O'O$ considered as a ray of the second pencil should make with the ray corresponding to it in the first pencil; therefore to $O'O$ or q' considered as a ray of the second pencil corresponds in the first pencil the tangent q to the circle at O.

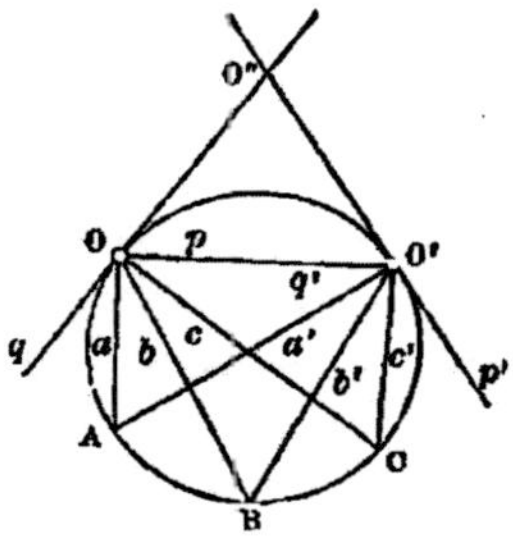

Fig. 89.

Imagine the circumference of the circle to be described by a moving point A; the rays AO, AO' or a, a' will trace out the two pencils. As A approaches O, the ray AO' will approach OO' or q' and the ray AO will approach q; and in the limit when A is indefinitely near to O, the ray AO will coincide with q or the tangent at O. This agrees with the definition of the tangent at O, as the straight line which joins two indefinitely near points of the circumference.

Similarly, to the ray OO' or p considered as belonging to the first pencil corresponds the ray p' of the second pencil, the tangent to the circle at O'.

144. Conversely, if any number of points A, B, C, D,... on a circle be joined to two points O and O' lying on the same

circle, the pencils $O(A, B, C, D, \ldots)$ and $O'(A, B, C, D, \ldots)$ so formed will be directly equal, since the angle AOB is equal to $AO'B$, AOC to $AO'C, \ldots$ BOC to $BO'C$, &c. But two equal pencils are always projective with one another (Art. 104). If then the points $A, B, C, \ldots$ remain fixed, while the centre of the pencil moves and assumes different positions on the circumference of the circle, the pencils so formed are all equal to one another, and consequently all projective with one another. The tangent at O is by definition the straight line which joins O to the point indefinitely near to it on the circle. It follows that in the projective pencils $O(A, B, C, \ldots)$ and $O'(A, B, C, \ldots)$, the ray of the first which corresponds to the ray $O'O$ of the second is the tangent at O.

145. It has been seen (Art. 73) that in two projective forms four harmonic elements of the one correspond to four harmonic elements of the other. If then the four rays $O(A, B, C, D)$ form a harmonic pencil, the same is the case with regard to the four rays $O'(A, B, C, D)$, whatever be the position of the point O' on the circle. By taking O' indefinitely near to A, we see that the pencil composed of the tangent at A and the chords AB, AC, AD will also be harmonic; so again the pencil composed of the chord BA, the tangent at B, and the chords BC, BD will be harmonic, &c.

When this is the case, *the four points A, B, C, D of the circle are said to be harmonic**.

146. *The tangents to a circle determine upon any pair of fixed tangents two ranges which are projective with one another.*

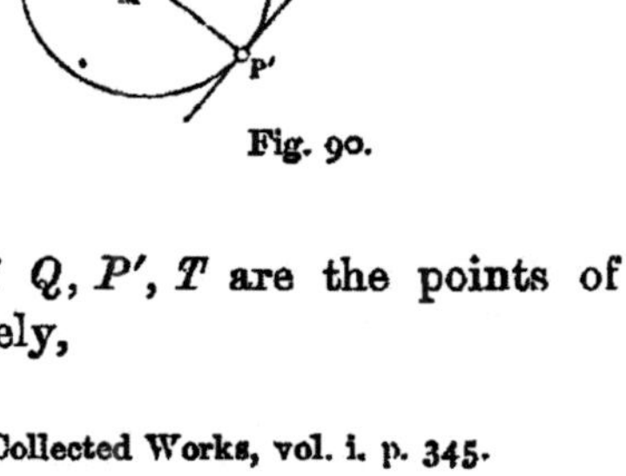

Fig. 90.

Let M (Fig. 90) be the centre of the circle, PQ and $P'Q'$ a pair of fixed tangents, and AA' a variable tangent. The part AA' of the variable tangent intercepted between the fixed tangents subtends a constant angle at M; for if Q, P', T are the points of contact of the tangents respectively,

* STEINER, *loc. cit.*, p. 157, § 43; Collected Works, vol. i. p. 345.

$$\begin{aligned} \text{angle } AMA' &= AMT + TMA' \\ &= \tfrac{1}{2} QMT + \tfrac{1}{2} TMP' \\ &= \tfrac{1}{2} QMP' \text{*}. \end{aligned}$$

Accordingly, as the tangent AA' moves, the rays MA, MA' will generate two projective pencils (Art. 108), and the points A, A' will trace out two projective ranges.

Since the angle AMA' is equal to the half of QMP', it is equal to either of the angles QMQ', PMP' (denoting by P and Q' the same point, according as it is regarded as belonging to the first or to the second tangent). Consequently Q and Q', P and P' are pairs of corresponding points of the two projective ranges; *i.e.* the points of contact of the two fixed tangents correspond respectively to the point of intersection of the tangents.

Imagine the circle to be generated, as an envelope, by the motion of the variable tangent; the points A, A' will trace out the two projective ranges. As the variable tangent approaches the position PQ, the point A' approaches Q', and A approaches the point which corresponds to Q', viz. Q; and in the limit when the variable tangent is indefinitely near to PQ, the point A will be indefinitely near to Q or the point of contact of the tangent PQ. The point of contact of a tangent must therefore be regarded as the point of intersection of the tangent with an indefinitely near tangent.

147. The preceding proposition shows that four tangents a, b, c, d to a circle are cut by a fifth in four points A, B, C, D whose anharmonic ratio is constant whatever be the position of the fifth tangent.

This tangent may be taken indefinitely near to one of the four fixed tangents, to a for example; in this case A will be the point of contact of a, and B, C, D the points of intersection ab, ac, ad respectively.

As a particular case, if a, b, c, d meet the tangent PQ in four harmonic points, they will meet every tangent in four harmonic points. The group constituted by the point of contact of a and the points of intersection ab, ac, ad will also be harmonic. In this case, *the four tangents a, b, c, d are said to be harmonic* †.

* PONCELET, *Propr. proj.*, Art. 462.

† STEINER, *loc. cit.*, p. 157, § 43; Collected Works, vol. i. p. 345.

148. *The range determined upon any given tangent to a circle by any number of fixed tangents is projective with the pencil formed by joining their points of contact to any arbitrary point on the circle.*

Let $A, B, C, \ldots X$ (Fig. 91) be points on the circle, and $a, b, c, \ldots x$ the tangents at these points respectively. If the points $A', B', C', \ldots$ in which the tangent x is cut by the tangents $a, b, c, \ldots$ be joined to the centre of the circle, the joining lines will be perpendicular respectively to the chords $XA, XB, XC, \ldots$ and will therefore (Art. 108) form a pencil equal to the pencil $X(A, B, C, \ldots)$. The range $A'B'C'\ldots$ is therefore projective with the pencil $X(A, B, C, \ldots)$.

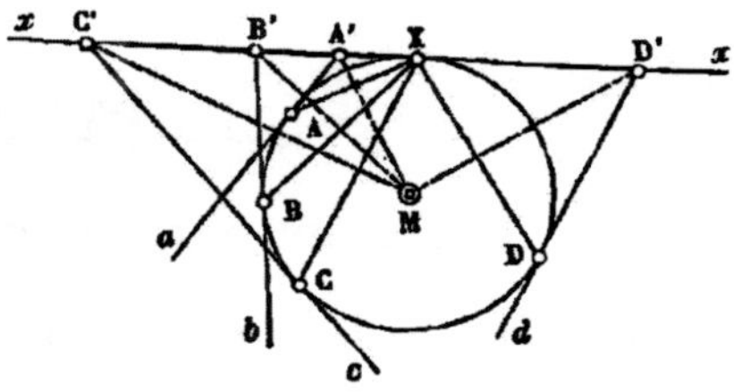

Fig. 91.

COROLLARY. *If four points on a circle are harmonic, then the tangents also at these points are harmonic; and conversely.*

For if, in what precedes, $X(ABCD)$ is a harmonic pencil, $A'B'C'D'$ will be a harmonic range; and conversely.

CHAPTER XIV.

PROJECTIVE FORMS IN RELATION TO THE CONIC SECTIONS.

149. LET the figures be constructed which are homological with those of Arts. 144, 146, 148. To the points and tangents of the circle will correspond the points and tangents of a conic section (Art. 23). A tangent to a conic is therefore a straight line which meets the curve in two points which are indefinitely near to one another; a point on the curve is the point of intersection of two tangents which are indefinitely near to one another. To two equal and therefore projective pencils will correspond two projective pencils, and to two projective ranges will correspond two projective ranges; for two pencils or ranges which correspond to one another in two homological figures are in perspective. We deduce therefore the following propositions:

(1). *If any number of points $A, B, C, D, \ldots$ on a conic are joined to two fixed points O and O' lying on the same conic* (Fig. 92), *the pencils $O(A, B, C, D, \ldots)$ and $O'(A, B, C, D, \ldots)$ so formed are projective with one another. To the ray OO' of the first pencil corresponds the tangent at O', and to the ray $O'O$ of the second pencil corresponds the tangent at O.*

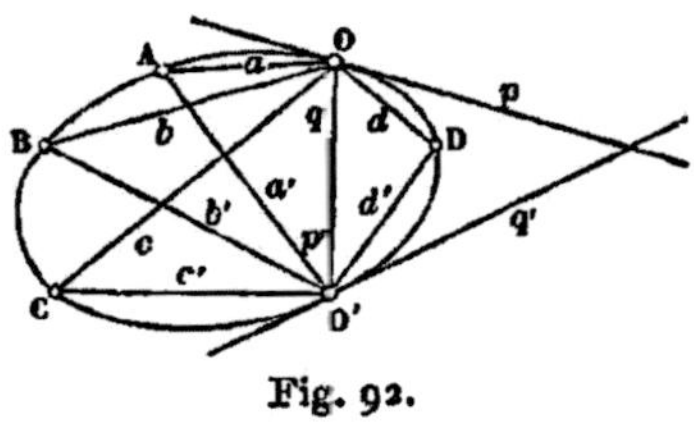

Fig. 92.

(2). *Any number of tangents $a, b, c, d, \ldots$ to a conic determine on a pair of fixed tangents o and o'* (Fig. 93) *two projective ranges. To the point oo' or Q of the first range corresponds the point of contact Q' of o', and to the same point $o'o$ or P' of the second range corresponds the point of contact P of o**.

* STEINER, *loc. cit.*, p. 139, § 38; Collected Works, vol. i. pp. 332, 333.

(3). *The range which a variable tangent to a conic determines upon a fixed tangent is projective with the pencil formed by joining the*

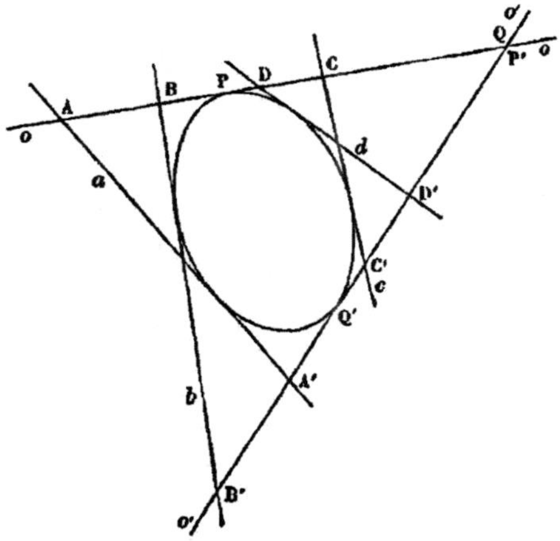

Fig. 93.

point of contact of the variable tangent to any fixed point of the conic. (Fig. 94.)

150. We proceed now to the theorems converse to those of Art. 149. The proofs here given are due to M. Ed. Dewulf.

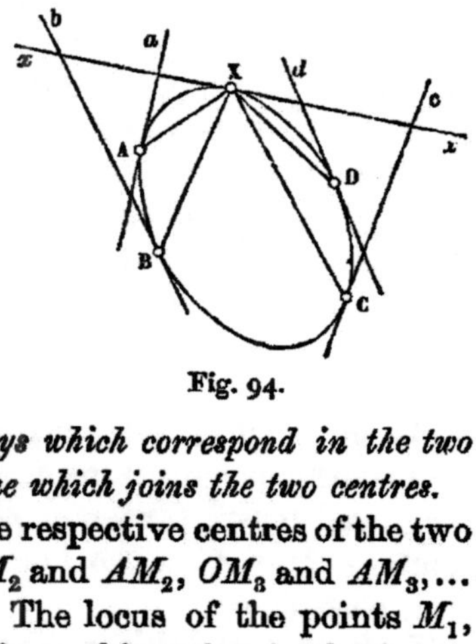

Fig. 94.

I. *If two (non-concentric) pencils lying in the same plane are projective with one another (but not in perspective), the locus of the points of intersection of pairs of corresponding rays is a conic passing through the centres of the two pencils; and the tangents to the locus at these points are the rays which correspond in the two pencils respectively to the straight line which joins the two centres.*

Let O and A (Fig. 95) be the respective centres of the two pencils, and let OM_1 and AM_1, OM_2 and AM_2, OM_3 and AM_3,... be pairs of corresponding rays. The locus of the points M_1, M_2, M_3,... will pass through O, since this point is the intersection of the ray AO of the pencil A with the corresponding ray of the pencil O. Similarly A will be a point on the locus.

Let o be that ray of the pencil O which corresponds to the ray AO of the pencil A. Describe a circle touching o at O,

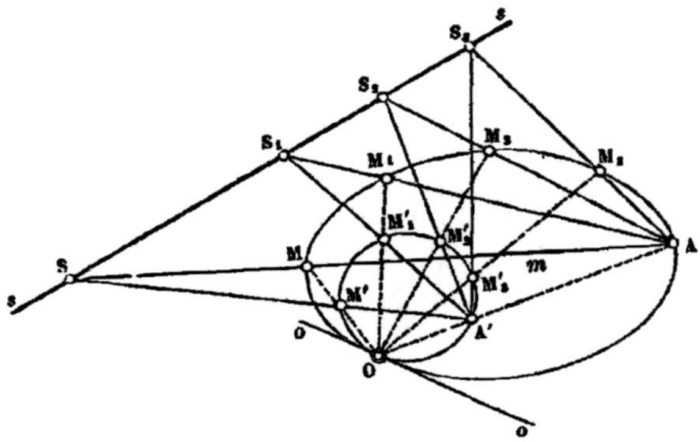

Fig. 95.

and let this circle cut OA in A', and OM_1, OM_2, OM_3, ... in the points M_1', M_2', M_3', ... respectively.

The pencils $O(M_1'\ M_2'\ M_3 ...)$ and $A'(M_1'\ M_2'\ M_3' ...)$ are directly equal to one another; and since by hypothesis the pencil $O(M_1'\ M_2'\ M_3' ...)$ or $O(M_1\ M_2\ M_3 ...)$ is projective with the pencil $A(M_1\ M_2\ M_3 ...)$, therefore the pencils $A'(M_1'\ M_2'\ M_3' ...)$ and $A(M_1\ M_2\ M_3 ...)$ are projective. But they are in perspective, since the ray $A'O$ in the one corresponds to the ray AO in the other (Art. 80); therefore pairs of corresponding rays will intersect in points S_1, S_2, S_3, ... lying on a straight line s. In order, then, to find that point of the locus which lies on any given ray m of the pencil A, it is only necessary to produce m to meet s in S, to join SA' cutting the circle in M', and to join OM'; this last line will cut m in the required point M. But this construction is precisely the same as that employed in Art. 23 (Fig. 11) in order to draw the curve homological with a circle, having given the axis s and centre O of homology, and a pair of corresponding points A and A'. The locus of the points M is therefore a conic section.

II. *If two (non-collinear) ranges lying in the same plane are projective with one another (but not in perspective), the envelope of the straight lines joining pairs of corresponding points is a conic, i. e. the straight lines all touch a conic. This conic touches the bases of the two ranges at the points which correspond in these respectively to the point of intersection of their bases.*

Let s and s' (Fig. 96) be the bases of the two ranges, and let A and A', B and B', C and C', ... be pairs of corresponding

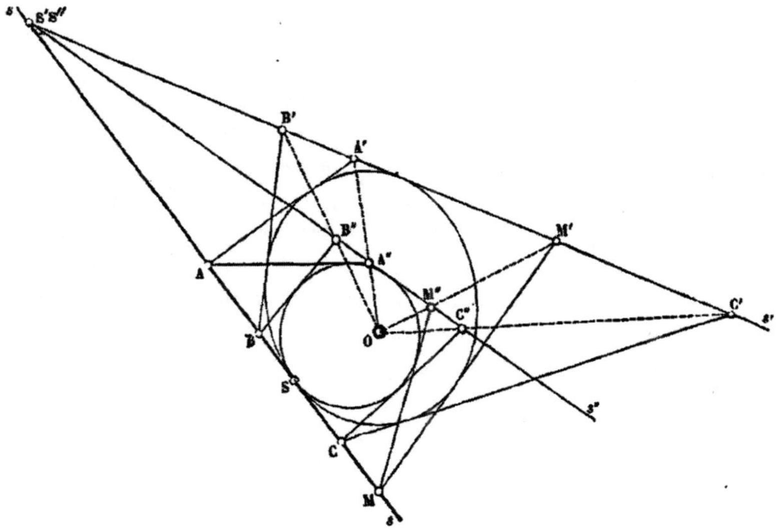

Fig. 96.

points. The curve enveloped by the straight lines AA', BB', CC',... will touch s, since this is the straight line joining the point ss' or S' of the second range with the corresponding point S of the first. Similarly, s' will be a tangent to the envelope.

Describe a circle touching s at S, and draw to it tangents a'', b'', c'',... s'' from the points A, B, C,... S' respectively. The tangents a'', b'', c'', ... will determine on s'' a range which is projective with s and therefore also with s'. But the point S' corresponds to itself in the two ranges s' and s''; these are therefore in perspective (Art. 80), and the straight lines $A''A'$, $B''B'$, $C''C'$,... will meet in one point O. In order then to draw a tangent to the envelope from any given point M lying on the line s, it is only necessary to draw from M a tangent m to the circle, meeting s'' in M'', and to join OM''; this last line will cut s' in that point M' of the range s' which corresponds to the point M of the range s, and MM' will be the required tangent to the envelope. But this construction is precisely the same as that made use of in Art. 23 (Fig. 12) in order to draw the curve homological with a circle, taking a given tangent to the circle as axis of homology, any given point O as centre of homology, and s', s'' as a pair of corresponding straight lines. The envelope of the lines MM' is therefore a conic section.

The theorems (I) and (II) of the present Article are correlative (Art. 33), since the figure formed by the points of intersection of corresponding rays of two projective pencils is correlative to that formed by the straight lines joining corresponding points of two projective ranges. Thus *in two figures which are correlative to one another (according to the law of duality in a plane), to points lying on a conic in one correspond tangents to a conic in the other.*

151. Having regard to Arts. 73 and 79, the propositions of Arts. 149, 150 may be enunciated as follows:

The anharmonic ratio of the four straight lines which connect four fixed points on a conic with a variable point on the same is constant.

The anharmonic ratio of the four points in which four fixed tangents to a conic are cut by a variable tangent to the same is constant *.

The anharmonic ratio of four points A, B, C, D lying on a conic is the anharmonic ratio of the pencil $O(A, B, C, D)$ formed by joining them to any point O on the conic. *The anharmonic ratio of four tangents a, b, c, d to a conic* is that of the four points $o(a, b, c, d)$, where o is an arbitrary tangent to the conic.

If this anharmonic ratio is equal to -1, the group of four points or tangents is termed *harmonic.*

The anharmonic ratio of four tangents to a conic is equal to that of their points of contact †.

Consequently the tangents at four harmonic points are harmonic, and *vice versa.*

The locus of a point such that the rays joining it to four given points $ABCD$ form a pencil having a given anharmonic ratio is a conic passing through the given points.

* STEINER, *loc. cit.*, p. 156, § 43; Collected Works, vol. i. p. 344.

† CHASLES, *Géométrie Supérieure*, Art. 663.

The tangent to the locus at one of these points, at A for example, is the straight line which forms with AB, AC, AD a pencil whose anharmonic ratio is equal to the given one.

The curve enveloped by a straight line which is cut by four given straight lines in four points whose anharmonic ratio is given is a conic touching the given straight lines.

The point of contact of one of these straight lines, a for example, forms with the points ab, ac, ad a range whose anharmonic ratio is equal to the given one*.

152. *Through five given points O, O', A, B, C in a plane* (Fig. 92), *no three of which lie in a straight line, a conic can be described.* For we have only to construct the two projective pencils which have their centres at two of the given points, O and O' for example, and in which three pairs of corresponding rays OA and $O'A$, OB and $O'B$, OC and $O'C$ intersect in the three other points. Any other pair OD and $O'D$ of corresponding rays will give a new point D of the curve.

To construct the tangent at any one of the given points, at O for example, we have only to determine that ray of the pencil O which corresponds to the ray $O'O$ of the pencil O'.

Through five given points *only one* conic can be drawn; for if there could be two such, they would have an infinite number of other points in common (the intersections of all the pairs of corresponding rays of the projective pencils); which is impossible.

Given five straight lines o, o', a, b, c in a plane (Fig. 93), *no three of which meet in a point, a conic can be described to touch them.* For we have only to construct the two projective ranges which are determined upon two of the given lines, o and o' for example, by the three others a, b, c, and of which three pairs of corresponding points oa and $o'a$, ob and $o'b$, oc and $o'c$ are given. The straight line d which joins any other pair of corresponding points of the two ranges will be a new tangent to the curve.

To construct the point of contact of any one of the given straight lines, that of o for example, we have only to determine that point of the range o which corresponds to the point $o'o$ of the range o'.

Only one conic can be drawn to touch five given straight lines; for if there could be two such, they would have an infinite number of common tangents (all the straight lines which join pairs of corresponding points of the projective ranges); which is impossible.

* STEINER, *loc. cit.*, pp. 156, 157, § 43; Collected Works, vol. i. pp. 344, 345.

From this we see also that:

Through four given points can be drawn an infinite number of conics; and two such conics have no common points beyond these four.

There can be drawn an infinite number of conics to touch four given straight lines; and two such conics have no common tangents beyond these four.

153. The theorems of Art. 88 may now be enunciated in the following manner:

If a hexagon $ab'ca'bc'$ is circumscribed to a conic (Figs. 97 and 61), *the straight lines p, q, r which join the three pairs of opposite vertices are concurrent.*

If a hexagon $AB'CA'BC'$ is inscribed in a conic (Figs. 98 and 60), *the three pairs of opposite sides intersect one another in three collinear points P, Q, R.*

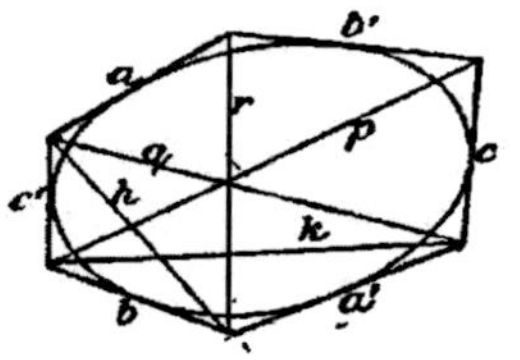

Fig. 97.

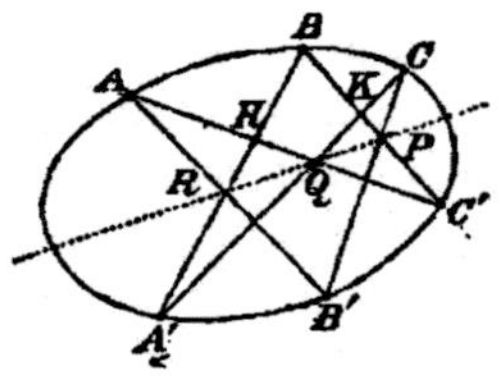

Fig. 98.

This is known as BRIANCHON'S theorem *.

This is known as PASCAL'S theorem †.

These results are of such importance in the theory of conics that they deserve independent proofs.

The ranges $a(ba'b'c')$ and $c(ba'b'c')$ are projective (Art. 149); the pencils formed by joining them to the points (ba'), (bc') respectively are therefore projective. If the line joining (ac'), $(a'b)$ be denoted by h, and that joining (bc'), $(a'c)$ by k, the pencils in

The pencils $A(BA'B'C')$ and $C(BA'B'C')$ are projective (Art. 149); the ranges in which they cut BA', BC' respectively are therefore projective. If AC', $A'B$ cut in H and BC', $A'C$ in K, ranges in question are $(BA'RH)$ and $(BKPC')$. Since they have

* This theorem was published for the first time by BRIANCHON in 1806, and repeated in his *Mémoire sur les lignes du second ordre* (Paris, 1817: p. 34).

† This theorem was given in PASCAL'S *Essai sur les Coniques*, a small work of six pages 8vo., published in 1640, when its author was only sixteen years old. It was republished in the *Œuvres de Pascal* (The Hague, 1779), and again by H. WEISSENBORN, in the preface to his book *Die Projection in der Ebene* (Berlin, 1862).

question are $(ba'rh)$ and $(bkpc')$. Since they have the ray b in common, they are in perspective; therefore $(a'k)$, (rp), (hc') are collinear, that is p, q, r are concurrent.

the point B in common, they are in perspective; therefore $A'K$, RP, HC' are concurrent, that is P, Q, R are collinear.

154. Pascal's theorem has reference to six points of a conic, Brianchon's theorem to six tangents; these six points or tangents may be chosen arbitrarily from among all the points on the curve and all the tangents to it. Now a conic is determined by five points or five tangents; in other words, five points or five tangents may be chosen at will from among all the points or lines of the plane, but as soon as these five elements have been fixed, the conic is determined. Pascal's theorem then expresses the condition which six points on a plane must satisfy if they lie on a conic; and Brianchon's theorem expresses similarly the condition which six straight lines lying in a plane must satisfy if they are all tangents to a conic. And the condition in each case is both necessary and sufficient.

That it is necessary is seen from the theorems themselves. For six points on a conic, taken in any order, may be regarded as the vertices of an inscribed hexagon *; but since Pascal's theorem is true for every inscribed hexagon, the three pairs of opposite sides must meet in three collinear points in whatever order the six points be taken.

The condition is also sufficient. For suppose (Fig. 98) that the hexagon $AB'CA'BC'$, formed by taking the six points in a certain order, possesses the property that the pairs of opposite sides BC' and $B'C$, CA' and $C'A$, AB' and $A'B$ intersect in three collinear points P, Q, R. Through the five points $AB'CA'B$ one conic (and one only) can be drawn; if X be the point where this conic cuts AC' again, then $AB'CA'BX$ is an inscribed hexagon, and its pairs of opposite sides $B'C$ and BX, XA (or $C'A$) and CA', $A'B$ and AB' will meet in three collinear points. But the second and third of these points are Q and

* It is perhaps hardly necessary to remind the reader that the hexagons to which Pascal's and Brianchon's theorems refer are not hexagons in Euclid's sense —*i. e.* they are not necessarily convex (non-reentrant) figures.

R; therefore BX must meet $B'C$ at the point of intersection of $B'C$ and QR, *i.e.* at P. Both BC' and BX thus pass through P, and they must therefore coincide. Since then the point X lies not only on AC' but also on BC', it must coincide with the point C' itself.

The condition is therefore sufficient; and it has already been shown to be necessary.

By taking the six points in all the different orders possible, sixty* simple hexagons can be made. From the reasoning above, it follows that if any one of these hexagons possesses the property that its three pairs of opposite sides intersect in three collinear points, the six points will lie on a conic, and consequently all the other hexagons will possess the same property †.

By analogous considerations having reference to Brianchon's theorem, properties correlative to those just established may be shown to be true of a system of six straight lines ‡.

155. Consider the two triangles which are formed, one by the first, third, and fifth sides, the other by the second, fourth, and sixth sides, of the inscribed hexagon $AB'CA'BC'$ (Fig. 98). Let BC' and $B'C$, CA' and $C'A$, AB' and $A'B$ be taken as corresponding sides of the triangles. By Pascal's theorem these sides intersect in pairs in three collinear points; and therefore (Art. 17) the two triangles are homological. Pascal's theorem may therefore be enunciated as follows:

If two triangles are in homology, the points of intersection of the sides of the one with the non-corresponding sides of the other lie on a conic.

Similarly, in a circumscribed hexagon $ab'ca'bc'$ (Fig. 97) let the vertices of even order and those of odd order respectively be regarded as the angular points of two triangles, and let bc' and $b'c$, ca' and $c'a$, ab' and $a'b$ be taken to be corresponding vertices. By Brianchon's theorem these vertices lie two and two on three straight lines which meet in a point; therefore

* In general, a complete n-gon includes in itself $\frac{1}{2}(n-1)(n-2)\ldots 1$ simple n-gons.

† STEINER, *loc. cit.*, p. 311, § 60, No. 54; Collected Works, vol. i. p. 450.

‡ A system of six points on a conic thus determines sixty different lines such as PQR in Fig. 98, or *Pascal lines* as they have been called. So too a system of six tangents to a conic determines sixty different *Brianchon points*.

(Art. 16) the two triangles are homological. Brianchon's theorem may therefore be enunciated as follows:

If two triangles are in homology, the straight lines joining the angular points of the one to the non-corresponding angular points of the other all touch a conic.

The two theorems may be included under the one enunciation:

If two triangles are in homology, the points of intersection of the sides of the one with the non-corresponding sides of the other lie on a conic, and the straight lines joining the angular points of the one to the non-corresponding angular points of the other all touch another conic *.

156. Returning to Fig. 98, let the points A, B', C, A', B be regarded as fixed, and C' as variable; Pascal's theorem may then be presented in the following form:

If a triangle $C'PQ$ move in such a way that its sides PQ, QC', $C'P$ turn round three fixed points R, A, B respectively, while two of its vertices P, Q slide along two fixed straight lines CB', CA' respectively, then the remaining vertex C' will describe a conic which passes through the following five points, viz. the two given points A and B, the point of intersection C of the given straight lines, the point of intersection B' of the straight lines AR and CB', and the point of intersection A' of the straight lines BR and CA' †.

So also Brianchon's theorem may be expressed in the following form:

If a triangle $c'pq$ (Fig. 99) *move in such a way that its vertices pq, qc', $c'p$ slide along three fixed straight lines r, a, b respectively, while two of its sides p, q turn round two fixed points cb', ca' respectively, then the remaining side c' will envelope a conic which touches the following five straight lines, viz. the two given straight lines a and b, the straight line c which joins the fixed points, the straight line b' which joins the points ar and cb', and the straight line a' which joins the points br and ca'.*

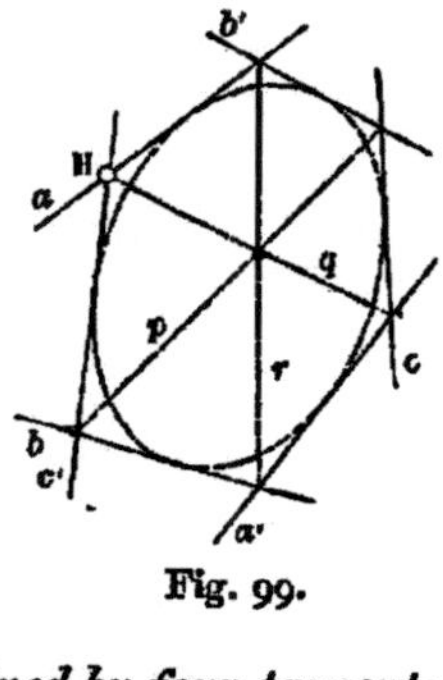

Fig. 99.

157. (1). If in the theorems of Art. 152 (right) one of the tangents is supposed to lie at infinity, the conic becomes a parabola (Art. 23). Thus *a parabola is determined by four tangents,*

* Möbius, *loc. cit.*, Art. 278.

† This theorem was given by Maclaurin, in 1721; cf. *Phil. Trans. of the Royal*

or (Art. 152, right) *only one parabola can be drawn to touch four given straight lines; and no two parallel tangents can be drawn to a parabola.*

(2). If the same supposition is made in theorem (2) of Art. 149, it is seen that the points at infinity on the two tangents o and o' are corresponding points of the projective ranges determined on these tangents; for the straight line which joins them is a tangent to the curve. It follows (Art. 100) that

The tangents to a parabola meet two fixed tangents to the same in points forming two similar ranges; or

Two fixed tangents to a parabola are cut proportionally by the other tangents.*

(3). Let A and A', B and B', C and C',... be the points in which the various tangents to the parabola meet the two fixed tangents (Fig. 100), and let P and Q' be the respective points of contact of the latter. The point of intersection of

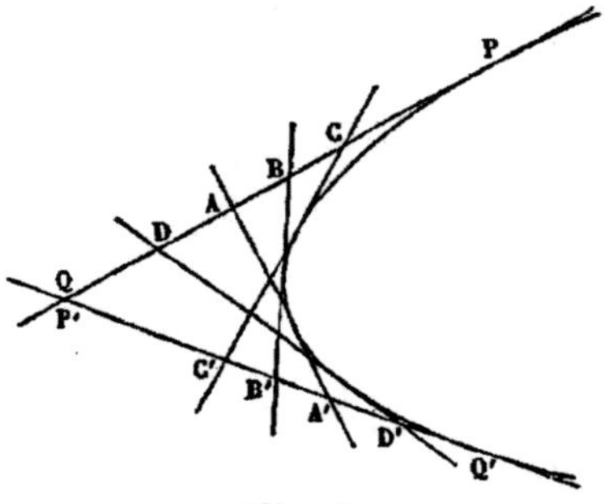

Fig. 100.

the two fixed tangents will be denoted by Q or P' according as it is regarded as a point of the first or of the second tangent. We have then

$$\frac{AB}{A'B'}=\frac{AC}{A'C'}=\ldots=\frac{BC}{B'C'}=\ldots=\frac{AP}{A'P'}=\frac{AQ}{A'Q'}=\ldots=\frac{PQ}{P'Q'}.$$

(4). Conversely, *given two straight lines in a plane, on which lie two similar ranges (which are not in perspective), the straight lines connecting pairs of corresponding points will envelope a parabola which*

Society of London for 1735, and CHASLES, *Aperçu historique sur l'origine et le développement des méthodes en Géométrie* (Brussels, 1837; second edition, Paris, 1875). If B lies at infinity, the theorem becomes identical with lemma 20, book i. of NEWTON'S *Principia.*

* APOLLONII PERGAEI *Conicorum* lib. iii. 41.

touches the given straight lines at the points which correspond in the two ranges respectively to their point of intersection.

For the points at infinity on the given straight lines being corresponding points (Art. 99), the straight line which joins them will be a tangent to the envelope; thus the envelope is a conic (Art. 150 (II.)) which has the line at infinity for a tangent, *i. e.* it is a parabola.

158. In theorem I of Art. 150 (Fig. 95) suppose that the point A lies at infinity, or, in other words, that the pencil A consists of parallel rays. To the straight line OA, considered as a ray a' of the pencil O (viz. that ray which is parallel to the rays of the other pencil), corresponds that ray a of the pencil A which is the tangent at the point A. This ray a may be at a finite, or it may be at an infinite distance.

In the first case (Fig. 101) the straight line at infinity is a ray j of the pencil A, and to it corresponds in the pencil O a ray j' different from a' and consequently not passing through A; the conic will therefore be a hyperbola (Art. 23) having $A (\equiv aa')$ and jj' for its points at infinity; the straight line a is one asymptote and j' is parallel to the other.

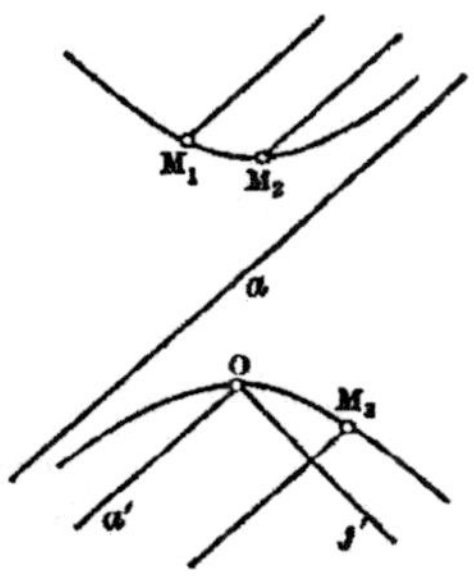

Fig. 101.

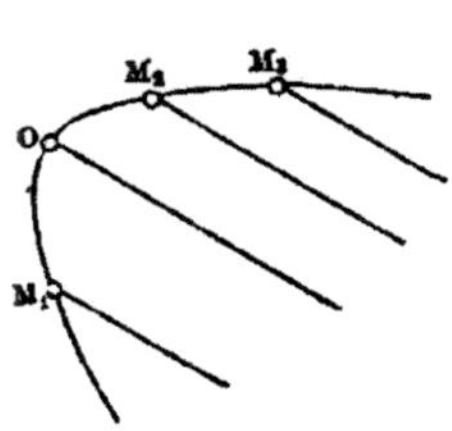

Fig. 102.

In the second case (Fig. 102) the line at infinity is the tangent at A to the conic, which is therefore a parabola.

159. If in this same theorem of Art. 150 the points A and O are supposed both to lie at infinity (Fig. 103), the two projective pencils will each consist of parallel rays; and since the conic which these pencils generate must pass through A and O it is a hyperbola (Art. 23). The asymptotes of the hyperbola are the tangents to the curve at its infinitely distant points*; they will therefore be the rays a and o' of the first

* Desargues, *loc. cit.*, p. 210; Newton, *Principia*, lib. i. prop. 27, Scholium.

and second pencil which correspond to the straight line at infinity considered as a ray of the second and first pencil respectively.

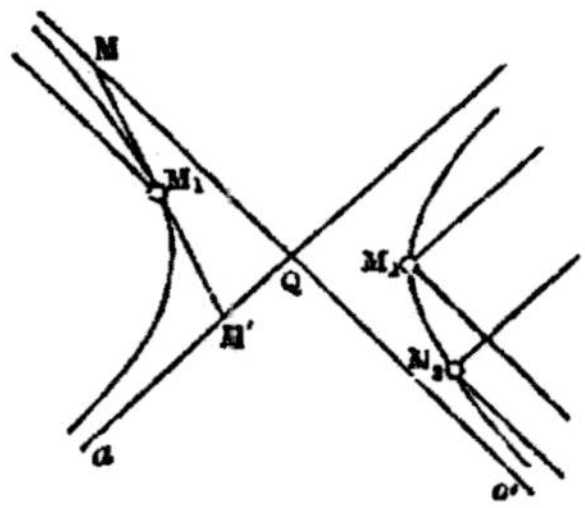

Fig. 103.

By the general theorem of Art. 149, the asymptotes of a hyperbola are cut by the other tangents in points forming two projective ranges, in which the points of contact (which are in this case at infinity) correspond respectively to the point of intersection Q of the asymptotes. The equation of Arts. 74 and 109 (1), viz.

$$JM \,.\, I'M' = \text{constant}$$

becomes therefore in this case

$$QM \,.\, QM' = \text{constant},$$

M and M' being the points of intersection of any tangent with the asymptotes. We conclude therefore that

The segments which are determined by any tangent to a hyperbola on the two asymptotes (measured from the point of intersection of the asymptotes), are such that the rectangle contained by them is constant.

This may be stated in a different form as follows:

The triangle formed by any tangent to a hyperbola and the asymptotes has a constant area.*

160. Again, let the theorem of Art. 149 be applied to the case of two fixed parallel tangents which are cut by a variable tangent in M and M'. In the projective ranges thus generated the points which correspond respectively to the infinitely distant point of intersection of the two fixed tangents are their points of contact; if these be denoted by J and I', we have by Art. 74 the equation

$$JM \,.\, I'M' = \text{constant}.$$

Therefore, *the segments which a variable tangent to a conic cuts off from two fixed parallel tangents (measured from the points of contact of these latter) are such that the rectangle contained by them is constant†.*

* Apollonius, *loc. cit.*, iii. 43.

† Ibid., iii. 42.

CHAPTER XV.

CONSTRUCTIONS AND EXERCISES.

161. By help of Pascal's and Brianchon's theorems may be solved the following problems:

Given five tangents a, b', c, a', b, to a conic, to draw from any given point H, lying on one of these tangents a, another tangent to the curve (Fig. 104).

Given five points A, B', C, A', B on a conic, to find the point of intersection of the curve with any given straight line r drawn through one of these points A (Fig. 105).

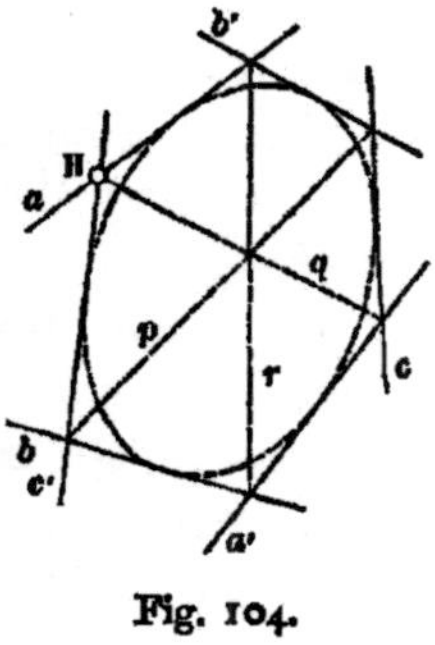

Fig. 104.

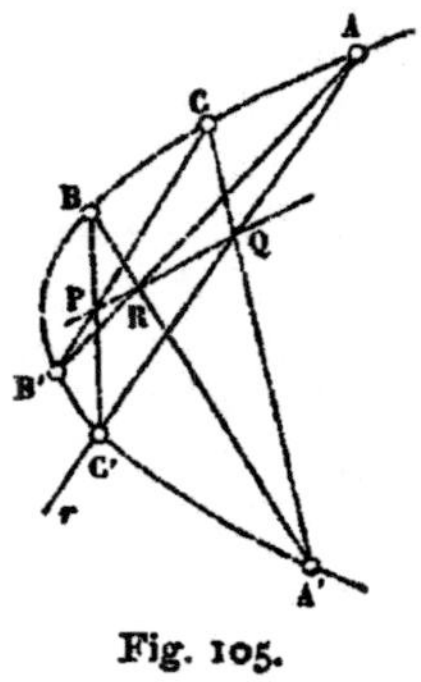

Fig. 105.

If c' be the required tangent, $ab'ca'bc'$ is a hexagon to which Brianchon's theorem applies. Let r be the diagonal connecting one pair ab' and $a'b$ of opposite vertices, and let q be the diagonal connecting another such pair ca' and $c'a$ (where $c'a$ is the given point H); then the diagonal which connects the remaining pair bc' and $b'c$ must pass through the point qr. If then p be the straight

If C' be the required point, $AB'CA'BC'$ is a hexagon to which Pascal's theorem applies. Let R be the point of intersection of one pair AB' and $A'B$ of opposite sides, and let Q be the point of intersection of another such pair CA' and r; then QR must pass through the point of intersection of the remaining pair BC' and $B'C$. If then PB be joined, it will cut the given line joining the points qr and $b'c$, the straight line which joins pb to the given point H is the required tangent.

By assuming different positions for the point H, all lying on one of the given tangents, and repeating in each case the above construction, any desired number of tangents to the conic may be drawn.

Brianchon's theorem therefore serves to construct, by means of its tangents, the conic which is determined by five given tangents *.

straight line r in the required point C'.

By assuming different positions for the given straight line r, all passing through one of the given points on the conic, and repeating in each case the above construction, any desired number of points on the conic may be found.

Pascal's theorem therefore serves to construct, by means of its points, the conic which is determined by five given points †.

162. Particular cases of the problem of Art. 161 (right).

I. Suppose the point B to lie at infinity; the problem then becomes the following:

Given four points A, B', C, A' on a hyperbola and the direction of one asymptote, to find the second point of intersection C' of the curve with a given straight line r drawn through A (Fig. 106).

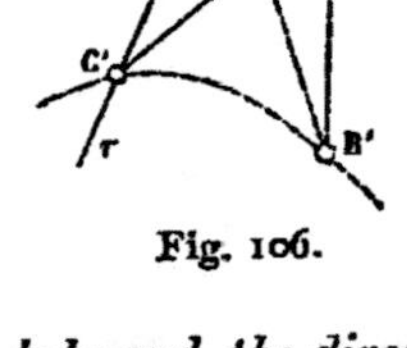

Fig. 106.

Solution. This is deduced from that of the general problem by taking the point B to lie at infinity in the given direction. We draw through A' a straight line m in this direction; if then AB' meets m in R, and $A'C$ meets r in Q, we join QR meeting $B'C$ in P, and draw through P a parallel to m; this parallel will cut r in the required point C'.

II. Suppose the point A to lie at infinity; the problem is then:

Given four points B', C, A', B on a hyperbola and the direction of one asymptote, to find the point of intersection of the curve with a given straight line r drawn parallel to this asymptote (Fig. 107).

Solution. Draw through B' a straight line parallel to the given direction. If this line meet $A'B$ in R, and if $A'C$ meet r

* Brianchon, *loc. cit.*, p. 38; Poncelet, *loc. cit.*, Art. 209.

† Newton, *Principia*, prop. 22; Maclaurin, *De linearum geometricarum proprietatibus generalibus* (London, 1748), § 44.

in Q, join QR cutting $B'C$ in P. Then if BP be joined, it will cut r in the required point C'.

III. Suppose the two points A' and B both to lie at infinity. The problem then becomes:

Given three points A, B', C on a hyperbola and the directions of both asymptotes, to find the second point of intersection of the curve with a given straight line r drawn through A (Fig. 108).

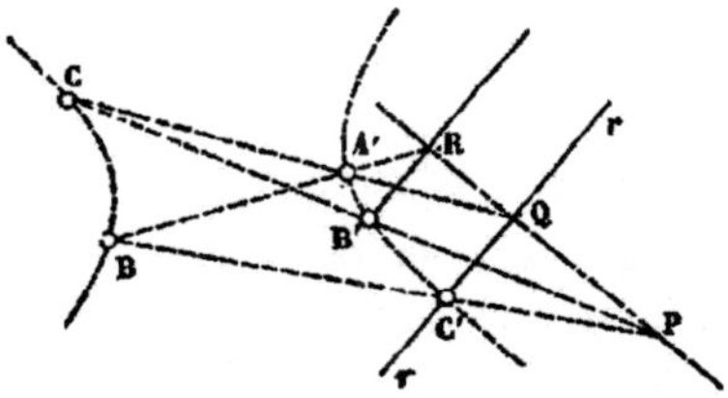

Fig. 107.

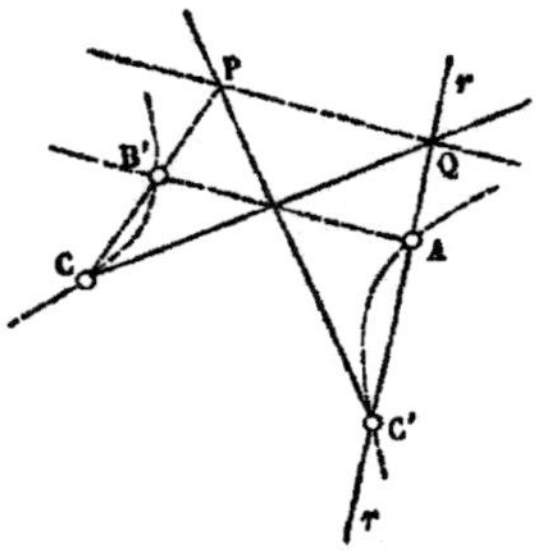

Fig. 108.

Solution. Through the point Q, where the given straight line r meets a straight line drawn through C parallel to the direction of the first asymptote, draw a parallel to AB'. Let P be the point where this parallel cuts $B'C$; then a parallel through P to the second asymptote will cut r in the required point C'.

IV. If the two points A and B' both lie at infinity, the problem is:

Given three points C, A', B of a hyperbola and the directions of both asymptotes, to find the point of intersection of the curve with a given straight line r drawn parallel to one of the asymptotes (Fig. 109).

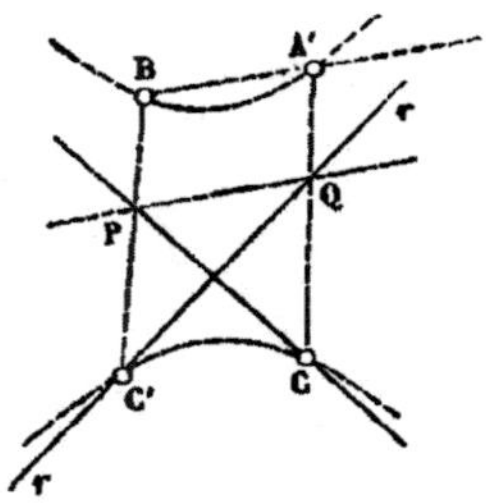

Fig. 109.

Solution. Through Q, the point of intersection of r and CA', draw a parallel to $A'B$; let P be the point where this parallel meets the straight line drawn through C parallel to the other asymptote. Then if BP be joined, it will cut r in the required point C'.

V. If, lastly, the points B', C, A', B are finite and the straight line AC' lies at infinity, the problem becomes the following:

Given four points B', C, A', B of a hyperbola and the direction of one asymptote, to find the direction of the other asymptote (Fig. 110).

Solution. Through the point R, in which $A'B$ meets the straight line drawn through B' in the given direction, draw a parallel to CA'; let P be the point where this parallel cuts $B'C$. Then if BP be joined, it will be parallel to the required direction.

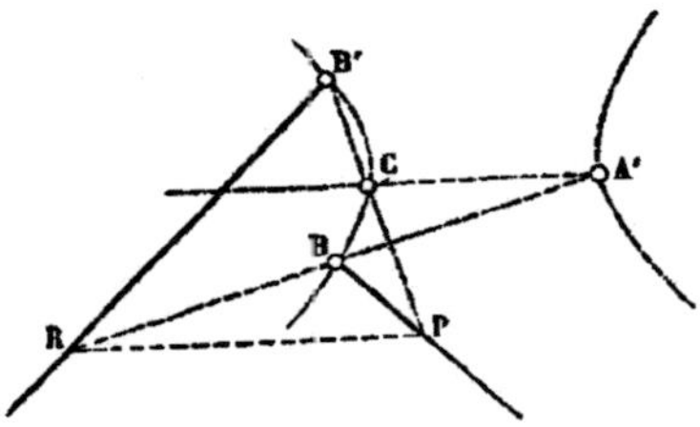

Fig. 110.

It will be a useful exercise for the student to deduce the constructions for these particular cases from the general construction; in order to do this it is only necessary to remember that to join a finite point to a point lying at infinity in a given direction we merely draw through the former point a parallel to the given direction.

163. Particular cases of the problem of Art. 161 (left).

I. Suppose the point ac' to lie at infinity; then the problem becomes the following:

Given five tangents a, b', c, a', b to a conic, to draw the tangent which is parallel to one of them, to a, for example (Fig. 111).

Solution. Draw through the point $a'c$ a straight line q parallel to a; join ab' and $a'b$ by the straight line r, and join the points qr and $b'c$ by the straight line p. Then if through the point pb a parallel be drawn to a, it will be the required tangent.

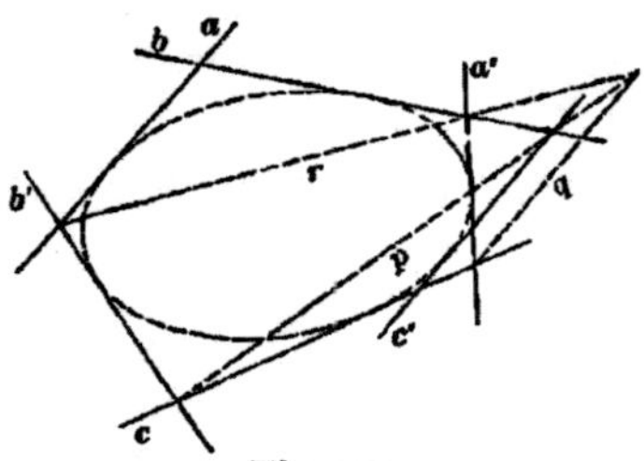

Fig. 111.

From a given point in the plane of a conic two tangents at most can be drawn to the curve (Art. 23); so that from a point lying on a given tangent only one other tangent can be drawn. If then the conic is a parabola, it cannot have a pair of parallel tangents. (This has already been seen in Art. 157 (1).)

II. Suppose the straight line b to lie at infinity; the problem is then:

Given four tangents a, b′, c, a′ to a parabola, to draw from a given point H lying on one of them, a, another tangent to the curve (Fig. 112).

Solution. Through the point ab' draw a straight line r parallel to a'; join the points H and $a'c$ by the straight line q, and the points qr and $b'c$ by the straight line p. The straight line drawn through H parallel to p will be the required tangent.

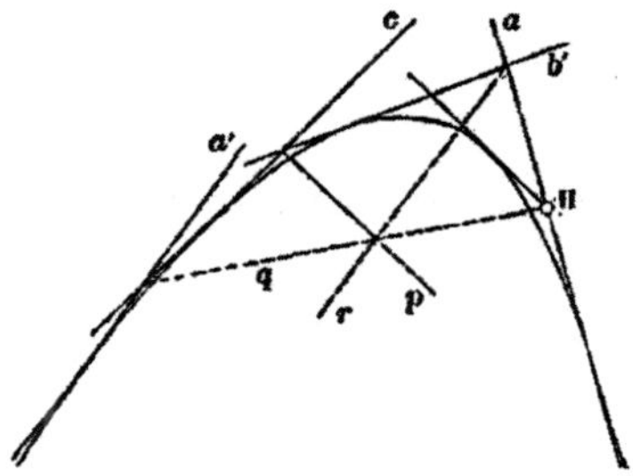

Fig. 112.

III. If the straight line a lies at infinity, we have the problem:

Given four tangents b′, c, a′, b to a parabola, to draw the tangent which is parallel to a given straight line (Fig. 113).

Solution. Through $a'b$ draw the straight line r parallel to b', and through $a'c$ draw the straight line q parallel to the given direction; join the points qr, $b'c$ by the straight line p. The straight line through pb parallel to the given direction is the tangent required.

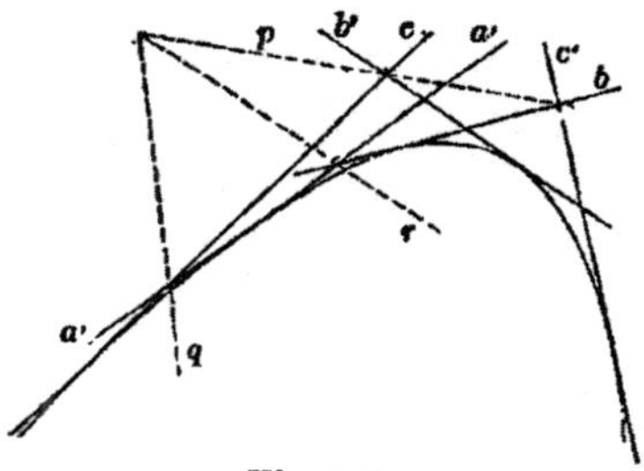

Fig. 113.

IV. If in problem II the point H assume different positions on a, or if in III the given straight line assume different directions, we arrive at the solution of the problem:

To construct by means of its tangents the parabola which is determined by four given tangents.

CHAPTER XVI.

DEDUCTIONS FROM THE THEOREMS OF PASCAL AND BRIANCHON.

164. We have already given some propositions and constructions (Arts. 161–163) which follow immediately from the theorems of Pascal and Brianchon, by supposing some of the elements to pass to infinity. Other corollaries may be deduced by assuming two of the six points or six tangents to approach indefinitely near to one another *.

If $AB'CA'BC'$ are six points on a conic, Pascal's theorem asserts that the pencils $A(A'B'CC')$ and $B(A'B'CC')$, for example, are projective with one another. To the ray AB of the first pencil corresponds in the second the tangent at B, so that we may say that the group of four lines

$$AA', AB', AC, AB$$

is projective with the group

$$BA', BB', BC, \text{tangent at } B.$$

But this amounts evidently to saying that the point C', which was at first taken to have any arbitrary position on the curve, has come to be indefinitely near to the point B. Instead then of the inscribed hexagon we have now the figure made up of the inscribed pentagon $AB'CA'B$ and the tangent b at the vertex B (Fig. 114); and Pascal's theorem becomes the following:

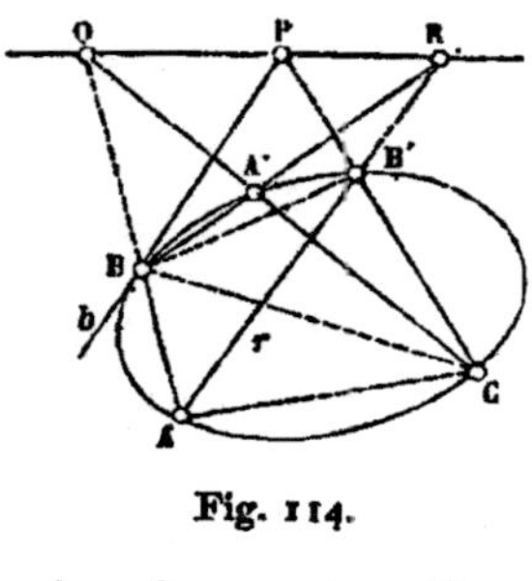

Fig. 114.

If a pentagon is inscribed in a conic, the points of intersection R, Q of two pairs of non-consecutive sides (AB' and $A'B$, AB and CA'), and the

* Carnot, *loc. cit.*, pp. 455, 456.

point P where the fifth side ($B'C$) meets the tangent at the opposite vertex (B), are collinear.

This corollary may also be deduced from the construction (Art. 84, right) for two projective pencils. Three pairs of corresponding rays are here given, viz. AA' and BA', AC and BC, AB' and BB'. We cut the two pencils by the transversals CA', CB' respectively; if R be the point of intersection of $A'B$ and AB', then any pair of corresponding rays of the two pencils must cut the transversals CA', CB' respectively in two points which are collinear with R. In order then to obtain that ray of the second pencil which corresponds to AB, viz. the tangent at B, we join R to the point of intersection Q of CA' and AB, and join QR meeting CB' in P; then BP is the required ray b. But this construction agrees exactly with the corollary enunciated above.

165. By help of this corollary the two following problems can be solved:

(1). *Given five points A, B', C, A', B of a conic, to draw the tangent at one of them B* (Fig. 114).

Solution. Join Q, the point of intersection of AB and CA', to R, the point of intersection of AB' and $A'B$; if P is the point where QR meets $B'C$, then BP will be the required tangent *.

Particular cases.

Given four points of a hyperbola and the direction of one asymptote, to draw the tangent at one of the given points. (This is obtained by taking one of the points A, B', C, A' to lie at infinity.)

Given four points of a hyperbola and the direction of one asymptote, to draw that asymptote. (B at infinity.)

Given three points of a hyperbola and the directions of both asymptotes, to draw the tangent at one of the given points. (Two of the four points A, B', C, A' at infinity.)

Given three points of a hyperbola and the directions of both asymptotes, to draw one of the asymptotes. (B and one of the other points at infinity.)

(2). *Given four points A, B, A', C of a conic and the tangent at one of them B, to construct the conic by points; for example, to find the point of the curve which lies on a given straight line r drawn through A* (Fig. 114).

Solution. Let R be the point where $A'B$ meets r, and Q the point where AB meets CA'; and let QR cut the given tangent in P. The point B' where CP cuts the given straight line r will be the one required.

By supposing one or more of the elements of the figure to lie at

* MACLAURIN, *loc. cit.*, § 40.

infinity, *e.g.* one of the points A, A', C; or two of these points; or the point A and the line r; or the point B; or the point B and one of the other points; or the point B and the given tangent; we obtain the following particular cases:

To construct by points a hyperbola, having given

three points of the curve, the tangent at one of these points, and the direction of one asymptote;

or: two points, the tangent at one of them, and the directions of both asymptotes;

or: three points and an asymptote;

or: two points, one asymptote, and the direction of the other asymptote.

Given three points of a hyperbola, the tangent at one of them, and the direction of an asymptote, to find the direction of the other asymptote.

To construct by points a parabola, having given three points of the curve (lying at a finite distance) and the direction of the point at infinity on it.

166. Returning to the hexagon $AB'CA'BC'$ inscribed in a conic, let not only C' be taken indefinitely near to B, but also C indefinitely near to B'. The figure will then be that of an inscribed quadrangle $AB'A'B$ together with the tangents at B and B' (Fig. 115), and Pascal's theorem becomes the following:

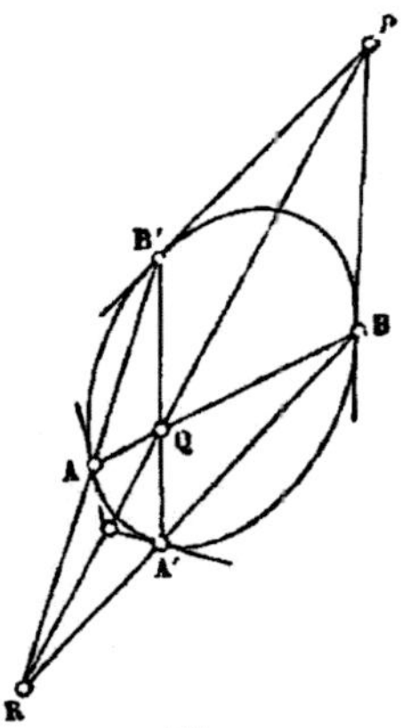

Fig. 115.

If a quadrangle is inscribed in a conic, the points of intersection of the two pairs of opposite sides, and the point of intersection of the tangents at a pair of opposite vertices, are three collinear points.

This property coincides with one already obtained elsewhere (Art. 85, right). For considering the projective pencils of which BA and $B'A$, BA' and $B'A'$, ... are corresponding rays, it is seen that the straight line which joins the point of intersection Q of BA and $B'A'$ to the point of intersection R of $B'A$ and BA' must pass through the point of intersection P of the rays which correspond in the two pencils respectively to the straight line joining their centres B and B'.

167. By help of the foregoing corollary the following problems can be solved:

(1). *Given four points A, B', A', B of a conic and the tangent BP at one of them B, to draw the tangent at another of the points B'* (Fig. 115).

Solution. Let AB and $A'B'$ meet in Q, and AB' and $A'B$ in R; and let QR meet the given tangent in P. Then $B'P$ will be the required tangent*.

By supposing one of the given points, or the given tangent, to lie at infinity, the solutions of the following particular cases are obtained:

To draw the tangent at a given point of a parabola, having given in addition two other points on the curve, the tangent at one of them, and the direction of one asymptote; *or*, one other point, the tangent at this, and the directions of both asymptotes; *or*, one other point, one asymptote, and the direction of the other asymptote.

To draw the asymptote of a hyperbola when its direction is known, having given in addition three points on the curve and the tangent at one of them; *or*, two points on the curve, the tangent at one of them, and the direction of the second asymptote; *or*, two points on the curve and the second asymptote.

To draw the tangent at a given point of a parabola, having given two other finite points on the curve, and the direction of the point at infinity on it.

(2). *To construct a conic by points, having given three points A, B, B' on the curve and the tangents BP, $B'P$ at two of them; i. e.* to determine, for example, the point A' in which an arbitrary straight line r drawn through B is cut by the conic (Fig. 116).

Solution. Join the point of intersection P of the given tangents to the point R where r cuts AB'; and let PR cut AB in Q. If $B'Q$ be joined, it will cut r in the required point A'.

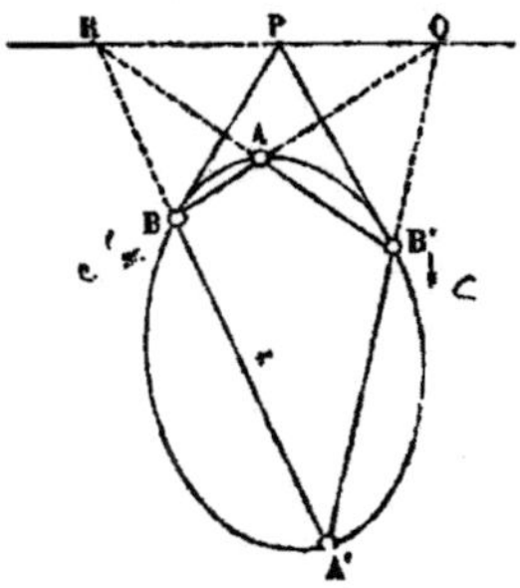

Fig. 116.

By supposing one of the points A, B, B' or one of the lines BP, $B'P$, r to lie at infinity, we shall obtain the solutions of the following particular cases:

To construct by points a hyperbola, having given two points on the curve, the tangents at these, and the direction of one asymptote; *or*, one point on the curve, the tangent there, one asymptote and the direction of the second asymptote; *or*, one point on the curve and both asymptotes.

To construct by points a parabola, having given two points on the

* MACLAURIN, *loc. cit.*, § 38.

curve, the tangent at one of them, and the direction of the point at infinity on the curve.

168. The tangents at the other vertices A and A' of the quadrangle $ABA'B'$ (Fig. 116) will also intersect on the straight line joining the points $(AB, A'B')$ and $(AB', A'B)$. Hence the theorem of Art. 166 may be enunciated in the following, its complete form:

If a quadrangle is inscribed in a conic, the points of intersection of the two pairs of opposite sides, and the points of intersection of the tangents at the two pairs of opposite vertices, are four collinear points.

If two opposite vertices of the quadrangle be taken to lie at infinity, this becomes the following:

If on a chord of a hyperbola, as diagonal, a parallelogram be constructed so as to have its sides parallel to the asymptotes, the other diagonal will pass through the point of intersection of the asymptotes.

169. THEOREM. *The complete quadrilateral formed by four tangents to a conic, and the complete quadrangle formed by their four points of contact, have the same diagonal triangle.*

In the last two figures write C, D, E, G in place of

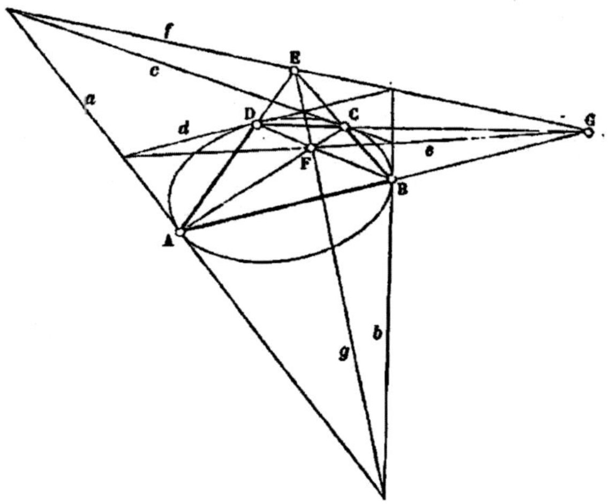

Fig. 117.

A', B', R, Q respectively. In the inscribed quadrangle $ABCD$ (Fig. 117) the point of intersection of the tangents at A and C, that of the tangents at B and D, the point of intersection of the sides AD, BC, and that of the sides AB, CD all lie on one straight line EG. If the same points A, B, C, D are taken in a different order, two other inscribed quadrangles $ACDB$ and $ACBD$ are obtained, to each of which the theorem of Art. 168 may be applied. Taking the quadrangle $ACDB$, it is seen that the point of intersection of the tangents at A and D, that of the tangents at C and B, the point of intersection of the sides AB, CD, and that of the sides AC, BD all lie on one straight line FG. So too the quadrangle $ACBD$ gives four points lying on one straight line EF; viz. the points of intersection of the tangents at A and B, of the tangents at C and D, of the sides AD, CB, and of the sides AC, BD*.

The three straight lines EG, GF, FE thus obtained are the sides of the diagonal triangle EFG (Art. 36, [2]) of the complete quadrangle whose vertices are the points A, B, C, D; and since the same straight lines contain also the points in which intersect two and two the tangents a, b, c, d at these points, they are also the diagonals of the complete quadrilateral formed by these four tangents. The theorem is therefore proved.

170. In the complete quadrilateral $abcd$ the diagonal f, whose extremities are the points ac, bd, cuts the other two diagonals g and e in E and G respectively; these two points are therefore harmonically conjugate with regard to ac and bd (Art. 56). The correlative theorem is: The two opposite sides of the complete quadrangle $ABCD$ which meet in F are harmonically conjugate with regard to the straight lines which connect F with the two other diagonal points E and G (Art. 57). Summing up the preceding, we may enunciate the following proposition (Fig. 117):

If at the vertices of a (simple) quadrangle $ABCD$, inscribed in a conic, tangents a, b, c, d be drawn, so as to form a (simple) quadrilateral circumscribed to the conic, then this quadrilateral possesses the following properties with regard to the quadrangle: (1) *the diagonals of the two pass through one point* (F) *and form a harmonic pencil;* (2) *the points of intersection of the pairs of opposite sides of the two lie on one straight line* (EG) *and form a harmonic range;* (3) *the*

* MACLAURIN, *loc. cit.*, § 50; CARNOT, *loc. cit.*, pp. 453, 454.

diagonals of the quadrilateral pass through the points of intersection of the pairs of opposite sides of the quadrangle *.

171. By help of the theorem of Art. 169, when we are given four tangents a, b, c, d to a conic and the point of contact A of one of them, we can at once find the points of contact of the three others; and when we are given four points A, B, C, D on a conic and the tangent a at one of them, we can draw the tangents at the three other points †.

Solution. Draw the diagonal triangle EFG of the complete quadrilateral $abcd$; then AG, AF, AE will cut b, c, d respectively in the required points of contact B, C, D.

Draw the diagonal triangle EFG of the complete quadrangle $ABCD$; then the straight lines joining ag, af, ae to B, C, D respectively will be the required tangents.

172. The theorem of Art. 169 may be enunciated with regard to the (simple) quadrilateral formed by the four straight lines a, b, c, d; it then takes the following form, under which it is seen to be already included in the theorem of Art. 170 ‡:

In a quadrilateral circumscribed to a conic, the straight lines which join the points of contact of the pairs of opposite sides pass through the point of intersection of the diagonals (Fig. 118).

This property coincides with one already proved with regard to two projective ranges (Art. 85, left). For consider the projective ranges on a and c as bases, in which ab and cb, ad and cd, ... are corresponding points; the straight lines which connect the pairs of points ab and cd, cb and ad respectively, must intersect on the straight line which connects the points corresponding in the two ranges respectively to ac; but this is the straight line joining the points of contact of a and c.

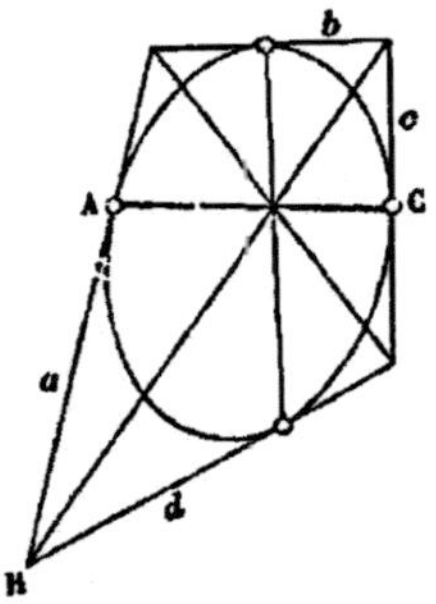

Fig. 118.

If the conic is a hyperbola, and we consider the quadrilateral which is formed by the asymptotes and any pair of tangents, the foregoing theorem expresses that the diagonals of such a quadrilateral are parallel to the chord which joins the points of contact of the two tangents §.

* CHASLES, *Sections coniques*, Art. 121.

† MACLAURIN, *loc. cit.*, §§ 38, 39.

‡ NEWTON, *loc. cit.*, Cor. ii. to lemma xxiv.

§ APOLLONIUS, *loc. cit.*, iii. 44.

173. The theorem of Art. 172 gives the solution of the problem: *To construct a conic by tangents, having given three tangents a, b, c and the points of contact A and C of two of them;* to draw, for example, through a given point H lying on a a second tangent to the curve (Fig. 118).

Solution. Join the point ab to the point of intersection of AC and $H(bc)$; the joining line will meet c in a point which when joined to H gives the required tangent d.

If one of the points A, C or one of the given tangents be supposed to lie at infinity, the solution of the following particular cases is obtained:

To construct by tangents a hyperbola, having given one asymptote, two tangents to the curve, and the point of contact of one of them; *or*, both asymptotes and one tangent.

To construct by tangents a parabola, having given the point at infinity on the curve, two tangents, and the point of contact of one of them; *or*, two tangents and the points of contact of both.

Given four tangents to a conic and the point of contact of one of them, to find the points of contact of the others.

174. If in Pascal's theorem the points A', B', C' be taken to lie indefinitely near to A, B, C respectively, the figure becomes that of an inscribed triangle ABC together with the tangents at its vertices (Fig. 119); and the theorem reduces to the following:

In a triangle inscribed in a conic, the tangents at the vertices meet the respectively opposite sides in three collinear points.

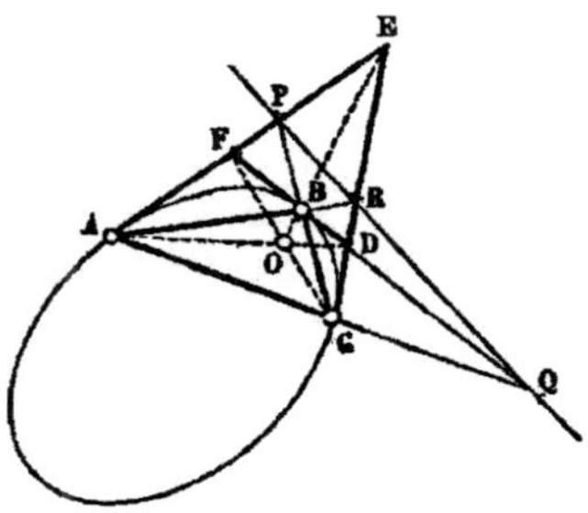

Fig. 119.

175. This gives the solution of the problem:

Given three points A, B, C of a conic and the tangents at two of them A and B, to draw the tangent at the third point C (Fig. 119).

Solution. Let P, Q be the points where the given tangents at A, B cut BC, CA respectively; if PQ cut AB in R, then CR is the tangent required.

The following are particular cases:

Given two points on a hyperbola, the tangents at these points, and the direction of one asymptote, to construct the asymptote itself.

Given one asymptote of a hyperbola, one point on the curve, the tangent at this point, and the direction of the second asymptote, to construct this second asymptote.

Given both asymptotes of a hyperbola and one point on the curve, to draw the tangent at this point.

(From the solution of this problem, it follows that the segment determined on any tangent by the asymptotes is bisected at the point of contact.)

Given two points on a parabola, the direction of the point at infinity on the curve, and the tangent at one of the given points, to draw the tangent at the other given point.

176. The inscribed triangle ABC and the triangle DEF formed by the tangents (Fig. 119) possess the property that their respective sides BC and EF, CA and FD, AB and DE intersect in pairs in three collinear points. The triangles are therefore homological, and consequently (Art. 18) the straight lines AD, BE, CF which connect their respective vertices pass through one point O. Thus we have the proposition:

In a triangle circumscribed to a conic, the straight lines which join the vertices to the points of contact of the respectively opposite sides are concurrent.

177. By help of this proposition the following problem can be solved:

Given three tangents to a conic and the points of contact of two of them, to determine the point of contact of the third.

Solution. Let DEF (Fig. 119) be the triangle formed by the three tangents, and let A, B be the points of contact of EF, FD respectively. If AD and BE intersect in O, then FO will cut the tangent DE in the required point of contact C.

Particular cases.

Given one asymptote of a hyperbola, two tangents, and the point of contact of one of them, to determine the point of contact of the other.

Given both asymptotes of a hyperbola, and one tangent, to determine the point of contact of the latter.

Given two tangents to a parabola and their points of contact, to determine the direction of the point at infinity on the curve.

Given two tangents to a parabola, the point of contact of one of them, and the direction of the point at infinity on the curve, to determine the point of contact of the other given tangent.

178. As a particular case of the theorem of Art. 176, consider a parabola and the circumscribing triangle formed by the tangents at any two points A, B, and the straight line at infinity, which is also

a tangent. If the tangents at A and B meet in C (Fig. 120), the straight line joining C to the middle point D of the chord AB will be parallel to the direction in which lies the point at infinity on the curve.

Again, if any point M be taken on AB, and parallels MP, MQ be drawn to BC, AC respectively to meet AC, BC in P, Q; and if MR be drawn parallel to DC to meet PQ in R; then PQ will be a tangent to the parabola, and R its point of contact.

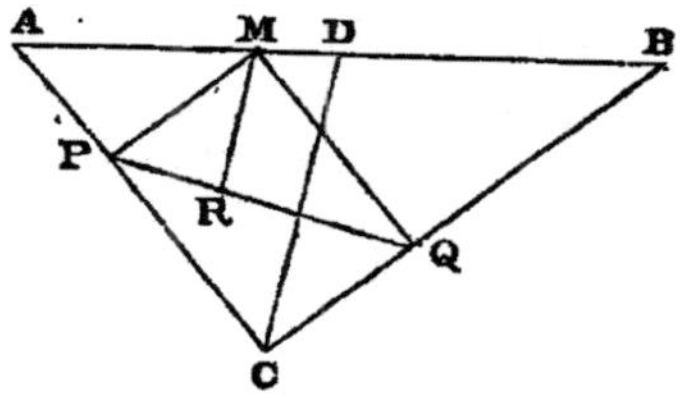

Fig. 120.

179. Just as from Pascal's theorem a series of special theorems have been derived, relating to the inscribed pentagon, quadrangle, and triangle, so also from Brianchon's theorem can be deduced a series of correlative theorems relating to the circumscribed pentagon, quadrilateral, and triangle.

Suppose *e.g.* that two of the six tangents a, b', c, a', b, c' which form the circumscribed hexagon (Art. 153, left), b and c' for example, lie indefinitely near to one another. Since a tangent intersects a tangent indefinitely near to it in its point of contact (Arts. 146, 149), the hexagon will be replaced by the figure made up of the circumscribed pentagon $ab'ca'b$ together with the point of contact of the side b (Fig. 121). Brianchon's theorem will then become the following:

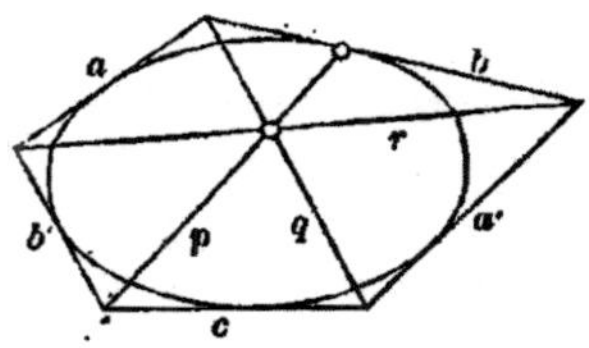

Fig. 121.

If a pentagon is circumscribed to a conic, the two diagonals which connect any two pairs of opposite vertices, and the straight line joining the fifth vertex to the point of contact of the opposite side, meet in the same point.

This theorem expresses a property of projective ranges which has already (Art. 85, left) been noticed.

For consider the two projective ranges determined by the other tangents on a and b as bases. Three pairs of corresponding points are given, *viz.* those determined by a', b', and c. Project the first range from the point ca' and the second from cb'; this gives two pencils in perspective of which corresponding pairs of rays intersect on the straight line r which joins the points ab', ba'. In order then to obtain that point of the second range which corresponds to the point ab of the first, *viz.* the point of contact of the tangent b, we draw the straight line q which joins the points ca' and ab, and then the straight line p which joins cb' and qr; then pb is the point required. But this construction agrees exactly with the theorem in question.

180. By means of the property of the circumscribed pentagon just established the following problems can be solved:

(1). *Given five tangents to a conic, to determine the point of contact of any one of them* *.

Particular case. Given four tangents to a parabola, to determine their points of contact, and also the direction of the point at infinity on the curve.

(2). *To construct by tangents a conic, having given four tangents and the point of contact of one of them.*

Particular cases.

To construct by tangents a hyperbola of which three tangents and one asymptote are given.

To construct by tangents a parabola, having given three tangents and the direction of the point at infinity on the curve; *or* three tangents and the point of contact of one of them.

181. The corollaries of Brianchon's theorem which relate to the circumscribed quadrilateral and triangle have already been given (they are the propositions of Arts. 172 and 176); they are correlative to the theorems of Arts. 166 and 174, just as those of Arts. 164 and 179 are correlative to one another.

It will be a very useful exercise for the student to solve for himself the problems enunciated in the present chapter: the constructions all depend upon two fundamental ones, correlative to one another, and following immediately from Pascal's and Brianchon's theorems.

182. The corollaries to the theorems of Pascal and Brianchon show that just as a conic is uniquely determined by five points or five tangents, so also it is uniquely determined by four points and the tangent at one of them, by four tangents and the point of contact of one of them, by three points and the tangents at two of them, or by three tangents and the points of contact of two of them. It follows that

(1). An infinite number of conics can be drawn to pass through three given points and to touch a given straight line at one of these points; *or* to pass through two given points and to touch at them two given straight lines; but no two of these conics can have another point in common.

* MACLAURIN, *loc. cit.*, § 41.

(2). An infinite number of conics can be drawn to touch a given straight line at a given point, and to touch two other given straight lines; *or* to touch two given straight lines at two given points; but no two of these conics can have another tangent in common.

If then two conics touch a given straight line at the same point (*i.e.* if the conics touch one another at this point), they cannot have in addition more than two common tangents or two common points; and if two conics touch two given straight lines at two given points (*i.e.* if two conics touch one another at two points) they cannot have any other common point or tangent.

Thus if two conics touch a straight line a at a point A, this point is equivalent to two points of intersection, and the straight line a is equivalent to two common tangents.

CHAPTER XVII.

DESARGUES' THEOREM.

183. THEOREM. *Any transversal whatever meets a conic and the opposite sides of an inscribed quadrangle in three conjugate pairs of points of an involution.*

CORRELATIVE THEOREM. *The tangents from an arbitrary point to a conic and the straight lines which join the same point to the opposite vertices of any circumscribed quadrilateral form three conjugate pairs of rays of an involution.*

This is known as DESARGUES' theorem *.

Let $QRST$ (Fig. 122) be a quadrangle inscribed in a conic,

Let $qrst$ (Fig. 123) be a quadrilateral circumscribed about a

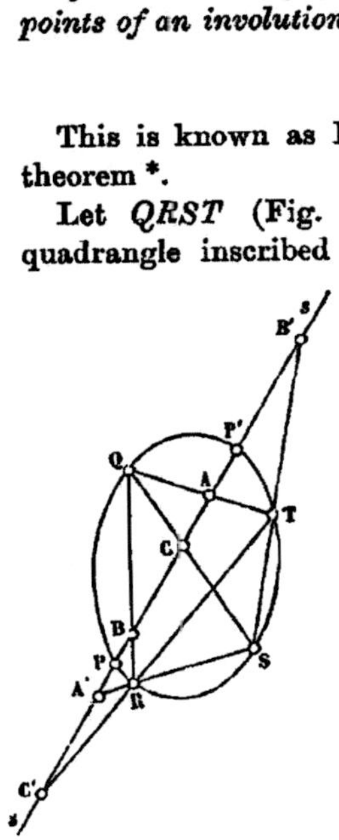

Fig. 122.

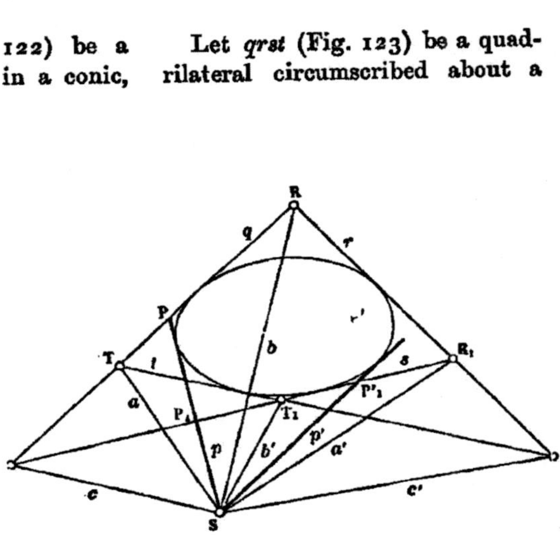

Fig. 123.

and let s be any transversal cutting the conic in P and P', and the sides QT, RS, QR, TS of the

conic; from any point S let tangents p, p' be drawn to the conic, and let the straight lines

* DESARGUES, *loc. cit.*, pp. 171, 176.

quadrangle in A, A', B, B' respectively.

The two pencils which join the points P, R, P', T of the conic to Q and S respectively are projective with one another (Art. 149), and the same is therefore true of the groups of points in which these pencils are cut by the transversal. That is, the group of points $PBP'A$ is projective with the group $PA'P'B'$, and therefore (Art. 45) with $P'B'PA'$; consequently (Art. 123) the three pairs of points

$$PP',\ AA',\ BB'$$

are in involution.

184. This theorem, like that of Pascal (Art. 153, right), enables us to construct by points a conic of which five points P, Q, R, S, T are given. For if (Fig. 122) an arbitrary transversal s be drawn through P, cutting QT, RS, QR, TS in A, A', B, B' respectively; and if (as in Art. 134) the point P' be found, conjugate to P in the involution determined by the pairs of points A, A' and B, B'; then will P' be another point on the conic to be constructed.

185. The pair of points C, C' in which the transversal cuts the diagonals QS and RT of the inscribed quadrangle belong also (Art. 131, left) to the involution determined by the points A, A' and B, B'.

Moreover, since the points A, A' and B, B' suffice to determine the involution, the points

a, a', b, b' be drawn which join S to the vertices qt, rs, qr, ts of the quadrilateral respectively.

The two groups of points in which q and s are cut by the tangents p, r, p', t are projective with one another (Art. 149), and the same is therefore true of the pencils formed by joining these points to S. That is, the group of rays $pbp'a$ is projective with the group $pa'p'b'$, and therefore (Art. 45) with $p'b'pa'$; consequently (Art. 123) the three pairs of rays

$$pp',\ aa',\ bb'$$

are in involution.

This theorem, like that of Brianchon (Art. 153, left), enables us to construct by tangents a conic of which five tangents p, q, r, s, t are given. For if (Fig. 123) an arbitrary point S be taken on p, and this point be joined to the points qt, rs, qr, ts respectively by the rays a, a', b, b'; and if (Art. 134) the ray p' be constructed, conjugate to p in the involution determined by the pairs of rays a, a' and b, b'; then will p' be another tangent to the conic to be constructed.

The pair of rays c, c' which connect S with the points of intersection qs and rt of the opposite sides of the circumscribed quadrilateral belong also (Art. 131, right) to the involution determined by the rays a, a' and b, b'.

Moreover, since the rays a, a' and b, b' suffice to determine the involution, the rays p, p' are a

P, P' are a conjugate pair of this involution for every conic, whatever be its nature, which circumscribes the quadrangle $QRST$.

Thus:

Any transversal meets the conics circumscribed about a given quadrangle in pairs of points forming an involution.

If the involution has double points, each of these is equivalent to two points of intersection P and P' lying indefinitely near to one another; and will therefore be the point of contact of the transversal with some conic circumscribing the quadrangle.

There are therefore either *two* conics which pass through four given points Q, R, S, T and touch a given straight line s (not passing through any of the given points), or there is *no* conic which satisfies these conditions.

186. If, from among the six points AA', BB', PP' of an involution, five are given, the sixth is determined (Art. 134). If then in Fig. 122 it is supposed that the conic is given, and that the quadrangle varies in such a way that the points A, A', B remain fixed, then also the point B' will remain invariable; consequently:

If a variable quadrangle move in such a way as to remain always inscribed in a given conic, while three of its sides turn each round one of three fixed collinear points, then the fourth side will turn round a fourth fixed point,

conjugate pair of this involution for every conic, whatever be its nature, which is inscribed in the quadrilateral $qrst$.

Thus:

The pairs of tangents drawn from any point to the conics inscribed in a given quadrilateral form an involution.

If the involution has double rays, each of these is equivalent to two tangents p and p' lying indefinitely near to one another; and will therefore be the tangent at S to some conic inscribed in the quadrilateral.

There are therefore either *two* conics which touch four given straight lines q, r, s, t and pass through a given point S (not lying on any of the given lines), or there is *no* conic which satisfies these conditions.

If, from among the six rays aa', bb', pp' of an involution, five are given, the sixth is determined (Art. 134). If then in Fig. 123 it is supposed that the conic is given, and that the quadrilateral varies in such a way that the rays a, a', b remain fixed, then also the ray b' will remain invariable; consequently:

If a variable quadrilateral move in such a way as to remain always circumscribed to a given conic, while three of its vertices slide each along one of three fixed concurrent straight lines, then the fourth vertex will slide along a

collinear with the three given ones. *fourth fixed straight line, concurrent with the three given ones.*

187. The theorem of the preceding Art. (left) may be extended to the case of any inscribed polygon having an even number of sides. Suppose such a polygon to have $2n$ sides, and to move in such a way that $2n-1$ of these pass respectively through as many fixed points all lying on a straight line s (Fig. 124). Draw the diagonals connecting the first of its vertices with the 4th, 6th, 8th, ... $2(n-1)^{th}$ vertex, thus dividing the polygon into $n-1$ simple quadrangles. In the first of these quadrangles the first three sides (which are the first three sides of the polygon) pass respectively through three fixed points on s; therefore also the fourth side (which is the first diagonal of the polygon) will pass through a fixed point on s. In the second quadrangle the first three sides (the first diagonal and the fourth and fifth side of the polygon) pass respectively through three fixed points on s; therefore the fourth side (the second diagonal of the polygon) will pass through a fixed point on s. Continuing in the same manner, we arrive at the last quadrangle and find that the fourth side of this (*i. e.* the $2n^{th}$ side of the polygon) passes through a fixed point on s. We may therefore enunciate the general theorem:

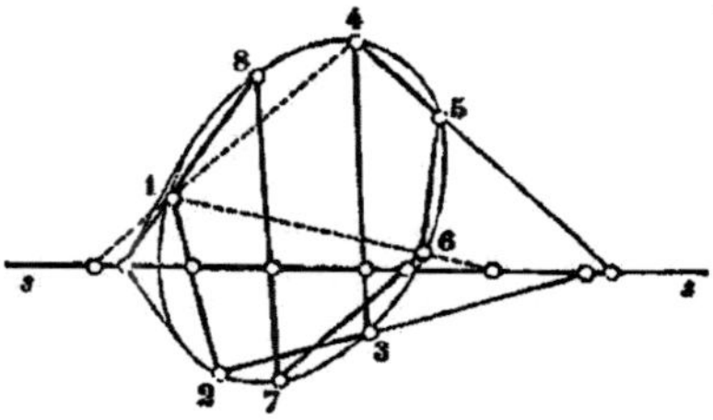

Fig. 124.

If a variable polygon of an even number of sides move in such a way as to remain always inscribed in a given conic, while all its sides but one pass respectively through as many fixed points lying on a straight line, then the last side also will pass through a fixed point collinear with the others.*

If tangents can be drawn to the conic from the fixed point round which the last side turns, and if each of these tangents is considered as a position of the last side, the two vertices which lie on this side will coincide and the polygon will have only $2n-1$ vertices. The point of contact of each of the two

* PONCELET, *loc. cit.*, Art. 513.

tangents will therefore be one position of one of the vertices of a polygon of $2n-1$ sides inscribed in the conic so that its sides pass respectively through the $2n-1$ given collinear points.

188. The solution of the correlative theorem is left as an exercise to the student: the enunciation is as follows:

If a variable polygon of an even number ($2n$) of sides moves so as to remain always circumscribed to a given conic, while all its vertices but one slide along as many fixed straight lines radiating from a centre, then the last vertex also will slide along a fixed straight line passing through the same centre (Fig. 125).

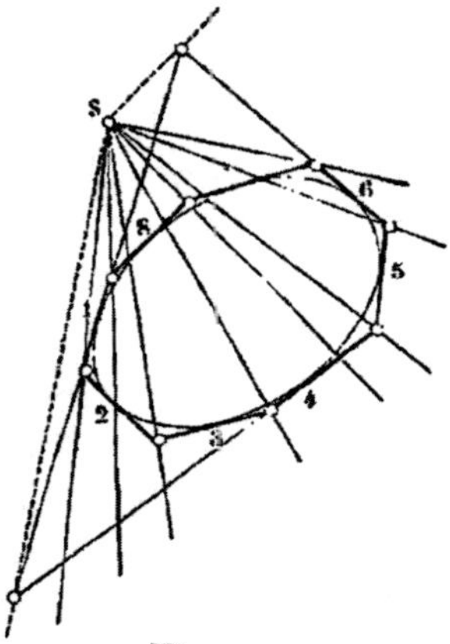

Fig. 125.

If the straight line on which this last vertex slides cut the conic in two points, and if the tangents at these be drawn, each of them will be one position of a side of a polygon of $2n-1$ sides circumscribed about the conic so that its vertices lie each on one of the $2n-1$ given concurrent straight lines.

189. If in Fig. 122 it be supposed that the points S and T lie indefinitely near to one another on the conic, or in other words that ST is the tangent at S, then the quadrangle $QRST$ reduces to the inscribed triangle QRS and the tangent at S (Fig. 126), so that Desargues' theorem becomes the following:

If a triangle QRS is inscribed in a conic, and if a transversal s meet two of its sides in A and A', the third side and the tangent at the opposite vertex in B and B', and the conic itself in P and P',

If in Fig. 123 the tangents s and t be supposed to lie indefinitely near to one another, so that st becomes the point of contact of the tangent s, then the quadrilateral $qrst$ reduces to the circumscribed triangle qrs and the point of contact of s (Fig. 127), so that the theorem correlative to that of Desargues becomes the following:

If a triangle qrs is circumscribed about a conic, and if from any point S there be drawn the straight lines a, a' to two of its vertices, the straight lines b, b' to the third vertex and the point of *these three pairs of points are in involution.*

190. This theorem gives a solution of the problem: *Given five*

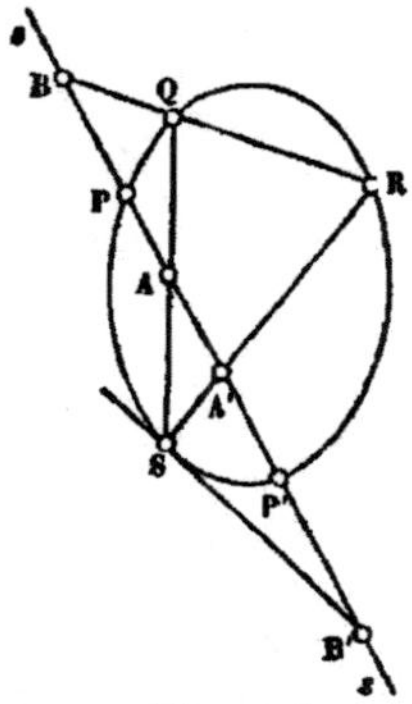

Fig. 126.

points P, P', Q, R, S on a conic, to draw the tangent at any one of them S.

For if A, A', B (Fig. 126) are the points in which the straight line PP' cuts the straight lines QS, SR, RQ respectively, we construct (as in Art. 134) the point B' conjugate to B in the involution determined by the two pairs of points A, A' and P, P'; then $B'S$ will be the required tangent.

191. If in Fig. 126 it be now supposed in addition that the points Q and R also lie indefinitely near to one another on the conic, *i.e.* that QR is the tangent at Q, then the inscribed quadrangle $QRST$ is replaced by the two tangents at Q and S and their chord of contact QS counted twice (Fig. 128).

Since the straight lines QT, RS now coincide, A and A' will

contact of the opposite side, and the tangents p, p' to the conic, then these three pairs of rays are in involution.

This theorem gives a solution of the problem: *Given five tangents*

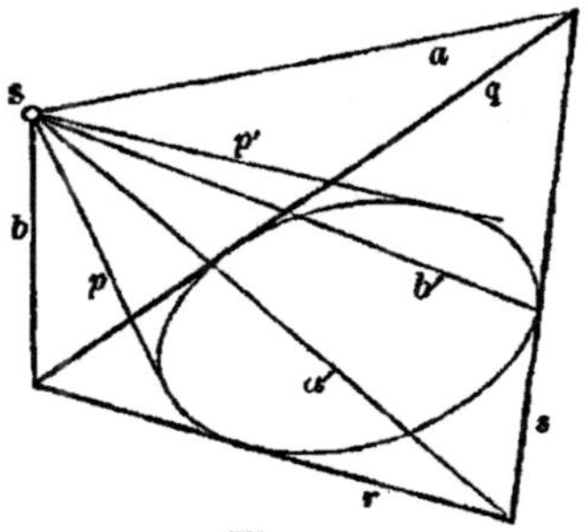

Fig. 127.

p, p', q, r, s to a conic, to find the point of contact of any one of them s.

For if a, a', b (Fig. 127) are the rays joining the point pp' to the points qs, sr, rq respectively, we construct (as in Art. 134) the ray b' conjugate to b in the involution determined by the two pairs of rays a, a' and p, p'; then $b's$ will be the required point of contact.

If in Fig. 127 it be now supposed in addition that the tangents q and r lie indefinitely near to one another, *i.e.* that qr is the point of contact of the tangent q, then the circumscribed quadrilateral $qrst$ is replaced by the points of contact of the tangents q and s and the point of intersection qs of these tangents counted twice (Fig. 129).

Since the points qt, rs now coincide in a single point qs, the also coincide in one point, which is consequently one of the double points of the involution determined by the pairs of conjugate

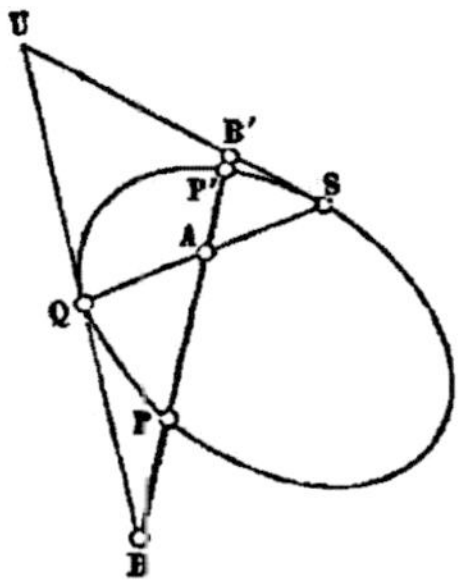

Fig. 128.

points P, P' and B, B'. In this case, then, Desargues' theorem becomes the following:

If a transversal cut two tangents to a conic in B and B', their chord of contact in A, and the conic itself in P and P', then the point A is a double point of the involution determined by the pairs of points P, P' and B, B'.

Or, differently stated:

If a variable conic pass through two given points P and P' and touch two given straight lines, the chord which joins the points of contact of these two straight lines will always pass through a fixed point on PP'.

If the tangents QU, SU vary at the same time with the conic, while the points P, P', B, B' remain fixed, the chord of contact

rays a and a' will also coincide in a single ray a, which is consequently one of the double rays of the involution determined by the

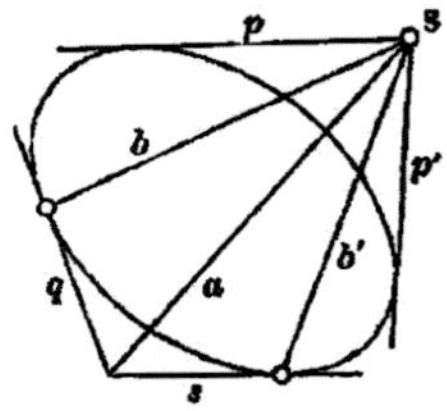

Fig. 129.

pairs of conjugate rays p, p' and b, b'. The theorem correlative to that of Desargues then becomes the following:

If a given point S be joined to two points on a conic by the straight lines b, b', and to the point of intersection of the tangents at these points by the straight line a; and if from the same point S there be drawn the two tangents p, p' to the conic; then a is a double ray of the involution determined by the pairs of rays p, p' and b, b'.

Or, differently stated:

If a variable conic touch two given straight lines p and p' and pass through two given points, the tangents at these two points will always intersect on a straight line passing through pp'.

If the points of contact of q and s vary at the same time with the conic, while the straight lines p, p', b, b' remain fixed, the point

QS must still always pass through one or other of the double points of the involution determined by the pairs of points P, P' and B, B'. If then four collinear points P, P', B, B' are given and any conic is drawn through P and P', and then the pairs of tangents from B and B' to this conic; then if each tangent from B is taken together with each tangent from B', four chords of contact will be obtained, which intersect one another two and two in the double points of the involution determined by P, P' and B, B' *.

of intersection qs must still always lie on one or other of the double rays of the involution determined by the pairs of rays p, p' and b, b'. If then four concurrent straight lines p, p', b, b' are given and any conic is drawn touching p and p', and then the two pairs of tangents to this conic at the points where it is cut by b and b'; then if the tangents at the two points on b are combined with the tangents at the two points on b', each with each, four points of intersection will be obtained, which lie two and two on the double rays of the involution determined by p, p' and b, b'.

192. From the theorem of the last Article (left) is derived a solution of the problem: *Given four points P, P', Q, S on a conic and the tangent at one of them Q, to draw the tangent at any other of the given points S* (Fig. 128).

For if A, B are the points in which PP' cuts QS and the given tangent respectively, and we construct the point B' conjugate to B in the involution determined by the pair of points P, P' and the double point A; then the straight line SB' will be the tangent required.

From the theorem of the last Article (right) is derived a solution of the problem: *Given four tangents p, p', q, s to a conic and the point of contact of one of them q, to determine the point of contact of any other of the given tangents s* (Fig. 129).

For if a, b are the rays which connect pp' with qs and with the given point of contact respectively, and we construct the ray b' conjugate to b in the involution determined by the pair of rays p, p' and the double ray a; then sb' will be the required point of contact.

193. Consider again the theorem of Art. 191; and suppose that the conic is a hyperbola, and that its asymptotes are the tangents given (Fig. 130). The chord of contact QS lies in this case entirely at infinity; so that the involution ($PP', BB', \ldots$) has one double point at infinity, and therefore (Arts. 59, 125) the other double point

* BRIANCHON, *loc. cit.*, pp. 20, 21.

is the common point of bisection of the segments $PP', BB', \ldots$ We conclude that:

If a hyperbola and its asymptotes be cut by a transversal, the segments intercepted by the curve and by the asymptotes respectively have the same middle point.

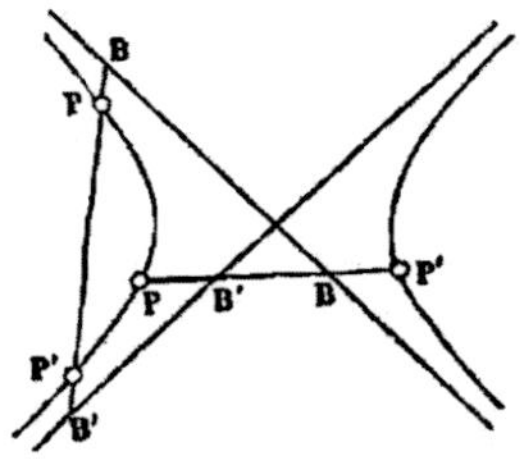

Fig. 130.

From this it follows that

$$PB = B'P' \text{ and } PB' = BP' \,^*,$$

which gives a rule for the construction of a hyperbola when the two asymptotes and a point on the curve are given †.

194. Consider once more the theorem of Art. 191 (left), and suppose now that the points P and P' are indefinitely near to one another, *i.e.* let the transversal be a tangent to the conic (Fig. 131). Its point of contact P will be the second double point of the involution determined by the pair of points B, B' and the double point A; consequently (Art. 125) P and A are harmonic conjugates

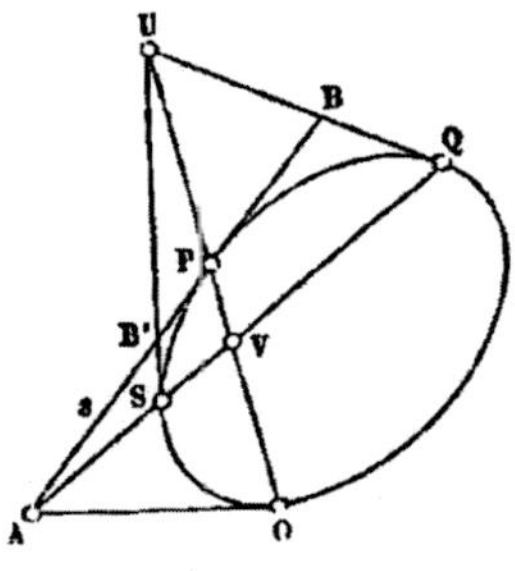

Fig. 131.

Consider once more the theorem of Art. 191 (right), and suppose now that the tangents p and p' lie indefinitely near to one another, *i.e.* let the point S lie on the conic itself (Fig. 132). The tangent to the conic at S will be the second double ray of the involution determined by the pair of rays b, b' and the double ray a; consequently (Art. 125) p and a are harmonic conjugates with

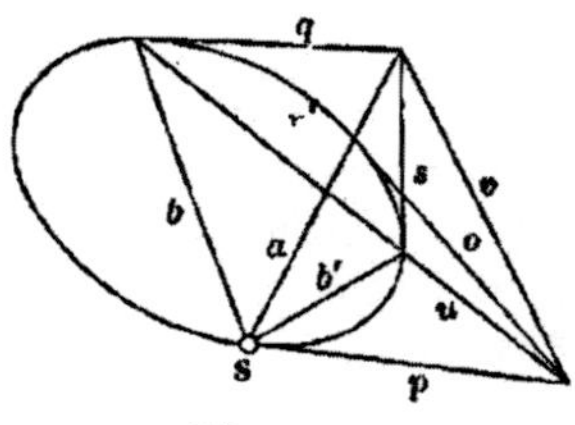

Fig. 132.

* Apollonius, *loc. cit.*, ii. 8, 16.

† Ibid., ii. 4.

with regard to B and B'; and we conclude that:

In a triangle UBB' circumscribed to a conic, any side BB' is divided harmonically by its point of contact P and the point where it meets the chord QS joining the points of contact of the other two sides.

195. From A a second tangent can be drawn to the conic; let its point of contact be O. Since the four points P, A, B, B', which have been shown to be harmonic, are respectively the point of contact of the tangent AB, and the three points where this tangent cuts three other tangents OA, QB, SB' respectively, it follows that the tangents AB, OA, QB, SB' will be cut by every other tangent in four harmonic points (Art. 149); *i.e.* they are four harmonic tangents (Art. 151). And since the chord of contact QS of the conjugate tangents QB, SB' passes through A the point of intersection of the tangents at P and O, we have the theorem:

If the chord of contact of one pair of tangents to a conic pass through the point of intersection of another pair of tangents, then each pair is harmonically conjugate with regard to the other.

And conversely:

If four tangents to a conic are harmonic, the chord of contact of each pair of conjugate tangents passes through the point of intersection of the other pair.

regard to b and b'; and we conclude that:

In a triangle ubb' inscribed in a conic, any two sides b and b' are harmonic conjugates with regard to the tangent p at the vertex in which they meet and the straight line joining this vertex to the point of intersection of the tangents q and s at the other two vertices.

The straight line a cuts the conic in a second point; let the tangent at this be o. Since the four rays p, a, b, b', which have been shown to be harmonic, are respectively the tangent at S, and the straight lines which join S to three other points on the conic (the points of contact of o, q, and s) it follows that the straight lines connecting these four points with any other point on the conic will form a harmonic pencil (Art. 149); *i.e.* the four points are harmonic (Art. 151). And since the point of intersection of the tangents q and s lies on the chord of contact of the tangents p and o, we have the theorem:

If the point of intersection of the tangents at one pair of points on a conic lie on the chord joining another such pair of points, then each pair is harmonically conjugate with regard to the other.

And conversely:

If four points on a conic are harmonic, the point of intersection of the tangents at each pair of conjugate points lies on the chord joining the other pair.

196. These two correlative propositions can be combined into one by virtue of the property already established (Arts. 148, 149) that the tangents at four harmonic points on a conic are themselves harmonic, and conversely. We may then enunciate as follows:

If a pair of tangents to a conic meet in a point lying on the chord of contact of another pair, then also the second pair will meet in a point lying on the chord of contact of the first; and the four tangents (and likewise their points of contact) will form a harmonic system.*

Thus in Fig. 131 QS passes through A, the point of intersection of PA and OA, and similarly OP passes through U, the point of intersection of QB and SB'; and the pencil $U(QSPA)$ is harmonic, and likewise the pencil $A(OPQU)$.

In Fig. 132 the point qs lies on a, the chord of contact of o and p, and similarly the point op lies on the straight line u which joins the points of contact of q and s; and the range $u(qsap)$ is harmonic, and the range $a(opqu)$ also.

197. Example. Suppose the conic to be a hyperbola (Fig. 133).

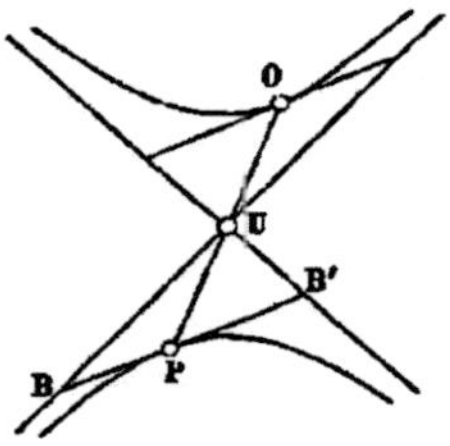

Fig. 133.

Its asymptotes are a pair of tangents whose chord of contact QS is the straight line at infinity; consequently the chord joining the points of contact of a pair of parallel tangents will pass through the point of intersection U of the asymptotes; and conversely, if through U a transversal be drawn, the tangents at the points P and O, where it cuts the curve, will be parallel. The point U will lie midway between P and O, since in general $UVPO$ (Fig. 131) is a harmonic range, and in this case V lies at infinity.

Any tangent to the curve cuts the asymptotes in two points B and B' which are harmonically conjugate with regard to the point of contact P and the point where the tangent meets the chord of contact of the asymptotes; but this last lies at infinity; therefore P is the middle point of BB'. Thus

The part of a tangent to a hyperbola which is intercepted between the asymptotes is bisected at its point of contact†.

This proposition is a particular case of that of Art. 193.

198. Theorem‡. *If a quadrangle is inscribed in a conic, the rectangle contained by the distances of any point on the curve from*

* De la Hire, *loc. cit.*, book i. prop. 30. Steiner, *loc. cit.*, p. 159, § 43; Collected Works, vol. i. p. 346.

† Apollonius, *loc. cit.*, ii. 319.

‡ To this Chasles has given the name of Pappus' theorem, since it corresponds to the celebrated '*problema ad quatuor lineas*' of this ancient geometer. Cf. *Aperçu historique*, pp. 37, 338.

one pair of opposite sides is to the rectangle contained by its distances from the other pair in a constant ratio.

In Fig. 122, the pairs of points P and P', A and A', B and B' being, by Desargues' theorem, in involution, the anharmonic ratios $(PP'AB)$ and $(P'PA'B')$ are equal to one another, or

$$\frac{PA}{P'A} : \frac{PB}{P'B} = \frac{P'A'}{PA'} : \frac{P'B'}{PB'}$$
$$= \frac{PB'}{P'B'} : \frac{PA'}{P'A'}.$$

But $PA : P'A$ is equal to the ratio of the distances (measured in any the same direction) of the points P and P' from the straight line QT, and the other ratios in the foregoing equation may be interpreted similarly; we have therefore

$$\frac{(A)}{(A)'} : \frac{(B)}{(B)'} = \frac{(B')}{(B')'} : \frac{(A')}{(A')'},$$

or

$$\frac{(A) \cdot (A')}{(B) \cdot (B')} = \frac{(A)' \cdot (A')'}{(B)' \cdot (B')'},$$

where (A), (A'), (B), (B') denote the distances of the point P from the sides QT, RS, QR, ST respectively of the inscribed quadrangle $QRST$, and $(A)'$, $(A')'$, $(B)'$, $(B')'$ denote similarly the distances of the point P' from these sides respectively. (These distances may be measured either perpendicularly or obliquely, so long as they are all measured parallel to one another.) The ratio

$$\frac{(A)\,(A')}{(B)\,(B')}$$

is therefore constant for all points P on the conic; which proves the theorem.

199. Theorem. *If a quadrilateral is circumscribed about a conic, the rectangle contained by the distances of one pair of opposite vertices from any tangent is to the rectangle contained by the distances of the other pair from the same tangent in a constant ratio*.*

In Fig. 123 let the vertices qr, qt, st, sr of the circumscribed quadrilateral $qrst$ be denoted by R, T, T_1, R_1 respectively; let the points where the tangents p, p' meet the side q be called P, P' respectively †, and let the points where these same tangents meet the side s be called P_1, P_1' respectively. Since by the theorem correlative to that of Desargues, the pairs of rays p and p', a and a', b and b', are in involution, the anharmonic ratios $(bapp')$ and $(b'a'p'p)$ are equal to one another. Hence by theorem (2) of Art. 149,

* Chasles, *Sections coniques*, Art. 26.

† P' is not shown in the figure.

$$(RTPP') = (T_1R_1P_1'P_1)$$
$$= (R_1T_1P_1P_1') \text{ by Art. 45};$$
$$\therefore \quad \frac{RP}{TP} : \frac{RP'}{TP'} = \frac{R_1P_1}{T_1P_1} : \frac{R_1P_1'}{T_1P_1'},$$

whence
$$\frac{RP \,.\, T_1P_1}{TP \,.\, R_1P_1} = \frac{RP' \,.\, T_1P_1'}{TP' \,.\, R_1P_1'}.$$

But $RP : TP$ is equal to the ratio of the distances (measured in any the same direction) of the points R and T from the straight line p; so also $T_1P_1 : R_1P_1$ is the ratio of the distances of the points T_1 and R_1 from the same straight line p. The foregoing equation therefore expresses that the ratio

$$\frac{RP \,.\, T_1P_1}{TP \,.\, R_1P_1}$$

is constant for every tangent p to the conic; which proves the theorem.

CHAPTER XVIII.

SELF-CORRESPONDING ELEMENTS AND DOUBLE ELEMENTS.

200. CONSIDER two projective flat pencils, concentric or non-concentric. Through their common centre or through their two centres O and O' draw a conic or a circle, and let this cut the rays of the first pencil in $A, B, C, \dots$ and those of the second in $A', B', C', \dots$. Project these two series of points from two new points O_1, O_1' (or from the same point) lying on the conic; the two projecting pencils $O_1(ABC \dots)$ and $O_1'(A'B'C' \dots)$ are by Art. 149 projective with the two given pencils $O(ABC \dots)$ and $O'(A'B'C' \dots)$ respectively; and are therefore projective with one another.

The two series of points $ABC \dots$ and $A'B'C' \dots$ are said to form two projective ranges on the conic.*

I. Now project these two ranges (Fig. 134) from two of their corresponding points, say from A' and A. The projecting pencils

$$A'(A, B, C, \dots) \text{ and } A(A', B', C', \dots)$$

will be projective with one another; and since they have the self-corresponding ray AA', they are in perspective. Corresponding pairs of rays will therefore (Art. 80) intersect on a fixed straight line, so that AB' and $A'B$, AC' and $A'C$, AD' and $A'D \dots$, will meet on one straight line s. If any point be taken on s, the straight lines joining it to A and A' will cut the conic again in another pair of corresponding points of the ranges $ABCD \dots$ and $A'B'C'D' \dots$.

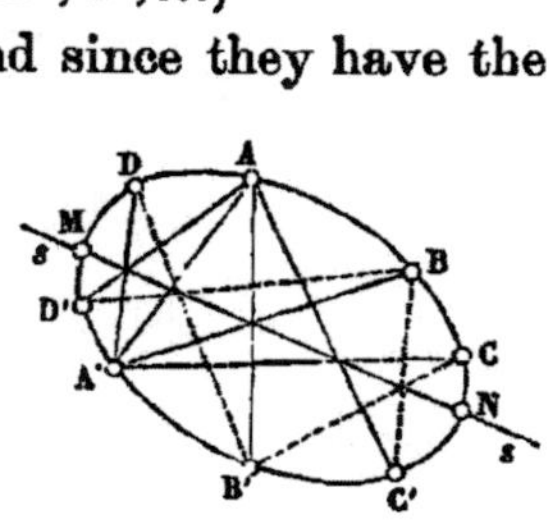

Fig. 134.

* BELLAVITIS, *Saggio di Geometria derivata* (Nuovi Saggi dell' Accademia di Padova, vol. iv. 1838, p. 270, note).

If instead of A' and A any other pair of corresponding points had been taken as centres of projection, say B' and B, the same straight line s would have been arrived at. For since $AB'CA'BC'$ is a hexagon inscribed in a conic, it follows by Pascal's theorem that the point of intersection of $B'C$ and BC' must lie on the straight line which joins the point of intersection of $A'B$ and AB' to that of $A'C$ and AC' (Art. 153, right).

II. Any point M in which the conic and the straight line s intersect is a *self-corresponding point* of the two ranges ABC... and $A'B'C'$.... For if M, M' be corresponding points

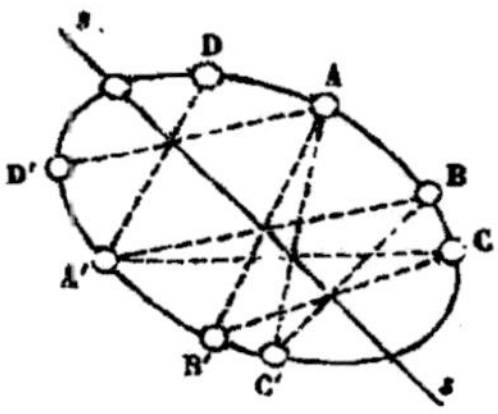

Fig. 135.

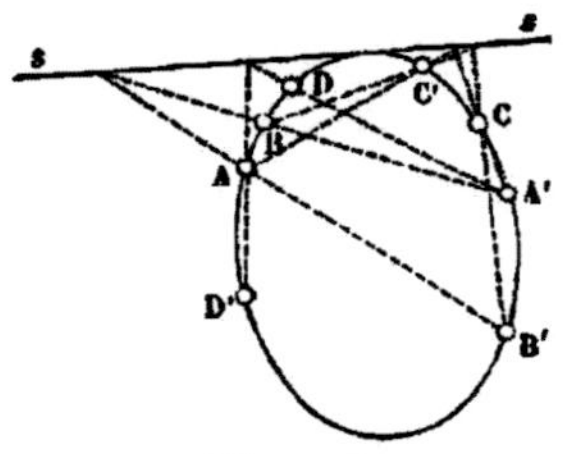

Fig. 136.

of the two ranges, it has been seen that $A'M$, AM' must intersect on s; if then M lie on s, M' must coincide with M; *i.e.* a pair of corresponding points of the two ranges are united at M.

The two ranges will therefore have two self-corresponding points, or only one, or none at all, according as the straight line s cuts the conic in two points (Fig. 135), *touches it* (Fig. 136), *or does not cut it* (Fig. 137).

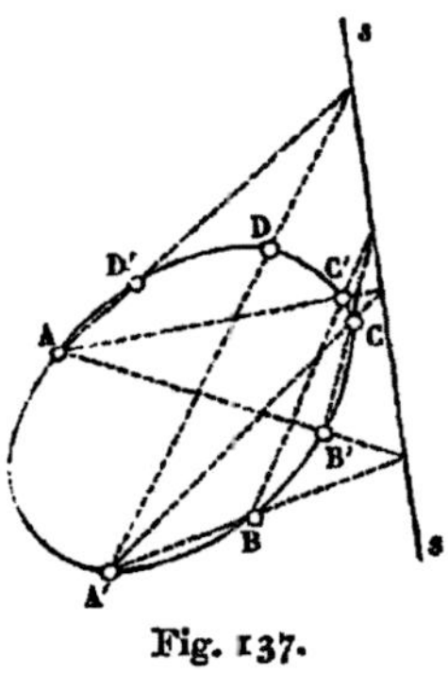

Fig. 137.

III. From what precedes it is clear that *two projective ranges of points on a conic are determined by three pairs of corresponding points A and A', B and B', C and C'*. For in order to find other pairs of corresponding points, and the self-corresponding points (when such exist), we have only to construct the straight line s which passes through the points of intersection of the three pairs of opposite sides of the hexagon $AB'CA'BC'$ (Figs. 98, 134, 135). The self-corresponding points will then be the points where s cuts the conic, and any number of pairs of corresponding points can be constructed by help of the property that any pair D and D' are such that the lines $A'D$ and AD' (or $B'D$ and BD', or $C'D$ and CD') intersect on s *.

201. Instead of *projective ranges of points on a conic* we may consider *projective series of tangents* to the same. Let o, o' be two projective ranges of points (either collinear or lying on different straight lines as bases). Describe a conic to touch o and o', and draw to this conic, from each pair of corresponding points A and A', B and B', C and C', ... the tangents a and a', b and b', c and c', ... If now these two series of tangents are cut by two other tangents o_1 and o_1', two new ranges of points will be obtained, which are projective with the given ranges respectively (Art. 149), and are therefore projective with one another.

Two series of tangents to a conic are said to be projective with one another when they are cut by any other tangent to the curve in two projective ranges.

I. Suppose the first series of tangents to be cut by the tangent a', and the second by the tangent a. The two projective ranges so formed are in perspective, since they have the self-corresponding point aa'; the straight lines which join the pairs of corresponding points $a'b$ and ab', $a'c$ and ac', ... will therefore pass through one point S. This point does not change if another pair of tangents b' and b are taken as transversals; for by Brianchon's theorem the straight lines which join the three pairs of opposite vertices $a'b$ and ab', $a'c$ and ac', $b'c$ and bc' of the circumscribed hexagon $ab'c\,a'bc'$ must meet in a point (Art. 153, left).

II. If the point S is such that tangents can be drawn from it to the conic, each of them will be a *self-corresponding line* of the two projective series of tangents abc ... and $a'b'c'$ ... .

[The proof of this is analogous to that of the corresponding property of two projective ranges of points on a conic (Art. 200, II).]

III. Two projective series of tangents to a conic are determined by three pairs of corresponding lines a and a', b and b', c and c'. For in order to find other pairs of corresponding lines, and the self-corresponding lines (when such exist), we have only to construct the point of intersection S of the diagonals which join two and two the opposite vertices of the circumscribed hexagon $ab'c\,a'bc'$. The self-corresponding lines will be the tangents from S to the conic, and any pair of corresponding lines d and d' may be constructed by means of the property that the points $a'd$ and ad' (or $b'd$ and bd', or $c'd$ and cd', . . .) are collinear with S.

* STEINER, *loc. cit.*, p. 174, § 46, iii.; Collected Works, vol. i. p. 357.

IV. A range of points $A, B, C, \ldots$ on a conic and a series of tangents $a, b, c, \ldots$ to the same are said to be projective with one another, when the pencil formed by joining $A, B, C, \ldots$ to any point on the conic is projective with the range determined by $a, b, c, \ldots$ on any tangent to the conic.

A range of points $A, B, C, \ldots$ on a conic, or a series of tangents $a, b, c, \ldots$ to the same, is said to be projective with a range of points on a straight line, or a pencil (flat or axial), when this last-mentioned range or pencil is projective with the pencil formed by joining $ABC \ldots$ to any point on the conic or with the range determined by $a, b, c, \ldots$ on any tangent to the conic.

V. These definitions premised, we may now include under the title of *one-dimensional geometric form* not only the range of collinear points, the flat pencil, and the axial pencil, but also the range of points on a conic and the series of tangents to a conic*; and with regard to these we may enunciate the general theorem: *Two one-dimensional forms which are each projective with a third (also of one dimension) are projective with one another* (cf. Art. 41).

VI. From these definitions it follows also that theorem (3) of Art. 149 may be enunciated in the following manner:

Any series of tangents to a conic is projective with the range formed by their points of contact.

VII. Let $A, B, C, \ldots$ and $A', B', C', \ldots$ be two projective ranges of points on a conic, and let $a, b, c, \ldots$ and $a', b', c', \ldots$ be the tangents at these points. The series of tangents $a, b, c, \ldots$ and $a', b', c', \ldots$ are projective with the series of points of contact $A, B, C, \ldots$ and $A', B', C', \ldots$ respectively, and are therefore projective with one another. Let s be the straight line on which the pairs of straight lines such as AB' and $A'B$, AC' and $A'C$, BC' and $B'C \ldots$ intersect; and let S be the point in which meet the straight lines joining pairs of points such as ab' and $a'b$, ac' and $a'c$, bc' and $b'c, \ldots$. If s cuts the conic in two points M and N, these must be the self-corresponding points of the ranges $ABC \ldots$ and $A'B'C' \ldots$; the tangents m and n at M and N respectively must therefore be the self-corresponding lines of the projective series $abc \ldots$ and $a'b'c' \ldots$; consequently the straight lines m and n will meet in S.

VIII. From the foregoing it follows that for the consideration of a

* The introduction of these new one-dimensional forms enables us now to add to the operations previously made use of (section by a transversal straight line and projection by straight lines radiating from a point) two others, *viz.* section of a flat pencil by a conic passing through the centre of the pencil, and projection of a range of collinear points by means of the tangents to a conic which touches the base of the range.

series of tangents can always be substituted that of their points of contact, and *vice versa*.

202. Instead of considering any two projective pencils as in Art. 200, take an involution of straight lines radiating from a point O. Suppose these to be cut by a conic passing through O in the pairs of points A and A', B and B', C and C', ..., and let these points be joined to any other point O_1 on the conic. Since by hypothesis (Arts. 122, 123) the pencils $O(AA'BC\ldots)$ and $O(A'AB'C'\ldots)$ are projective with one another, the pencils $O_1(AA'BC\ldots)$ and $O_1(A'AB'C'\ldots)$ are so too (Art. 149); and therefore the rays issuing from O_1 form an involution also. In this case we say that *the two projective ranges of points ABC ... and $A'B'C'$ ... on the conic form an involution;* or that *there is on the conic an involution formed by the pairs of conjugate points AA', BB', CC', ...**.

I. Similarly, if there is given an involution of points on a straight line o and if from the pairs of conjugate points there be drawn tangents a and a', b and b', c and c', ... to a conic touching o, these will be cut by any other tangent to the conic in an involution of points; in this case we say that aa', bb', cc', ... form *an involution of tangents to the conic* (cf. Art. 201).

II. If several pairs of tangents aa', bb', cc' ... to a conic form an involution, their points of contact AA', BB', CC', ... form an involution also, and conversely (Art. 201, VI).

203. Of the six points A, B', C, A', B, C' on a conic considered in Art. 200, let C' lie indefinitely near to A, and C indefinitely near to A'. The projective ranges $(ABC\ldots)$ or $(ABA'\ldots)$ and $(A'B'C'\ldots)$ or $(A'B'A\ldots)$ will then form an involution $(AA', BB', \ldots)$ and the inscribed hexagon is replaced by the figure made up of the inscribed quadrangle $AB'A'B$ and the tangents at the opposite vertices A and A' (Figs. 115, 138). We conclude that

An involution of points on a conic is determined by two pairs AA', BB'.

I. In order to find other pairs of conjugate points, it is only necessary to construct the straight line s which joins the point of intersection of AB' and $A'B$ to that of AB and $A'B'$; *i.e.* to

* STAUDT, *Beiträge zur Geometrie der Lage* (Nürnberg, 1856-57-60), Arts. 70 sqq.

draw the straight line joining the points of intersection of the pairs of opposite sides of the inscribed quadrangle $AB'A'B$. The points where s cuts the conic are the double points. Pairs of conjugate points will be constructed by remembering that any pair C and C' are such that the straight lines AC and $A'C'$ (or AC' and $A'C$, or BC and $B'C'$, or $B'C$ and BC') intersect on s.

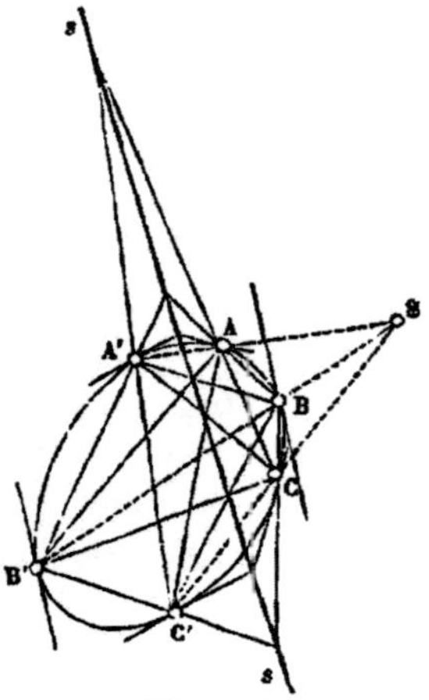

Fig. 138.

II. The tangents at a pair of conjugate points, such as A and A', B and B', ... likewise intersect on the straight line s (Art. 166).

III. Since the pairs of sides BC and $B'C'$, CA and $C'A'$, AB and $A'B'$ of the triangles ABC, $A'B'C'$ intersect in three points lying on a straight line s, the triangles are homological (Art. 17)*, and the straight lines AA', BB', CC' will meet in one point S. But AA' and BB' suffice to determine this point; accordingly:

Any pair of conjugate points of the involution are collinear with a fixed point S; or

Every straight line drawn through S to cut the conic determines on it a pair of conjugate points of the involution.

IV. It has been seen that if s cuts the conic in two points M and N, these are the double points of the involution. The tangents at M and N will therefore meet in S.

V. Conversely, *the pairs of points in which a conic is cut by the rays of a pencil whose centre S does not lie on the curve form an involution.*

For if A and A', B and B' are the points of intersection of the curve with two of the rays, these two pairs AA' and BB' determine an involution such that the straight line joining any pair of corresponding points always passes through a fixed point, viz. S. If the involution has double points, these are the intersections of the conic with the

* The triangles $A'BC$ and $AB'C'$, $AB'C$ and $A'BC'$, ABC' and $A'B'C$ are likewise homological in pairs.

straight line s which joins the point of intersection of AB and $A'B'$ to that of AB' and $A'B$.

VI. If from different points of a straight line s pairs of tangents a and a', b and b', c and c', ... be drawn to the conic, these form an involution. For if A and A', B and B', C and C', ... are the points of contact of the tangents a and a', b and b', c and c', ... respectively, and S is the point of intersection of the chords AA' and BB', then in the involution determined by the pairs A, A' and B, B' the straight line joining any other pair of conjugate points will pass through S. The point C and its conjugate lie therefore on a straight line passing through S, and the tangents at these points must meet on the straight line joining the points aa' and bb', *i.e.* on s; the conjugate of C is therefore C'. This shows that A and A', B and B', C and C' form a range of points in involution, and that consequently a and a', b and b', c and c' form a series of tangents in involution.

VII. If M and N are the double points of an involution AA', BB', CC', ... of points on a conic, it has been seen that AB, $A'B'$, MN are three concurrent straight lines (the same is the case with regard to AB', $A'B$, MN). In consequence then of theorem V, above, we conclude that:

If AA' and BB' are two pairs of conjugate elements of an involution, and MN the double elements, then MN, AB, and $A'B'$ (and similarly MN, AB', and $A'B$) are three pairs of conjugate elements of another involution.

VIII. The straight line s cuts the conic (see below, Art. 254) when the point S lies outside the conic (Fig. 138), that is,

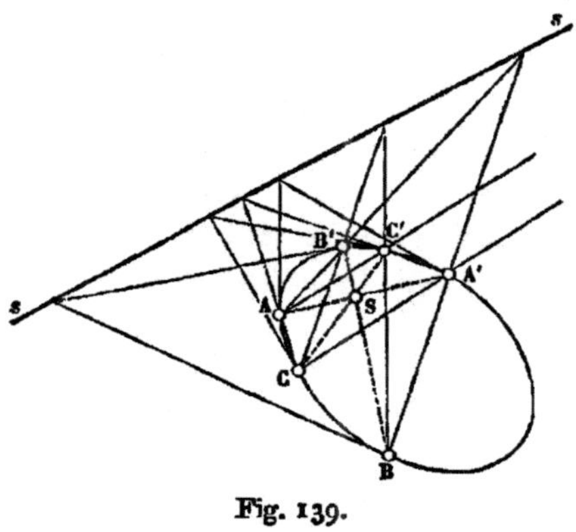

Fig. 139.

when the arcs AA' and BB' do not overlap one another; when these arcs overlap, the point S lies within the conic and the straight line s does not cut the latter (Fig. 139). We therefore arrive again at the property already proved in Art. 128, viz. that

An involution has two double elements when any two pairs of conjugate elements are such that they do not overlap; and it has no double elements when they are such that they do overlap.

In no case can an involution, properly so called, have only one double element. For if s were a tangent to the conic, S would be its point of contact, and of every pair of conjugate points one would coincide with S (cf. Art. 125).

204. If $(MNAB...)$ and $(MNA'B'...)$ are two projective ranges of points on a conic, M and N will be the self-corresponding points, and the straight line MN will pass through the point of intersection of AB' and $A'B$ (Art. 200). Now let B' be supposed to lie indefinitely near to A and similarly B to A', so that the straight lines AB' and $A'B$ become in the limit the tangents at A and A' respectively (Fig. 140). Since now $MNAA'$ and $MNA'A$ are groups of corresponding points of two projective ranges, the two pencils $mnaa'$ and $mna'a$ formed by joining them to any point O on the conic will be projective; and therefore $mnaa'$ is a harmonic pencil (Art. 83). We thus arrive again at the second theorem of Art. 195 (right); viz.

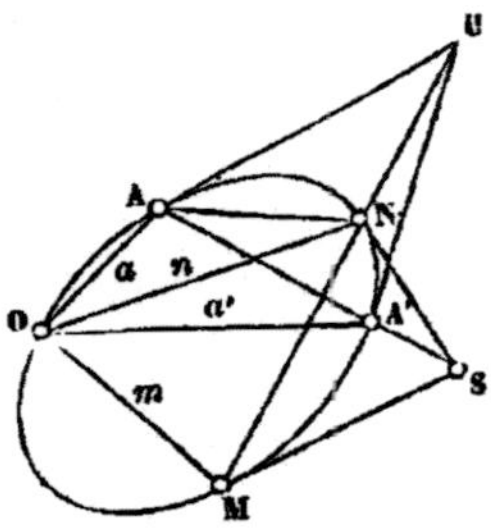

Fig. 140.

If four points M, N, A, A' on a conic are harmonic, the tangents at one pair of conjugate points, say A and A', intersect on the chord MN joining the other pair;

and its correlative (Art. 195, left),

If four tangents to a conic are harmonic, the point of intersection of one pair of conjugates lies on the chord of contact of the other pair.

From the former of these it follows that if through the point of intersection S of the tangents at M and N straight lines be drawn cutting the conic in A and A', B and B', C and C', ... respectively, any of these pairs of points will be harmonically conjugate with regard to M and N. The tangents at A and A', B and B', C and C', ... will therefore intersect in pairs on the straight line MN.

In other words:

If from any point there be drawn to a conic two tangents and a secant, the two points of contact and the two points of intersection form a harmonic system.

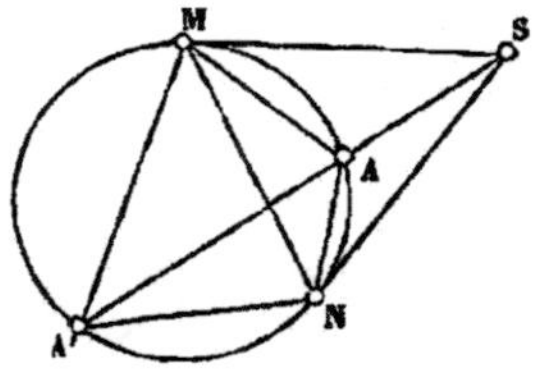

Fig. 141.

The points (AA'), (BB'), (CC'), ... form an involution of which M and N are the double points (Art. 203, III, IV). We therefore arrive again at the property of an involution that if it has two double elements these are separated harmonically by any pair of conjugate elements (Art. 125).

205. Suppose now that the conic is a circle (Fig. 141). From the similar triangles SAM, SMA',

$$AM : MA' :: SM : SA',$$

and from the similar triangles SAN, SNA'

$$AN : NA' :: SN : SA';$$

$$\therefore \quad \frac{AM}{AN} = \frac{A'M}{A'N}, \text{ (since } SM = SN\text{)},$$

or

$$AM . A'N = AN . A'M.$$

But by Ptolemy's theorem (Euc. vi. D),

$$AA'. MN = AM . A'N + AN . A'M.$$

If then M, N, A, A' are four harmonic points on a circle,

$$\tfrac{1}{2} AA'. MN = AM. A'N = AN. A'M.$$

206. The properties established in Art. 200 and the following Articles lead at once to the solution of the important problem:

To construct the self-corresponding elements of two superposed projective forms, and the double elements of an involution.

I. *Let two concentric projective pencils be given, which are determined by three pairs of corresponding rays* (Fig. 142); *it is required to construct their self-corresponding rays.*

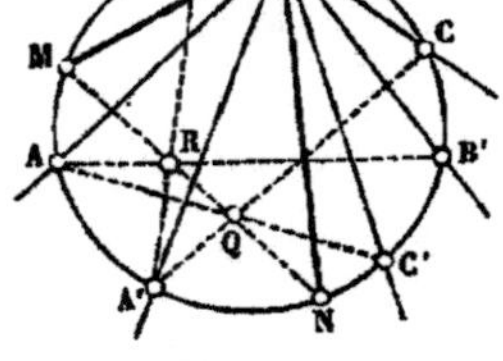

Fig. 142.

Through the common centre O describe any circle, cutting the three given pairs of rays in A and A', B and B', C and C' respectively. Let AB', $A'B$ meet in R, and AC', $A'C$ in Q; if the straight line QR cut the circle in two points M and N, then OM, ON will be the required self-corresponding rays.

II. *Let A and A′, B and B′, C and C′* (Fig. 143) *be three pairs of corresponding points of two collinear ranges; it is required to construct the self-corresponding points.*

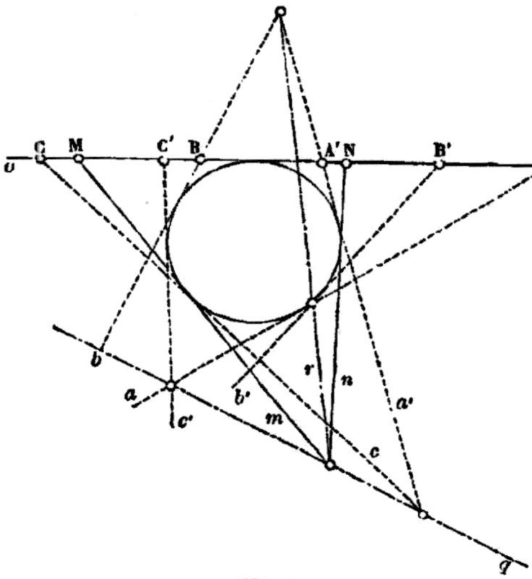

Fig. 143.

Describe any circle touching the common base o of the two ranges, and to this circle draw from the given points the tangents a and a', b and b', c and c'. Let r be the straight line which joins the points ab', $a'b$, and q that which joins the points ac', $a'c$. If the point qr lies outside the circle and from it the tangents m and n be drawn to the circle, then the points om, on in which these meet the base will be the required self-corresponding points of the two ranges.

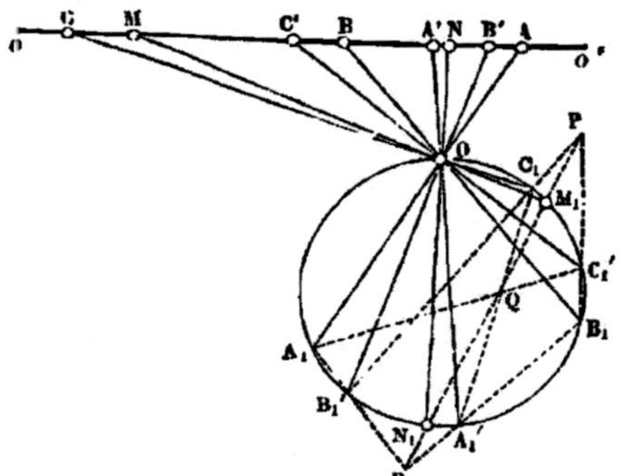

Fig. 144.

Otherwise (Fig. 144):

Draw any circle whatever in the plane and take on it any point

O. From O project the given points upon the circumference of the circle, and let A_1 and A_1', B_1 and B_1', C_1 and C_1' be the projections of A and A', B and B', C and C' respectively. Join A_1B_1', $A_1'B_1$ meeting in R, and A_1C_1', $A_1'C_1$ meeting in Q (or B_1C_1', $B_1'C_1$ meeting in P). If the straight line PQR cut the circle in two points M_1, N_1, and these be projected from the point O back upon the given base o, then their projections M, N will be the required self-corresponding points of the given ranges *.

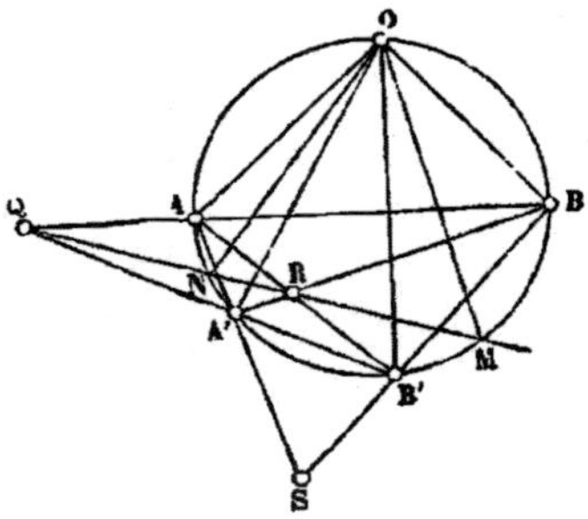

Fig. 145.

III. *In* (I) *let the two pencils be in involution* (Fig. 145), *and let it be required to find the double rays.*

Two pairs of conjugate rays suffice now to determine the pencils. Draw through the centre O any circle cutting the given rays in A and A', B and B' respectively. Let AB', $A'B$ meet in R, and AB, $A'B'$ in Q; if the straight line QR cut the circle in two points M and N, then OM, ON will be the required double rays of the involution.

IV. *Let A and A', B and B' be two given pairs of conjugates of an*

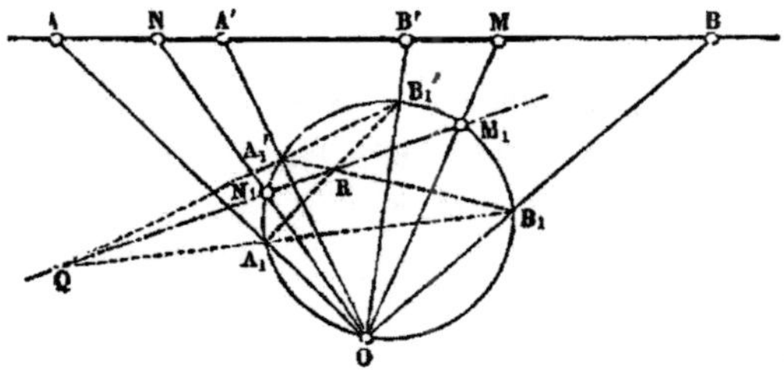

Fig. 146.

involution of points on a straight line; it is required to find the double points (Fig. 146).

Draw any circle in the plane and take on it any point O. From O project the given points upon the circumference of the circle, and let A_1 and A_1', B_1 and B_1' be the projections of A and A', B and B' respectively. Let A_1B_1', $A_1'B_1$ meet in R, and A_1B_1, $A_1'B_1'$ in Q. If QR cuts the circle in M_1, N_1, and these points be projected from O back upon the given straight line, then their projections M, N will be the required double points.

* Steiner, *loc. cit.*, pp. 68 and 174, §§ 17 and 46; Collected Works, vol. i. pp. 285, 356.

Otherwise:

Describe a circle touching the base AB... (Fig. 147), and draw to this circle from the points A and A', B and B', the tangents a and a',

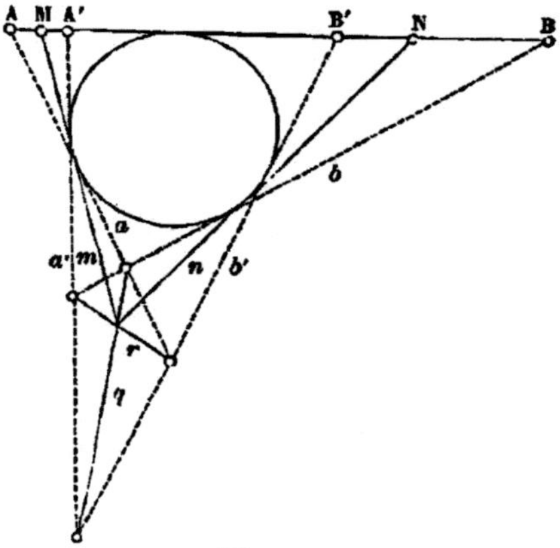

Fig. 147.

b and b', respectively. Let r be the straight line which joins the points ab', $a'b$, and q that which joins the points ab, $a'b'$. If the point qr lies outside the circle, the tangents m and n from this point to the circle will cut the base line of the involution in the required double points.

207. THEOREM. *A pencil in involution is either such that every ray is at right angles to its conjugate, or else it contains one and only one pair of conjugate rays including a right angle.*

Consider again Art. 206, III; if the point of intersection S of the straight lines AA', BB', ... is the centre of the circle (Fig. 148) then AA', BB', ... are all diameters, and therefore

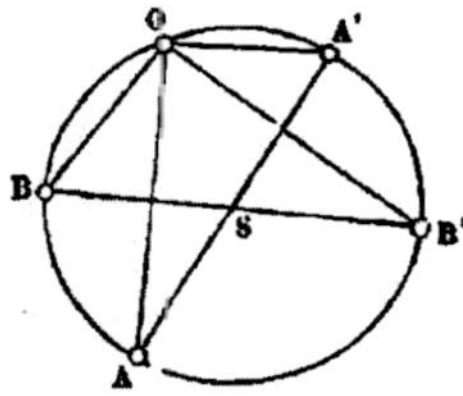

Fig. 148.

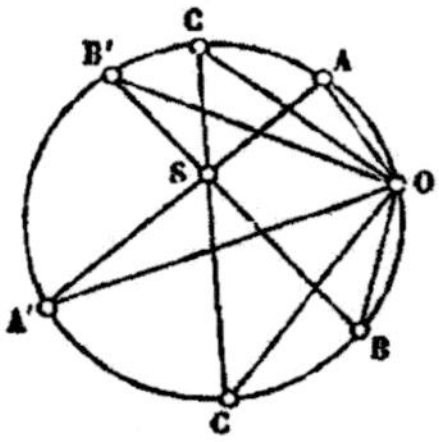

Fig. 149.

each ray OA, OB, ... will be at right angles to its conjugate OA', OB', ... In this case then the involution is formed by a series of right angles which have their common vertex at O.

But if S is not the centre of the circle (Fig. 149), draw the diameter through it; if C and C' are the extremities of this diameter, the rays OC, OC' will include a right angle. But these will be the only pair of conjugate rays which possess this property, since through S only one diameter can be drawn.

208. This proposition is only a particular case of the following one:

Two superposed involutions (or such as are contained in the same one-dimensional form) have always a pair of conjugate elements in common, except in the case where the involutions have double elements and the double elements of the one overlap those of the other.

Take two involutions of rays having a common centre O, and let a circle drawn through O cut the pairs of conjugate rays of the first involution in the pairs of points $(AA', BB', \ldots)$ and those of the second in $(GG', HH', \ldots)$. Let S be the point of intersection of $AA', BB', \ldots$ and T that of $GG', HH', \ldots$. If the straight line ST cut the circle in two points E and E', these will be a conjugate pair of each involution, since they are collinear with S and with T also. Let us now examine in what cases ST will cut the circle.

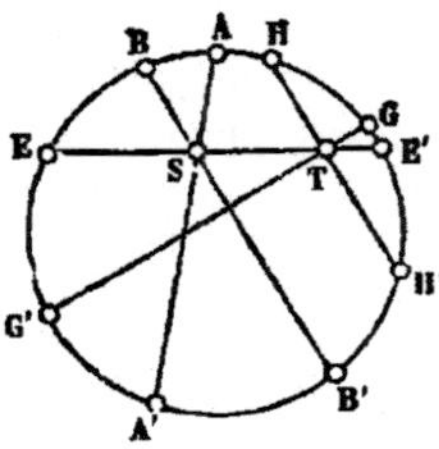

Fig. 150.

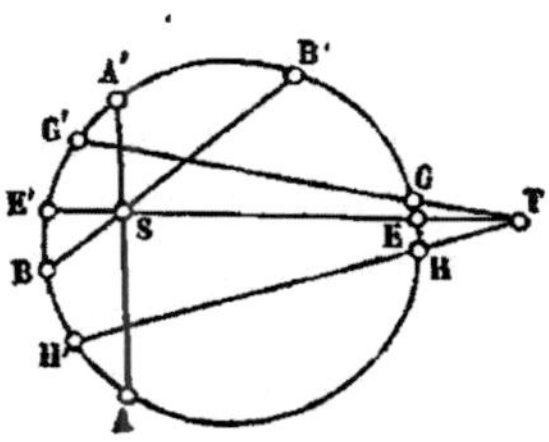

Fig. 151.

In the first place, it will certainly do so if one at least of the points S, T lies within the circle (Art. 203, VIII), *i.e.* if one at least of the involutions has no double elements (Figs. 150, 151).

Secondly, if both the points S, T lie outside the circle, *i.e.* if both the involutions have double elements, then the straight line ST may or may not cut the circle. If OM, ON are the double elements of the first involution, OU, OV those of the second, the rays OE, OE' must be harmonically conjugate both with regard to OM, ON and with regard to OU, OV; but (Art. 70) in order that there should exist a pair of elements which are at the same time harmonically conjugate with regard to each of the two pairs OM, ON and OU, OV, it is necessary and sufficient that these two pairs should not overlap. If then these pairs do not overlap, ST will cut the circle (Fig. 152);

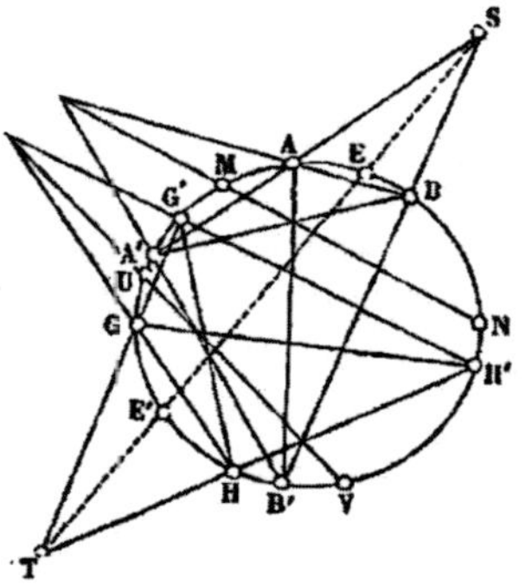

Fig. 152.

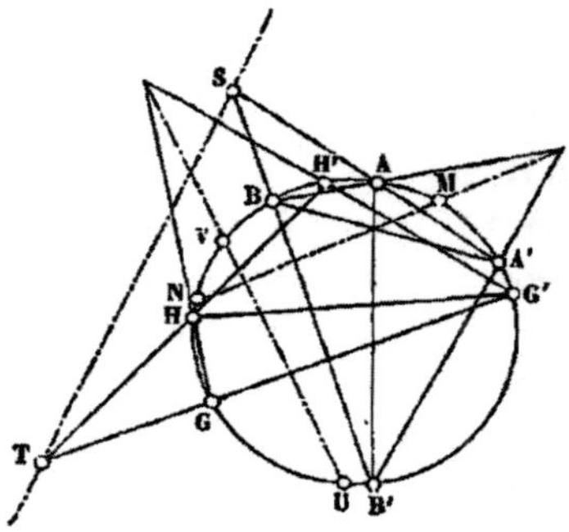

Fig. 153.

whereas if they do overlap, ST will not cut the circle (Fig. 153). The two involutions have therefore a common pair of conjugate elèments in all cases except this last, viz. when they both have double elements and these overlap.

[In Figs. 150, 151 and 152, are shown cases of two involutions having a common pair of conjugate elements E and E'; Fig. 153 on the other hand illustrates the case where no such pair exists.]

209. The preceding problem, viz. that of determining the common pair of conjugate elements of two involutions superposed one upon the other, depends upon the following, *viz. to determine* (in a range, in a pencil, or on a conic) *a pair of elements which are harmonically conjugate with regard to each of two given pairs.* This problem has already been solved, for the case of a range, in Art. 70; the following is another solution:

Suppose that we have to deal with a range of points lying on a straight line. Take any circle and a point O on it, and project the given points from O upon the circumference; let M, N and U, V be their projections (Fig. 152). Let the tangents at M and N to the circle meet in S, and the tangents at U and V in T. If the pair MN does not overlap the pair UV, then ST will cut the circle in two points E and E', which when projected back from O upon the given straight line will give the points required.

210. The double points of the involution determined by the pairs A, A' and B, B' are the common pair of conjugate elements of two other involutions; one of these is determined by the pairs A, B

and A', B', the other by the pairs A, B' and A', B (Art. 203, VII.)

From this follows a *construction for the double points of an involution of collinear points which is determined by the pairs* A, A' *and* B, B'. Take any point G outside the base of the involution and describe the circles GAB, $GA'B'$; they will meet in another point, say in H. Similarly let K be the second point of intersection of the circles GAB', $GA'B$. Every circle passing through G and H meets the base in a pair of conjugate points of the involution AB, $A'B'$ (Art. 127); so too every circle passing through G and K gives a pair of conjugate points of the involution AB', $A'B$. If then the circle GHK be described and it meet the base, the two points of intersection will be the double elements of the involution AA', BB'*.

211. It follows from the foregoing that the determination of the self-corresponding points of two projective ranges ABC ... and $A'B'C'$ ... on a conic (and consequently of the self-corresponding points of any two superposed projective forms) reduces to the construction of the straight line s on which intersect the pairs of straight lines AB' and $A'B$, AC' and $A'C$, BC' and $B'C$, ... Similarly the determination of the double points of an involution AA', BB', ... depends on the construction of the straight line s on which intersect the pairs of straight lines AB and $A'B'$, AB' and $A'B$, ... or the pairs of tangents at A and A', B and B', ...

Conversely, if any straight line s (which does not touch the conic) is given, an involution of points on the conic is thereby determined; for it is only necessary to draw, from different points of s, pairs of tangents to the conic, and the points of contact will be pairs of conjugate points of an involution.

But, on the other hand, in order that two projective ranges of points ABC ... and $A'B'C'$ ... may be determined, there must be given, in addition to the straight line s, a pair of conjugate points A and A' also; then the straight lines joining A and A' to any point on s will cut the conic in a pair of corresponding points B' and B.

Two projective ranges of points determine an involution; for they determine the straight line s, which determines the involution. If the two ranges have two self-corresponding points, these will also be the double points of the involution.

* CHASLES, *Géométrie supérieure*, Art. 263.

CHAPTER XIX.

PROBLEMS OF THE SECOND DEGREE.

212. Problem. *Given five points O, O', A, B, C on a conic, to determine the points of intersection of the curve with a given straight line s.*

Solution. Join any two of the points O, O' to each of the others A, B, C (Fig. 154); the

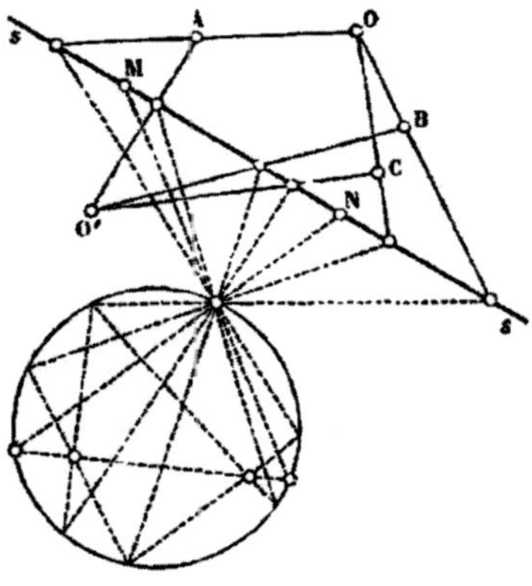

Fig. 154.

pencils $O\ (A, B, C, \ldots)$ and $O'\ (A, B, C, \ldots)$ will be projective, and will cut the transversal s in points forming two collinear projective ranges.

A point M which corresponds to itself in these two ranges will

Problem. *Given five tangents o, o', a, b, c to a conic, to draw a pair of tangents to the curve from a given point S.*

Consider the points where two of the tangents o, o' are met by the others a, b, c (Fig. 155); the

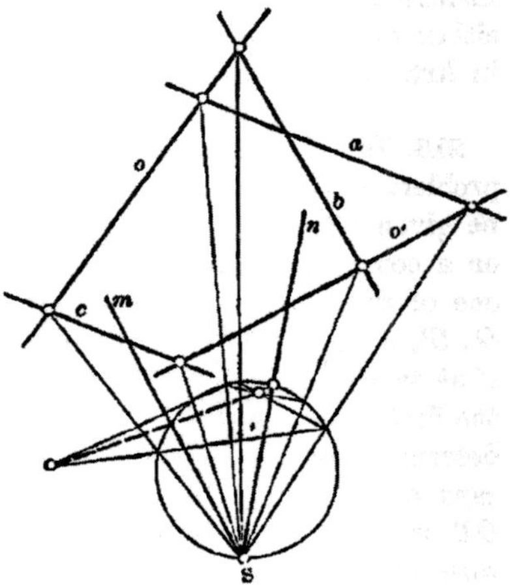

Fig. 155.

ranges $o\ (a, b, c, \ldots)$ and $o'\ (a, b, c, \ldots)$ will be projective, and if projected from S as centre will give two concentric projective pencils.

Any ray m which corresponds to itself in these two pencils will also be a point on the conic, since a pair of corresponding rays of the two pencils must meet in M. The points of intersection of the conic with the straight line s are therefore found as the self-corresponding points of the two collinear ranges which are determined on s by the three pairs of corresponding rays OA and $O'A$, OB and $O'B$, OC and $O'C$. There may be two such self-corresponding points, or only one, or none at all; consequently the straight line s may cut the conic in two points, or it may touch it, or it may not meet it at all. The construction of the self-corresponding points themselves may be effected by either of the methods explained in Art. 206, II.

also be a tangent to the conic, since a pair of corresponding points of the two ranges o and o' must lie on m. The tangents from S to the conic are therefore found as the self-corresponding rays of the two concentric pencils which are determined by the rays joining S to the three pairs of corresponding points oa and $o'a$, ob and $o'b$, oc and $o'c$. There may be two such self-corresponding rays, or only one, or none at all; consequently there can either be drawn from the point S two tangents to the conic, or S is a point on the conic, or else from S no tangent at all can be drawn. The construction of the self-corresponding rays themselves may be effected by the method explained in Art. 206, I.

213. In a similar manner the problem may be solved if there be given four points O, O', A, B on a conic and the tangent o at one of them O; *or* three points O, O', A and the tangents o and o' at two of them O and O'. In the first case the two pencils are determined by the three pairs of rays o and $O'O$, OA and $O'A$, OB and $O'B$; and in the second case by the three pairs o and $O'O$, OO' and o', OA and $O'A$.

In a similar manner the problem may be solved if there be given four tangents o, o', a, b to a conic and the point of contact O of one of them o; *or* three tangents o, o', a and the points of contact O and O' of two of them o and o'. In the former case the three pairs of points which determine the two ranges are O and $o'o$, oa and $o'a$, ob and $o'b$; in the latter case they are O and $o'o$, oo' and O', oa and $o'a$.

If however there be given five tangents, *or* four tangents and the point of contact of one of them, *or* three tangents and the points of contact of two of them, we may begin by first constructing such of the points of contact of the tangents as are not

If however there be given five points on the conic, *or* four points and the tangent at one of them, *or* three points and the tangents at two of them, we may begin by first constructing such of the tangents at the points as are not already given (Arts. 165,

already given (Arts. 180, 171, 177); the problem will then reduce to one of the cases given above.

171, 175); the problem will then reduce to one of the cases given above.

214. In the construction given in Art. 212 (left) suppose that the conic is a hyperbola and that the given straight line s is one of the asymptotes (Fig. 156). The collinear projective ranges determined on s by the pencils $O(A, B, C, \ldots)$ and $O'(A, B, C, \ldots)$ will have in this case one self-corresponding point, and this (being the point of contact of the hyperbola and the asymptote) will lie at an infinite distance. But in two collinear ranges whose self-corresponding points coincide in a single one at infinity, the segment intercepted between any pair of corresponding points is of constant length (Art. 103). We therefore conclude that

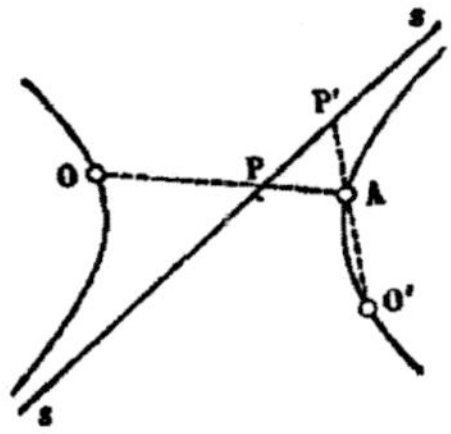

Fig. 156.

If from two fixed points O and O' on a hyperbola there be drawn two rays to cut one another on the curve, the segment PP' which these intercept on either of the asymptotes is of constant length *.

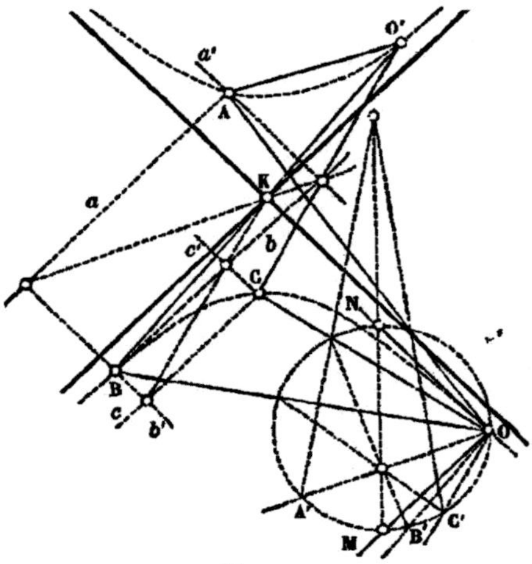

Fig. 157.

215. If in Art. 212 (left) the straight line s be taken to lie at infinity, the problem becomes the following:

Given five points O, O', A, B, C on a conic, to determine the points at infinity on it (Fig. 157).

* BRIANCHON, *loc. cit.*, p. 36.

Consider again the projective pencils $O(A, B, C, \ldots)$ and $O'(A, B, C, \ldots)$, which determine on the straight line at infinity s two collinear ranges whose self-corresponding points are the required points at infinity on the conic. Since each of these self-corresponding points must lie not only at the intersection of a pair of corresponding rays of the two pencils but also on the line at infinity s, the corresponding rays which meet in such a point must be parallel to one another; the problem therefore reduces to the determination of the pairs of corresponding rays of the two pencils which are parallel to one another.

In order then to solve the problem we draw through O the parallels OA', OB', OC' to $O'A$, $O'B$, $O'C$ respectively, and then construct (Art. 206, I) the self-corresponding rays of the two concentric pencils which are determined by the three corresponding pairs OA and OA', OB and OB', OC and OC'. If there are two self-corresponding rays OM and ON, the conic determined by the five given points is a hyperbola whose points at infinity lie in the directions OM, ON; *i.e.* whose asymptotes are parallel to OM and ON respectively.

If there is only one self-corresponding ray OM, the conic determined by the five given points is a parabola whose point at infinity lies in the direction OM.

If there is no self-corresponding ray, the conic determined by the five given points is an ellipse, since it does not cut the straight line at infinity.

If in the first case (Fig. 157) it is desired to construct the asymptotes themselves of the hyperbola, we consider this latter as determined by the two points at infinity and three other points, say A, B, and C; in other words, we regard the hyperbola as generated by the two projective pencils, one of which consists of rays all parallel to OM, and the other of rays all parallel to ON, and which are such that one pair of corresponding rays meet in A, a second pair in B, and a third pair in C. The rays which correspond in the two pencils respectively to the straight line at infinity (the line joining the centres of the pencils) will be the asymptotes required.

Let then a, b, c (Fig. 157) be the rays parallel to OM which pass through A, B, C respectively, and let a', b', c' be the rays parallel to ON which pass through the same points respectively. Join the points ab' and $a'b$ and the points bc' and $b'c$, and let K be the point of intersection of the joining lines; the straight lines drawn through K parallel to OM and ON will be the required asymptotes.

216. Problem. *Given five points A, B, C, D, E on a conic, to draw the tangents from a given point S to the conic.*

This problem also can be made to depend on that of Art. 212

(left), by making use of the properties of the involution (Art. 203) obtained by cutting the conic by transversals drawn through S.

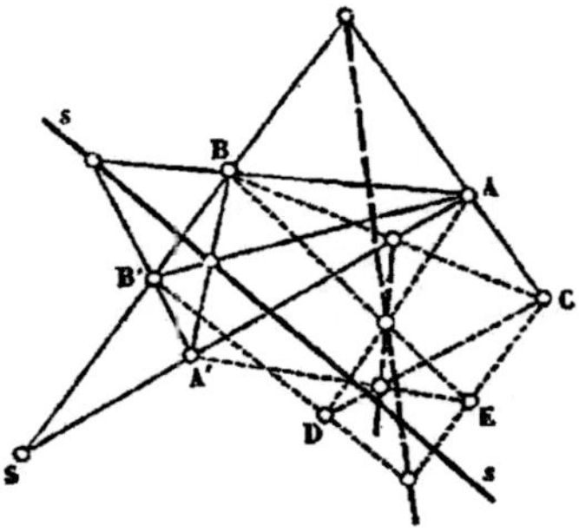

Fig. 158.

Join SA , SB (Fig. 158); these straight lines will cut the conic again in two new points A' and B', which can be determined (making use of the ruler only, and without drawing the curve) by means of Pascal's theorem (Art. 161, right). (In the figure the points A' and B' have been constructed by means of the hexagons $ADCBEA'$ and $BECADB'$ respectively). Now let the point of intersection of AB and $A'B'$ be joined to that of AB' and $A'B$; the joining line s will pass through the points of contact of the tangents from S (Art. 203). The problem therefore reduces to that of determining the points of intersection of the conic and the straight line s (Art. 212, left).

217. The problem, *To find the points of intersection of a given straight line s and a conic which is determined by five given tangents*, may similarly be made to depend on that of Art. 212 (right), by making a construction (Fig. 159) analogous to the foregoing one.

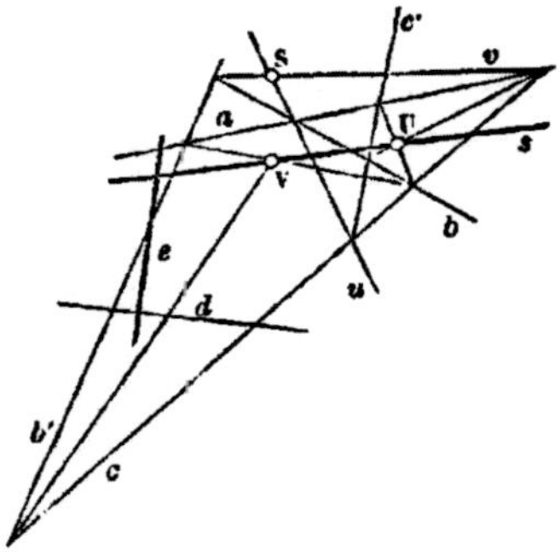

Fig. 159.

And the problem, *To draw through a given point a straight line which shall divide a given triangle into two parts having to one another a given ratio*, may be solved by reducing it to the following construction: To draw from the given point a tangent to a hyperbola of which the asymptotes and a tangent are known.

These are left as exercises to the student.

218. Problem. *To construct a conic which shall pass through four given points Q , R , S , T , and shall touch a given straight line s which does not pass through any of the given points.*

Solution. Let A , A', B , B'

To construct a conic which shall touch four given straight lines q, r, s, t, and shall pass through a given point S which does not lie on any of the given lines.

Let a , a' , b , b' be the rays

be the points where the sides QT, RS, QR, ST respectively of the quadrangle $QRST$ cut the straight line s (Fig. 160).

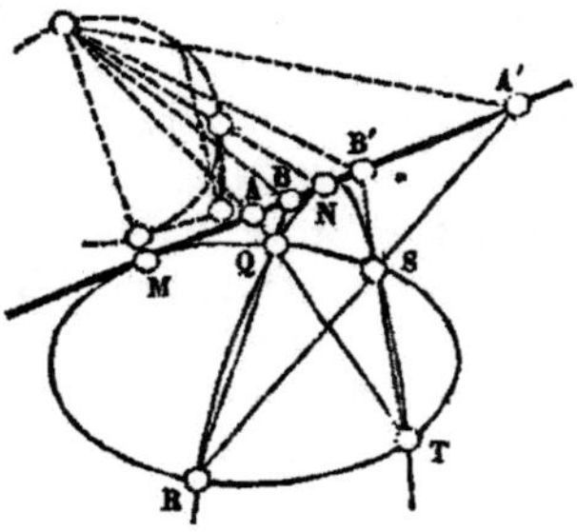

Fig. 160.

Construct the double points (if such exist) of the involution determined by the pairs of points A and A', B and B'.

If there are two double points M and N, each of them will be (Art. 185, left) the point of contact with s of some conic circumscribed about the quadrangle $QRST$. Each of the conics $QRSTM$, $QRSTN$ therefore gives a solution of the problem; and these conics can be constructed by points by help of Pascal's theorem (Art. 161, right).

If however there are no double points, there is no conic which satisfies the conditions of the problem.

joining the point S to the vertices qt, rs, qr, st respectively of the quadrilateral $qrst$ (Fig. 161). Construct the double rays (if

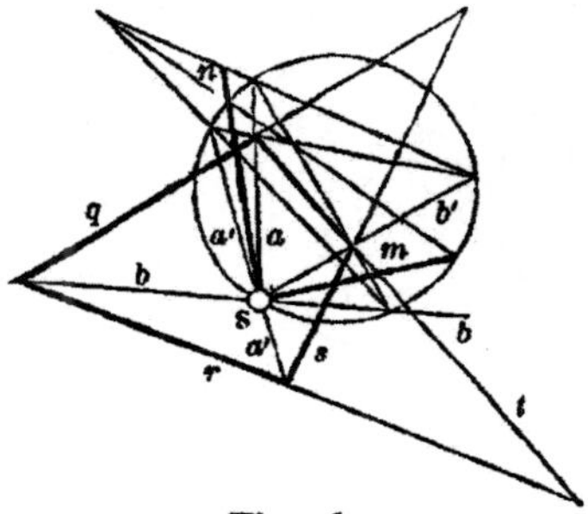

Fig. 161.

such exist) of the involution determined by the pairs of rays a and a', b and b'.

If there are two double rays m and n, each of them will be (Art. 185, right) a tangent at S to some conic inscribed in the quadrilateral $qrst$. Each of the conics $qrstm$, $qrstn$ therefore gives a solution of the problem; and these conics can be constructed by tangents by help of Brianchon's theorem (Art. 161, left).

If however there are no double rays, there is no conic which satisfies the conditions of the problem.

219. If in the foregoing Art. (left) the straight line s be taken to lie at infinity, the problem becomes the following:

To construct a parabola which shall pass through four given points Q, R, S, T.

To solve it, take any point O (Fig. 162), and through it draw the rays a, a', b, b' parallel respectively to the straight lines QT, RS, QR, ST; and construct the double rays (if such exist) of the involution determined by the pairs of rays a and a', b and b'.

Each of these double rays will determine the direction in which lies the point at infinity on a parabola passing through the four given points; the problem therefore reduces to the last problem of Art. 165. If however the involution has no double rays, no parabola can be found which satisfies the conditions of the problem.

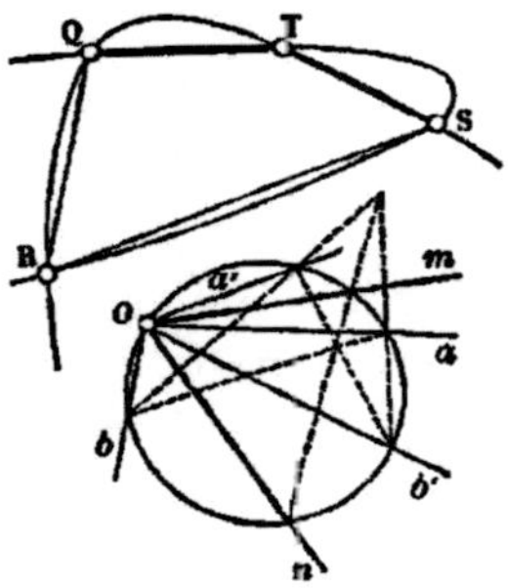

Fig. 162.

Through four given points therefore can be drawn either two parabolas or none; in the first case the other conics which pass through the given points are ellipses and hyperbolas; in the second case they are all hyperbolas. The first case occurs when each of the four points lies outside the triangle formed by the other three (*i. e.* when the quadrangle formed by the four points is *non-reentrant*); the second case when one of the four points lies within the triangle formed by the other three (*i. e.* when the quadrangle formed by the four points is *reentrant*).

220. If in Art. 218 (right) one of the straight lines q, r, s, t lies at infinity, the problem becomes the following:

To construct a parabola which shall touch three given straight lines and shall pass through a given point.

221. PROBLEM. *To construct a conic which shall pass through three given points P, P', P'' and shall touch two given straight lines q and s, neither of which passes through any of the given points.*

Solution. This depends on the theorem of Art. 191 (left). Join PP', and consider it as a transversal which cuts the conic in P and P', and the pair of tangents q and s in the two points B and B' (Fig. 163). If A and A_1 are the double points of the involution determined by the two pairs of points P and P', B and B', the chord of contact of the conic and the tangents q and s must pass through one of these points, by the theorem quoted above.

To construct a conic which shall touch three given straight lines p, p', p'' and shall pass through two given points Q and S, neither of which lies on any of the given straight lines.

The solution depends on the theorem of Art. 191 (right). Consider pp' as a point from which the tangents p and p' have been drawn to the conic, and the rays b and b' to the two points Q and S (Fig. 164). If a and a_1 are the double rays of the involution determined by the two pairs of rays p and p', b and b', the point of intersection of the tangents at Q and S to the conic must lie on one of these rays, by the theorem quoted above. Repeat the same

Repeat the same reasoning for the case of the transversal PP'', which cuts q and s in D and D'';

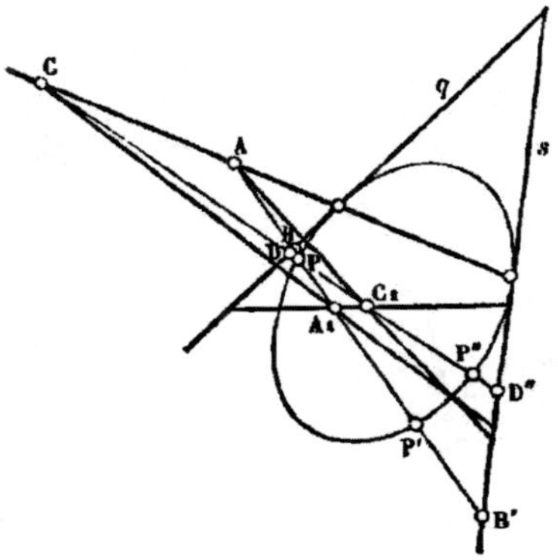

Fig. 163.

if C and C_1 are the double points of the involution determined by the two pairs of points P and P'', D and D'', the chord of contact must similarly pass through C or C_1. The problem admits therefore of four solutions; *viz.* when the two involutions (PP', BB') and (PP'', DD'') both have double points, there are four conics which satisfy the given conditions. If the double points are A, A_1 and C, C_1 respectively, the chords of contact of the four conics and the tangents q and s are AC, A_1C, AC_1, and A_1C_1. Of each of these conics five points are known, *viz.* P, P', P'', and the two points of intersection of AC (or of A_1C, or AC_1, or A_1C_1, as the case may be) with q and s; they can accordingly be constructed by points by means of Pascal's theorem (Art. 161, right).

reasoning for the case of the point pp'', from which are drawn the rays d and d'' to the points

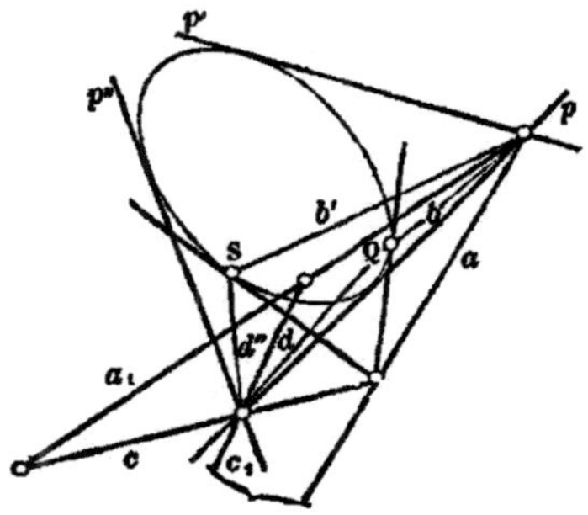

Fig. 164.

Q and S; if c and c_1 are the double rays of the involution determined by the two pairs of rays p and p'', d and d'', the point of intersection of the tangents must similarly lie on c or c_1. The problem admits therefore of four solutions; *viz.* when the two involutions (pp', bb') and (pp'', dd'') both have double rays, there are four conics which satisfy the given conditions. If the double rays are a, a_1 and c, c_1 respectively, the points of intersection of the tangents at Q and S to the four conics are ac, a_1c, ac_1, and a_1c_1. Of each of these conics five tangents are known, viz. p, p', p'', and the two straight lines which join ac (or a_1c, or ac_1, or a_1c_1, as the case may be) to Q and S; they can accordingly be constructed by tangents by means of Brianchon's theorem (Art. 161, left).

222. Problem. *To construct a polygon whose vertices shall lie on given straight lines (each on each), and whose sides shall pass through given points (each through each*).*

Solution. For the sake of simplicity suppose that it is required to construct a quadrilateral, whose vertices 1, 2, 3, 4 shall lie respectively on four given straight lines s_1, s_2, s_3, s_4, and whose sides 12, 23, 34, 41 shall pass respectively through four given points $S_{12}, S_{23}, S_{34}, S_{41}$ (Fig. 165). The method and reasoning will be the

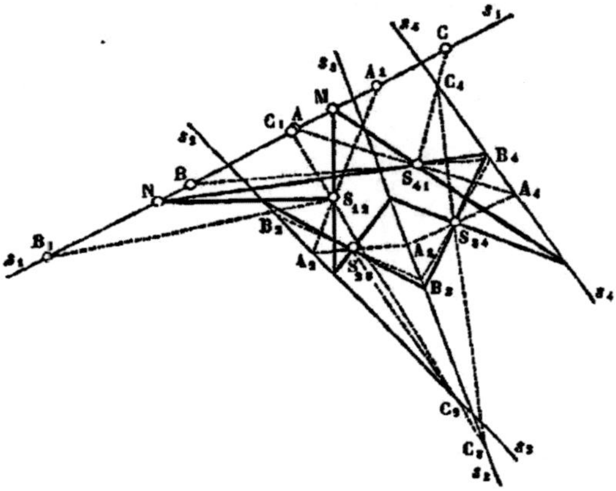

Fig. 165.

same as for a polygon of any number of sides. Take any points $A_1, B_1, C_1, \ldots$ on s_1 and project them from S_{12} as centre upon s_2; and let $A_2, B_2, C_2, \ldots$ be their projections. Project $A_2, B_2, C_2, \ldots$ from S_{23} as centre upon s_3, and let $A_3, B_3, C_3, \ldots$ be their projections. Project $A_3, B_3, C_3, \ldots$ from S_{34} as centre upon s_4, and let $A_4, B_4, C_4, \ldots$ be their projections. Finally project $A_4, B_4, C_4, \ldots$ from S_{41} as centre upon s_1, and let $A, B, C, \ldots$ be their projections.

The points $S_{12}, S_{23}, S_{34}, S_{41}$ are the centres of four projectively related pencils; for the first and second are in perspective (since their pairs of corresponding rays $A_1A_2, B_1B_2, \ldots$ and $A_2A_3, B_2B_3, \ldots$ intersect on s_2), the second and third are in perspective (pairs of corresponding rays intersect on s_3), and similarly the third and fourth are in perspective (pairs of corresponding rays intersect on s_4). Consequently (Art. 150) pairs of corresponding rays of the first and fourth pencils (such as A_1A_2 and A_4A) will intersect on a conic; or in other words the locus of the first vertex of the variable quadrilateral whose second, third, and fourth vertices (A_2, A_3, A_4) slide respectively on three given straight lines (s_2, s_3, s_4) and whose sides $(A_1A_2, A_2A_3, A_3A_4, A_4A)$ pass respectively through four given points,

* Poncelet, *loc. cit.*, p. 345.

is a conic*. This conic passes through the points S_{12}, S_{41}, the centres of the pencils which generate it; in order therefore to determine it, three other points on it must be known; the intersections of the three pairs of corresponding rays A_1A_2 and A_4A, B_1B_2 and B_4B, C_1C_2 and C_4C will suffice. It is then only necessary further to construct (Art. 212) the points of intersection M and N of the straight line s_1 with the conic determined by these five points; either M or N can then be taken as the first vertex of the required quadrilateral.

This construction may be looked at from another point of view. The broken lines $A_1A_2A_3A_4A$, $B_1B_2B_3B_4B$, and $C_1C_2C_3C_4C$ may be regarded as the results of so many attempts made to construct the required quadrilateral; these attempts however give polygons which are not closed, for A does not in general coincide with A_1, nor B with B_1, nor C with C_1. These attempts and all other conceivable ones which might similarly be made, but which it is not necessary to perform, give on the straight line s_1 two ranges $A_1B_1C_1\ldots$ and $ABC\ldots$; one being traced out by the first vertex and the other by the last vertex of the open polygon. These ranges are projective with one another, since the second has been derived from the first by means of projections from S_{12}, S_{23}, S_{34}, S_{41} as centres, and sections by the transversals s_2, s_3, s_4, s_1. Each of the self-corresponding points therefore of the two ranges will give a solution of the problem; for, if the first vertex of the polygon be taken there, the last vertex will also fall on the same point, and the polygon will be closed.

In the following examples also the method remains the same whatever be the number of sides of the polygon which it is required to construct.

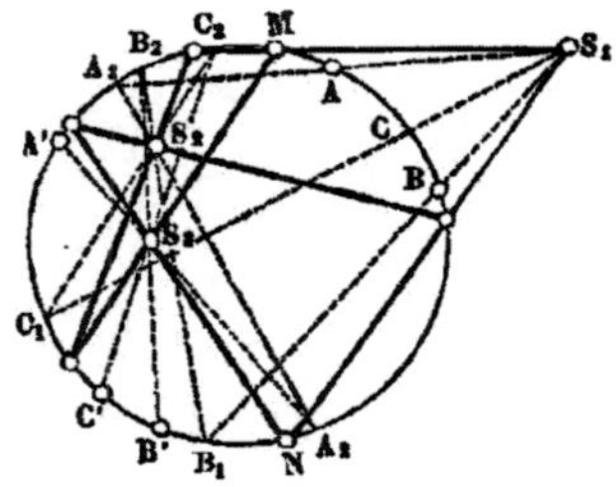

Fig. 166.

223. Problem. *To inscribe in a given† conic a polygon whose sides pass respectively through given points.*

Solution. Suppose that it is required to inscribe in the conic a triangle whose sides pass respectively through three given points S_1, S_2, S_3 (Fig. 166). Let us make three trials. Take then any three points A, B, C on the conic; join them to S_1 and let the joining lines cut the conic again in A_1, B_1, C_1; join these points to S_2 and let

* This theorem, *viz.* that 'if a simple polygon move in such a way that its sides pass respectively through given points and all its vertices except one slide respectively along given straight lines, then the remaining vertex will describe a conic,' is due to Maclaurin (Phil. Trans., London, 1735). Cf. Chasles, *Aperçu historique*, p. 150.

† *i.e.* either completely traced or determined by five given points.

the joining lines cut the conic again in A_2, B_2, C_2; finally join these points to S_3 and let the joining lines cut the conic again in A', B', C'. Since the point finally arrived at, A' or B' or C', does not in general coincide with the corresponding starting-point A or B or C, we shall have, instead of an inscribed triangle as required by the problem, three polygons AA_1A_2A', BB_1B_2B', CC_1C_2C' which are not closed. But since, by a series of projections from S_1, S_2, S_3 in succession as centres, we have passed from the range A, B, C, ... to the range A_1, B_1, C_1, ..., from this last to A_2, B_2, C_2, ..., and from this to A', B', C', ..., it follows that the range of points A, B, C, ..., with which we started is projective with the range of points A', B', C', ..., with which we ended (Arts. 200, 201, 203). The problem would be solved if one of the points in the latter range coincided with its correspondent in the former. If then the two projective ranges ABC ... and $A'B'C'$ ... have self-corresponding points, each of these may be taken as the first vertex of a triangle which satisfies the given conditions. We have therefore only to determine (Art. 200, II) the straight line on which intersect the three pairs of opposite sides of the inscribed hexagon $AB'CA'BC'$, and to construct (Art. 212) the points of intersection M and N of this straight line with the conic; each of them will give a solution of the problem*.

224. By a similar method may be solved the correlative problem: *To circumscribe about a given conic (i.e. one which is either completely drawn or determined by five tangents) a polygon whose vertices lie respectively on given straight lines.*

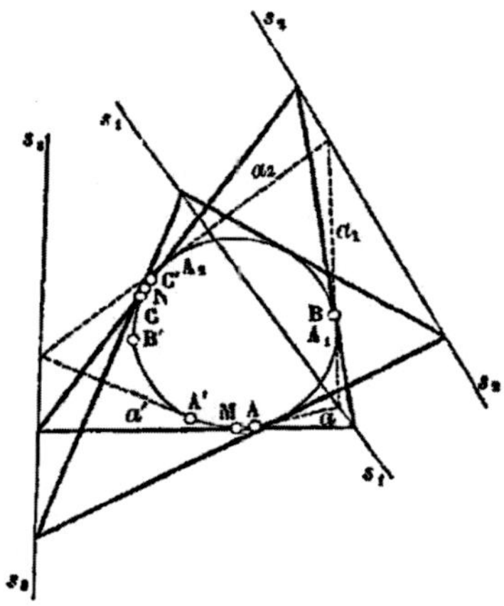

Fig. 167.

Suppose that it is required to circumscribe about the conic a triangle whose vertices lie respectively on the straight lines s_1, s_2, s_3 (Fig. 167). Take any point A on the conic and draw the tangent a at it; from the point where this tangent cuts s_1 draw another tangent a_1 (let its point of contact be A_1); from the point where a_1 cuts s_2 draw a third tangent a_2 (let its point of contact be A_2); finally, from the point where a_2 cuts s_3 draw the tangent a', and let its point of contact be A'. The problem would be solved if the point A'

* Poncelet, *loc. cit.*, p. 352.

coincided with A, *i.e.* if the tangents a' and a coincided with one another. Suppose that other similar trials have been made, taking other arbitrary points B, C, ... on the conic to begin with; then we shall arrive in succession at the ranges of points A, B, C, ..., A_1, B_1, C_1, ..., A_2, B_2, C_2, ..., and A', B', C', ..., which are all projectively related to one another. For the first range is projective with the second (Art. 203), since the tangents at A and A_1, B and B_1, C and C_1, ... always intersect on s_1; and for similar reasons the second and third, and the third and fourth, are projective with one another; consequently (Art. 201) the same is true of the fourth and the first. Since the problem would be solved if A' coincided with A, or B' with B, ..., each of the self-corresponding points of the projective ranges ABC ... and $A'B'C'$ ... may be taken as the point of contact of the first side of a triangle which satisfies the given conditions. We have therefore only to make three trials (Art. 200), *i.e.* to take any three points A, B, C on the conic and to derive from them the corresponding points A', B', C'; and then to construct the points of intersection of the conic with the straight line which joins the points of intersection of the three pairs of opposite sides (the Pascal line) of the inscribed hexagon $AB'CA'BC'$*.

225. The particular case of the problem of Art. 223 in which the given points S_1, S_2, ... lie all upon one straight line s must be considered separately. If the number of sides of the required polygon is even, the theorem of Art. 187 may be applied; in this case the problem has either no solution at all, or it has an infinite number of solutions. Suppose it required, for example, to inscribe in the conic an octagon of which the first seven sides pass respectively through the points S_1, S_2, ... S_7, then by the theorem just quoted the last side will pass through a fixed point S on s: this point S is not arbitrary, but its position is determined by those of the points S_1, S_2, ... S_7. If then the last of the given points S_8 coincides with S, there are an infinite number of octagons which satisfy the given conditions. But if S_8 does not coincide with S, there is no solution.

If the number of sides of the required polygon is odd, the problem becomes determinate. Suppose it is required to inscribe in the conic a heptagon (Fig. 124) whose sides pass respectively through the given collinear points S_1, S_2, S_3, ... S_7. By the theorem of Art. 187 there exist an infinite number of octagons whose first seven sides pass through seven given collinear points and whose eighth side passes through a fixed point S collinear with the others. If among these octagons there is one such that its eighth side touches the conic, the problem will be solved; for this octagon, having two of its vertices indefinitely

* PONCELET, *loc. cit.*, p. 354.

near to one another, will reduce to an inscribed heptagon, whose sides pass respectively through seven given points. If then tangents can be drawn from the point S to the conic, the point of contact of each of them will give a solution (Art. 187). According therefore to the position of the point S with reference to the conic, there will be two solutions, or only one, or none.

In Fig. 126 is shown the case of this problem where the polygon to be inscribed is a triangle *.

The solution of the correlative problem, *to circumscribe about a given conic a polygon whose vertices lie respectively on given rays of a pencil,* is left as an exercise to the student. This problem also is either indeterminate or impossible if the polygon is one of an even number of sides; it is determinate and of the second degree if the polygon is one having an odd number of sides (Figs. 125, 127).

226. Lemma. *If two conics cut one another in the points A, B, C, C', and if from A and B respectively two straight lines AFF', BGG' be drawn cutting the first conic in F and G, and the second in F' and G', then the chords FG, $F'G'$ will intersect in a point H lying on the chord CC'* (Fig. 168).

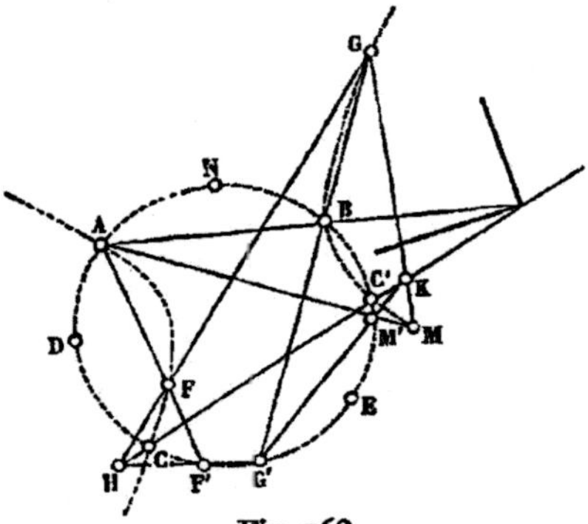

Fig. 168.

The transversal CC' cuts the first conic and the opposite sides of the inscribed quadrangle $ABGF$ in six points of an involution (Art. 183, left); and the same is true with regard to the second conic and the inscribed quadrangle $ABG'F'$. But the two involutions must coincide (Art. 127), since they have two pairs of conjugate points in common, *viz.* the points C, C' in which the transversal cuts both the conics, and the points in which it cuts the pair of opposite sides AFF', BGG', which belong to both quadrangles. The involutions will therefore have every pair of conjugate points in common, and therefore the transversal CC' will meet FG and $F'G'$ in the same point H, the conjugate of the point in which it meets AB†.

227. The preceding lemma, which is merely a corollary of Desargues' theorem, leads at once to the solution of the two following problems, one of which is of the first, and the other of the second degree.

* Pappus, *loc. cit.*, book vii. prop. 117.

† This may also be proved very simply by applying Pascal's theorem to each of the hexagons $AFGBCC'$, $AF'G'BCC'$ in turn.

I. PROBLEM. *Given three of the points of intersection A, B, C of two conics, and in addition two other points D, E of the first, and two other points F, G of the second, to determine the fourth point of intersection of the two conics* (Fig. 168).

Take two of the given points of intersection A and B, and join AF, BG. These straight lines will cut the first conic again in points F', G' respectively which can be determined by the method of Art. 161 (right). Join FG, $F'G'$, and let them meet in H. By the foregoing lemma H will lie on the chord joining the other two points of intersection of the conics. This chord will therefore be HC, and it remains only to determine the point C' where HC cuts either of the conics; C' will be the required fourth point of intersection of the conics.

II. PROBLEM. *Given two of the points of intersection, A, B, of two conics, and in addition the three points D, E, N of the first and the three points F, G, M of the second, to determine the other two points of intersection of the conics* (Fig. 168).

Join AF and BG, and let them meet the first conic again in F', G' respectively; join FG, $F'G'$, and let them meet in H. The point H will lie on the chord joining the two required points. Again, join AM, and let it meet the first conic again in M'; join GM, $G'M'$, and let these meet in K; then the point K also will lie on the same chord. The required points therefore lie on HK, and the problem reduces to the determination (Art. 212) of the points of intersection C, C' of the conics with HK *.

228. The solution just given of problem II holds good equally when the points A and B lie indefinitely near to one another, *i. e.* when the two conics touch a given straight line at the same given point.

In this case two conics are given which touch one another at a point A, and the straight line HK is constructed which joins their remaining points of intersection C and C'. If HK passes through A, one of the points C or C' must coincide with A, since a conic cannot cut a straight line in three points. When this is the case, three of the four points of intersection of the conics lie indefinitely near to one another, and may be said to coincide in the point A; and the conics are said to *osculate* at the point A. The construction gives a point H of the chord which joins A to the fourth point of intersection C of the conics. It may happen that this chord coincides with the tangent at A; in this case A represents four coincident points of intersection of the two conics (or rather, four such points lying indefinitely near to one another).

* GASKIN, *The geometrical construction of a conic section, &c.* (Cambridge, 1852), pp. 26, 40.

229. Let now the lemma of Art. 226 be applied to the case of a conic and a circle touching it at a point A. At A draw the normal to the conic (the perpendicular to the tangent at A), and let it cut the conic again in F and the circle again in F'. On AF as diameter describe a circle; this circle, which touches the conic at A and cuts it at F, will cut it again at another point G such that AGF is a right angle. Join AG and let G' be the point where it cuts the first circle. Join FG, $F'G'$; by the lemma they will intersect on the chord HK; but they are parallel to one another, since $AG'F'$ also is a right angle. Thus *for any circle whatever which touches the conic at A, the chord of intersection HK with the conic has a constant direction, viz. that parallel to FG.*

If HK passes through A, the conic and the circle osculate at this point. If then a parallel through A to FG cut the conic again in C, the circle which touches the conic at A and cuts it at C will be the osculating circle (circle of curvature) at A *.

[In the particular case where A is a vertex (Art. 297) of the conic, F will be the other vertex, FG the tangent at F, AC the tangent at A, and C will coincide with A. It is seen then that the osculating circle at a vertex of a conic has not only three but *four* indefinitely near points in common with the conic.]

Conversely, the conic can be constructed which passes through three given points A, P, Q and has a given circle for its osculating circle at one of these points A.

For join AP, AQ, and let them cut the given circle in P', Q' respectively; and join PQ, $P'Q'$, meeting in U. If AU be joined and cut the circle again in C, the required conic will pass through C. It is therefore determined by the four points A, P, Q, C and the tangent at A (which is the same as the tangent to the circle there).

230. The proposition correlative to the lemma of Art. 226 may be enunciated as follows:

If a and b are a pair of common tangents to two conics, and if from two points taken on a and b respectively the tangents f, g be drawn to the first conic and the tangents f', g' to the second, then the points fg and $f'g'$ will be collinear with the point of intersection of the second pair of common tangents to the conics.

This proposition enables us to solve the problems which are correlative to I and II of Art. 227; *viz.* given three (or two) of the common tangents to two conics, and in addition two (or three) tangents to the first and two (or three) tangents to the second, to determine the remaining common tangent (or the two remaining common tangents) to the conics.

231. Problem. *Given eleven points A, B, C, D, E; A_1, B_1, C_1, D_1, E_1; P;*

* Poncelet, *loc. cit.*, Arts. 334–337.

to construct by points the conic which passes through P and through the four points of intersection of the two conics which are determined by the points A, B, C, D, E and A_1, B_1, C_1, D_1, E_1 respectively. The conics are supposed not to be traced, nor are their points of intersection given.*

Solution. Draw through P any transversal, and construct (Art. 212, left) the points M and M' in which it cuts the conic $ABCDE$ and the points N and N' in which it cuts the conic $A_1B_1C_1D_1E_1$. Since these two conics and the required one all pass through the same four points, Desargues' theorem may be applied to them. If therefore (Art. 134, left) the point P' be constructed, conjugate to P in the involution determined by the pairs of points M and M', N and N', this point P' will lie on the required conic. By causing the transversal to turn about the point P, other points on the required conic may be obtained.

232. PROBLEM. *Given ten points A, B, C, D, E; A_1, B_1, C_1, D_1, E_1 and a straight line s; to construct a conic which shall touch s and shall pass through the four points of intersection of the two conics which are determined by the points A, B, C, D, E and A_1, B_1, C_1, D_1, E_1 respectively. The conics are supposed not to be traced, nor are their points of intersection given.*

Solution. Construct (Art. 212) the points of intersection M and M' of s with the conic $ABCDE$, and the points of intersection N and N' of s with the conic $A_1B_1C_1D_1E_1$, and then (Art. 134) the double points of the involution determined by the two pairs of points M and M', N and N'. If P is one of these double points, it will be the point of contact (Art. 185) of s with a conic drawn through the four points of intersection of the conics $ABCDE$ and $A_1B_1C_1D_1E_1$ to touch s. The problem thus reduces to that of the preceding Article.

233. The correlative constructions give the solutions of the correlative problems: *viz.* to construct a conic which passes through a given point (or which touches a given straight line), and which is inscribed in the quadrilateral formed by the four common tangents to two conics; the conics being supposed each to be determined by five given tangents, but not to be completely traced; and their four common tangents being supposed not to be given.

234. PROBLEM. *Through a given point S to draw a straight line which shall be cut by four given straight lines a, b, c, d in four points having a given anharmonic ratio.*

Solution. It has been seen (Art. 151) that the straight lines which are cut by four given straight lines in four points having a given anharmonic ratio are all tangents to one and the same conic

* PONCELET, *loc. cit.*, Art. 389.

touching the given straight lines; and that if A, B, C are the points where d cuts a, b, c respectively, and D is the point of contact of d, the anharmonic ratio $(ABCD)$ is equal to that of the four points in which the straight lines a, b, c, d are cut by any other tangent to the conic. Accordingly, if on the straight line d that point D be constructed (Art. 65) which gives with the points

$$ad\,(\equiv A)\,,\ bd\,(\equiv B)\,,\ cd\,(\equiv C)$$

an anharmonic ratio $(ABCD)$ equal to the given one, and if then the straight lines be constructed (Art. 213, right) which pass through S and touch the conic determined by the four tangents a, b, c, d and the point of contact D of d, each of these straight lines will give a solution of the proposed problem.

If one of the straight lines a, b, c, d lie at infinity, the problem becomes the following:

Given three straight lines a, b, c and a point S, to draw through S a straight line such that the segment intercepted on it between a and b may be to that intercepted on it between a and c in a given ratio.

To solve this, construct on the straight line a that point A which is so related to the points $ab\,(\equiv B)$ and $ac\,(\equiv C)$ that the ratio $AB:AC$ has the given value; and draw from S the tangents to the parabola which is determined by the tangents a, b, c and the point of contact A of a.

The correlative construction gives the solution of the following problem: On a given straight line s to find a point such that the rays joining it to four given points A, B, C, D form a pencil having a given anharmonic ratio.

235. PROBLEM. *Given two projective ranges of points lying on the straight lines u, u' respectively; to find two corresponding segments MP, $M'P'$ such that the angles MOP, $M'O'P'$ which they subtend at two fixed points O, O' respectively may be given in sign and magnitude.*

Solution. Take on u' two points A' and D' such that the angle $A'O'D'$ may be equal to the second of the given angles; let A and D be the points on u which correspond respectively to A' and D', and let A_1 be a point on u such that the angle A_1OD is equal to the first of the given angles. The problem would evidently be solved if OA_1 coincided with OA, since in this case the angles AOD and $A'O'D'$ would be equal to the given angles respectively. If the rays $O'A'$, OA, $O'D'$, OD, OA_1 be made to vary simultaneously, they will trace out pencils which are projectively related. For those traced out by $O'A'$ and $O'D'$ respectively are projective, and similarly those traced out by OA_1 and OD respectively, since the angles $A'O'D'$ and A_1OD are constant (Art. 108); and the pencils traced out by

OA and $O'A'$ respectively, and by OD, $O'D'$ respectively, are projective since the given ranges on u and u' are so. Consequently the pencils generated by OA and OA_1 respectively are projective, and their self-corresponding rays give the solutions of the problem. If three trials be made of a similar kind to the foregoing one, three pairs of corresponding rays OA and OA_1, OB and OB_1, OC and OC_1 will be obtained; let the self-corresponding rays of the concentric projective pencils determined by these three pairs be constructed (Art. 206, I). If one of these self-corresponding rays meets u in M, and if the point P be taken on u such that the angle MOP is equal to the first of the given ones, and if then on u' the points M', P' be found which correspond to M, P respectively, the angle $M'O'P'$ will be equal to the second of the given angles, and the problem will be solved.

236. PROBLEM. *Given two projective ranges of points A, B, C,... and A', B', C',... lying on the straight lines u and u' respectively, to find two corresponding segments which shall be equal, in sign and magnitude, to two given segments.*

Solution. Take on u' a segment $A'D'$ equal to the second of the given ones, and let AD be the segment on u which corresponds to $A'D'$. Take on u the point A_1 such that A_1D is equal to the first of the given segments; then the problem would be solved if A_1 coincided with A. If the points A, A', D', D, A_1 be made to vary simultaneously, the ranges traced out by A and A' respectively will be projective with one another, as also those traced out by D and D' respectively (by reason of the projective relation existing between ABC... and $A'B'C'$...); and the ranges traced out by A and D respectively, and similarly those traced out by A' and D' respectively, will be projective with one another, since they are generated by segments of constant length sliding along straight lines (Art. 103). Consequently also the ranges traced out by A and A_1 are projectively related, and their self-corresponding points give the solutions of the problem. It is therefore only necessary to obtain three pairs of corresponding points A and A', B and B', C and C', by making three trials, and then to construct the self-corresponding points of the ranges determined by these three pairs (Art. 206, II).

237. The student cannot have failed to remark that the method employed in the solution of the preceding problems has been in all cases substantially the same. This method is general, uniform, and direct; and it may be applied in a more or less simple manner to all problems of the second degree, *i.e.* to all questions which when treated algebraically would depend on a quadratic equation. It consists in making three trials, which give three pairs of corresponding elements of two superposed projective forms; the self-corresponding elements of these systems give the solutions of the problem. This method is precisely analogous to that known in Arithmetic as the 'rule of false position,' and it has on that account been termed a *geometric method of false position* *.

238. Problems of the second degree (and those which are reducible to such) are solved, like all those occurring in elementary Geometry, by means of the ruler and compasses only, that is to say by means of the intersections of straight lines and circles †. But again, the solution of any such problem can be made to depend on the determination of the self-corresponding elements of two superposed projective forms, which determination depends (Art. 206) on the construction of the self-corresponding points of two projective ranges lying on a circle whose position and size is entirely arbitrary. It follows that a single circle, described once for all, will enable us to solve all problems of the second degree which can be proposed with reference to any given elements lying in one plane (the plane in which the circle is drawn) ‡. This circle once described, any such problem will reduce to that of constructing three pairs of points of the two projective systems whose self-corresponding elements give the solution of the problem. This done, we proceed to transfer to the circumference of the circle, by means of projections and sections, these three pairs of points. This will give three pairs of points on the circle; taking these as the pairs of opposite vertices of an inscribed hexagon, we have only further to draw the straight line which joins the points of intersection of the three pairs of opposite sides (the Pascal line) of this hexagon.

It is hardly necessary to remark that instead of the solution of such a problem being made to depend on the common elements of two superposed projective forms, it may always be reduced to the determination of the double elements of an involution (Art. 211).

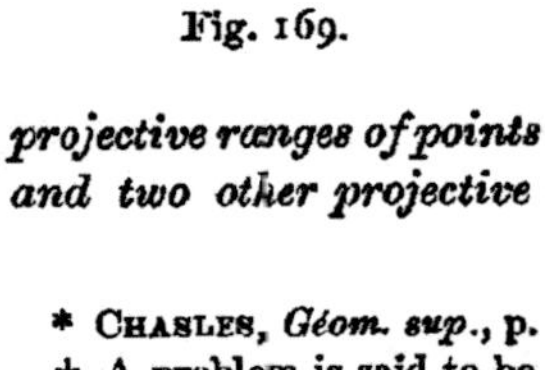

Fig. 169.

The following Articles (239 to 249) contain examples of problems solved by means of the method just explained.

239. Problem. *Given* (Fig. 169) *two projective ranges of points lying on the straight lines u and u′ respectively, and two other projective ranges of points lying on the straight lines v*

* Chasles, *Géom. sup.*, p. 212.

† A problem is said to be *of the first degree* when it can be solved with help of the ruler only, *i. e.* by the intersections of straight lines. See Lambert, *loc. cit.*, p. 161; Brianchon, *loc. cit.*, p. 6; Poncelet, *loc. cit.*, p. 76.

‡ Poncelet, *loc. cit.*, p. 187; Steiner, *Die geometrischen Constructionen ausgeführt mittelst der geraden Linie und eines festen Kreises* (Berlin, 1833), p. 67; Collected Works, vol. i. pp. 461–522; Staudt, *Geometrie der Lage* (Nürnberg, 1847), § 23.

and v' respectively; it is required to draw through a given point O two straight lines s and s', which shall cut u and u' in a pair of corresponding points and also v and v' in a pair of corresponding points.

Through O draw any straight line cutting u', v' in A', P' respectively; let A be the point on u which corresponds to A', and let P be the point on v which corresponds to P'. The problem would be solved if the straight lines OA and OP coincided with one another. If these straight lines be made to change their positions simultaneously, they will trace out two concentric projective pencils (determined by three trials of a similar kind to the one just made); and the self-corresponding rays of these pencils will give the solutions of the problem.

240. In the preceding problem the straight lines u and u' might be taken to coincide, and similarly v and v'. If all four straight lines coincided with one another, the problem would become the following:

Given two projective ranges u, u' and two other projective ranges v, v' all lying on one straight line, to find a pair of points which shall correspond to one another when regarded as points of the ranges u, u' respectively, and likewise when regarded as points of the ranges v, v' respectively.

241. Problem. *Between two given straight lines u and u_1 to place a segment such that it shall subtend given angles at two given points O and S* (Fig. 170).

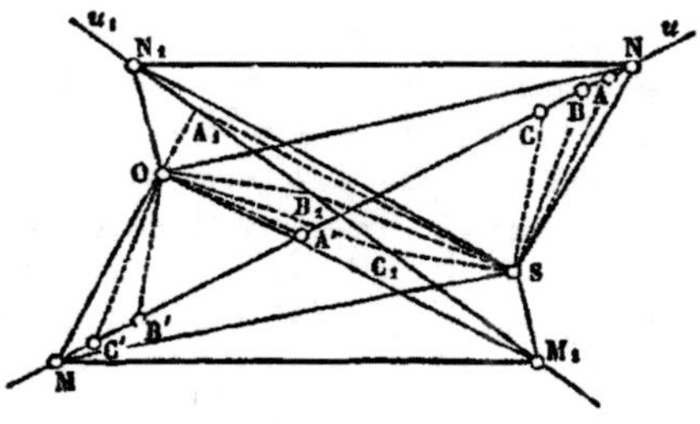

Fig. 170.

Draw any ray SA to meet u in A; draw SA_1 to meet u_1 in A_1 so that ASA_1 may be equal to the second of the given angles; join OA_1, and draw OA' to meet u in A' so that $A'OA_1$ may be equal to the first of the given angles. Then the problem would be solved if OA coincided with OA'. Three trials of a similar kind to the one just made will give three pairs of corresponding rays (OA and OA', OB and OB', OC and OC') of the two projective pencils which would be traced out by causing OA and OA' to change their positions simultaneously; the self-corresponding rays OM and ON of these pencils will give the solutions (MM_1 and NN_1) of the problem.

242. PROBLEM. *Given two projective ranges u and u′; if a pair of corresponding points A and A′ of these ranges be taken, it is required to find another pair of corresponding points M and M′ such that the ratio of the length of the segment AM to that of the segment A′M′ may be equal to a given number λ.*

Let A and A', B and B', C and C' be three pairs of corresponding points of the two ranges. On u take two new points B'', C'' such that $AB''=\lambda.A'B'$ and $AC''=\lambda.A'C'$. The points A, B'', C'' determine a range which is similar (Art. 99) to the range $A', B', C', \ldots$ and therefore projective with $A, B, C, \ldots$. The collinear ranges $A, B'', C'', \ldots$ and $A, B, C, \ldots$ have already one self-corresponding point in A; their other self-corresponding point M (Art. 90) will give the solution of the problem, since $AM=AM''=\lambda . A'M'$. This problem is therefore of the first degree.

243. PROBLEM. *Given two collinear projective ranges ABC ... and A′B′C′ ..., to find a pair of corresponding points M and M′ such that the segment MM′ shall be bisected at a given point O.*

Take three points A'', B'', C'' such that O is the middle point of each of the segments AA'', BB'', CC''; the points A'', B'', C'' determine a range which is equal to the range $ABC \ldots$, and therefore projective with the range $A'B'C' \ldots$. Construct the self-corresponding points of the collinear projective ranges $A'B'C' \ldots$ and $A''B''C'' \ldots$; if M' or M'' is one of them, then MM' will have its middle point at O, and will be a segment such as is required.

244. PROBLEM. *Given a straight line and two points E, F on it; to determine on the straight line two points M and M′ such that the segment MM′ may be equal in length to a given segment, and the anharmonic ratio (EFMM′) equal to a given number.*

Take on the given straight line any three points A, B, C; then find on it three points A', B', C' such that the anharmonic ratios $(EFAA')$, $(EFBB')$, $(EFCC')$ may each be equal to the given number; and again three points A'', B'', C'' such that the segments AA'', BB'', CC'' may each be equal in length to the given segment. The ranges $ABC \ldots$ and $A'B'C' \ldots$ will be projectively related (Arts. 79, 109), and the same will be the case with regard to the ranges $ABC \ldots$ and $A''B''C'' \ldots$ (Art. 103); therefore $A'B'C' \ldots$ and $A''B''C'' \ldots$ will be projective with one another. If these ranges have self-corresponding points, and if M' or M'' is one of them, the segment MM' and the anharmonic ratio $(EFMM')$ will have the given values, and the problem is solved.

245. PROBLEM. *To inscribe in a given triangle PQR a rectangle of given area* (Fig. 171).

Suppose $MSTU$ to be the rectangle required; if MS' be drawn parallel to PR, a parallelogram $MSPS'$ will be formed which is equal in area to the rectangle; so that for the given problem may be substituted the following equivalent one:

To find on the base QR of a given triangle PQR a point M such that if MS, MS' be drawn parallel to the sides PQ, PR to meet PR, PQ in S, S' respectively, the rectangle contained by PS and PS' shall be equal to a given square k^2.

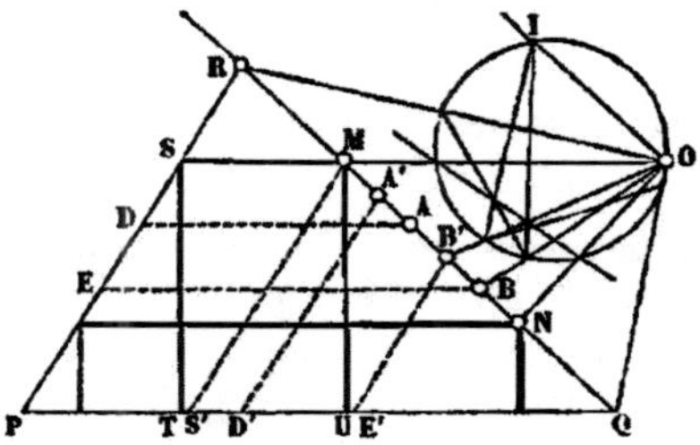

Fig. 171.

Take any point A on QR, draw AD parallel to PQ to meet PR in D, and take on PQ a point D' such that the rectangle contained by PD and PD' may be equal to k^2; then draw $D'A'$ parallel to PR to meet QR in A'. If the points A and A' coincided with one another, the problem would be solved.

Now let the points A, D, D', A' be made to vary simultaneously; they will trace out ranges which are all projective with one another. For since D is the projection of A made from the point at infinity on PQ, and A' the projection of D' made from the point at infinity on PR, the first and second ranges are in perspective, and the third and fourth likewise. But the second and third ranges are projective with one another, since the relation $PD \cdot PD' = k^2$ shows (Art. 74) that the points D and D', in moving simultaneously, describe two projective ranges such that the point P, regarded as belonging to either range, corresponds to the point at infinity regarded as belonging to the other*.

Three similar trials give three pairs of points similar to A and A'; if the self-corresponding points of the ranges determined by these pairs be constructed, they will give the solutions of the problem.

Instead of taking the point A quite arbitrarily in the three trials, any particular positions may be chosen for it, and by this means the construction may often be simplified. This remark applies to all the problems which we have discussed. With regard to the present one, it is clear that if A be taken at infinity, its projection D will also lie at infinity; consequently D' will coincide with P, and therefore A' with R. Again, if A be taken coincident with Q, its projection D will coincide with P, and consequently D', and therefore also A', will pass off to infinity. We have thus two trials, neither of which requires

* If the two ranges be called u and u', and the construction of Art. 85 (left) be referred back to, it will be seen that the auxiliary range u'' lies in this case entirely at infinity. If then a pair of corresponding points D and D' have been found, and we wish to find the point E' which corresponds to any other point E of PR ($\equiv u$), we have only to join $D'E$, and to draw DE' parallel to $D'E$ to meet PQ ($\equiv u'$) in E'.

any construction; the pairs which result from them are composed respectively of the point at infinity and R, and of Q and the point at infinity. If the pair given by the third trial be called B, B', and if A, A' stand for any pair whatever, we have (Art. 74)

$$QA \cdot RA' = QB \cdot RB',$$

and therefore, if M is a self-corresponding point,

$$QM \cdot RM = QB \cdot RB',$$

from which the self-corresponding points could be found. But it is better in all cases to go back to the general construction of Art. 206. In this case the three pairs of conjugate points of the two ranges which are given are: B and B'; the point at infinity and R; Q and the point at infinity. Let then any circle be taken, and a point O on its circumference; from O draw the straight lines OB, OB', OR, OQ, and a parallel to QR, and let these cut the circle again in B_1, B_1', R_1, Q_1, and I respectively *. Join the point of intersection of $B_1 R_1$ and $B_1' I$ with that of $B_1 I$ and $B_1' Q_1$; if the joining line cut the circle in two points M_1 and N_1, the straight lines which join these to O will meet QR in the self-corresponding points M and N, and these give the solutions of the problem.

246. PROBLEM. *To construct a polygon, whose sides shall pass respectively through given points, and all whose vertices except one shall lie respectively on given straight lines; and which shall be such that the angle included by the sides which meet in the last vertex is equal to a given angle.*

Suppose, for example, that it is required to construct a triangle LMN (Fig. 172) whose sides MN, NL, LM shall pass through the given points O, V, U respectively, and whose vertices M, N shall lie on the given straight lines u, v respectively; and which shall be such that the angle MLN is equal to a given angle.

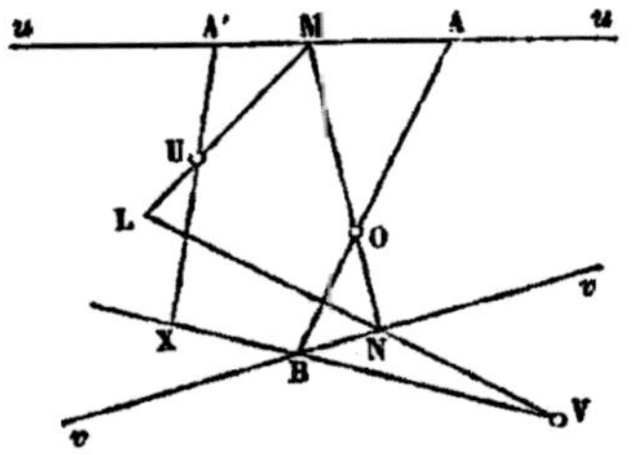

Fig. 172.

Through O draw any straight line to cut u in A and v in B; join BV, and through U draw the straight line UX making with BV an angle equal to the given one. Let UX meet u in A'; the problem would be solved if the point A' coincided with A. If the rays OA, UA' be made to vary simultaneously, they will determine on u two projective ranges; the solutions of the problem will be found by constructing the self-corresponding points of these ranges.

* Of these points only I is marked in the figure.

247. The following problem is included in the foregoing one:

A ray of light emanating from a given point O is reflected from n given straight lines in succession; to determine the original direction which the ray must have, in order that this may make with its direction after the last reflexion a given angle.

Let $u_1, u_2, \ldots u_n$ be the given straight lines (Fig. 173). If the ray OA_1 strike u_1 at A_1, then by the law of reflexion the incident and reflected rays will make equal angles with u_1; but the incident ray passes through the fixed point O; therefore the reflected ray will always pass through the point O_1 which is symmetrical to O with regard to u_1*. So again, if the ray after one reflexion strikes u_2 at A_2, it will be reflected according to the same law; consequently the ray after two reflexions will pass through a fixed point O_2 which is symmetrical to O_1 with regard to u_2; and so on. The paths of the ray before reflexion, and after one, two, ... n reflexions form therefore a polygon $OA_1A_2A_3 \ldots$, whose $n+1$ sides pass respectively through $n+1$ fixed points $O, O_1, O_2, \ldots O_n$, and which is such that n of its vertices lie respectively on n given straight lines $u_1, u_2, \ldots u_n$; while the angle included by the sides which meet in the last vertex is to be equal to a given angle. Thus the problem reduces, as was stated, to that of Art. 246.

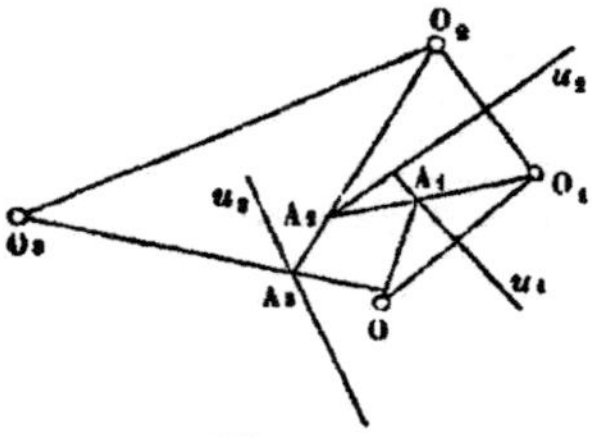

Fig. 173.

248. PROBLEM. *To construct a polygon whose vertices shall lie respectively on given straight lines, and whose sides shall subtend given angles at given points respectively.*

Suppose it required to construct a triangle whose vertices 1, 2, 3 shall lie on the given straight lines u_1, u_2, u_3 respectively, and whose sides 23, 31, 12 shall subtend at the given points S_1, S_2, S_3 respectively the angles $\omega_1, \omega_2, \omega_3$ which are given in sign and magnitude (Fig. 174). On u_1 take any point A; join AS_3, and make the angle AS_3B equal to ω_3; let S_3B cut u_2 in B. Join BS_1; make the angle BS_1C equal to ω_1, and let S_1C cut u_3 in C. Join CS_2; make the angle CS_2A' equal to ω_2, and let S_2A' cut u_1 in A'. The problem would be solved if S_2A' coincided

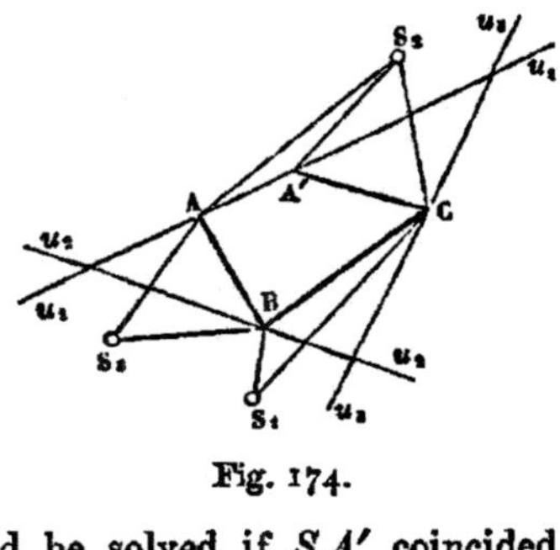

Fig. 174.

* *i.e.* a point O_1 such that OO_1 is bisected at right angles by u_1.

with S_2A. If S_2A be made to turn about S_2, the other rays S_3A, S_3B, S_1B, S_1C, S_2C, and S_2A' will change their positions simultaneously, and will trace out pencils which are all projectively related. For the ranges traced out by S_3A and S_3B respectively will be projective (Art. 108) since the angle AS_3B is constant; the ranges traced out by S_3B and S_1B respectively are projective since they are in perspective; and so on. The solutions of the problem will therefore be given by the self-corresponding rays of the concentric projective pencils which are generated by S_2A and S_2A' respectively.

In the same manner is solved the more general problem in which the straight lines joining $S_1, S_2, \ldots$ to the vertices of the polygon are no longer to include given angles, but are to be such that together with pairs of given straight lines meeting in $S_1, S_2, \ldots$ respectively they form at each of these points a pencil of four rays having a given anharmonic ratio. If at each of the points the pencil is to be harmonic, and the given straight lines such as to include a right angle, the problem can be enunciated as follows (Art. 60):

To construct a polygon whose vertices shall lie respectively on given straight lines, and whose sides shall subtend at given points angles whose bisectors are given.

249. The same method gives the solution of the problem:

To construct a polygon whose sides shall pass respectively through given points, and which shall be such that the pairs of adjacent sides divide given segments respectively in given anharmonic ratios *.

Particular cases of this problem may be obtained by supposing that each pair of adjacent sides is to intercept on a given straight line a segment given in magnitude and direction; or a segment which is divided by a given point into two parts having a given ratio to one another †.

* That is to say, two adjacent sides are to cut a given straight line, on which are two given points A, B, in two other points C, D such that the anharmonic ratio $(ABCD)$ may be equal to a given number.

† CHASLES, *Géom. sup.*, pp. 219–223; and TOWNSEND, *Modern Geometry* (Dublin, 1865), vol. ii. pp. 257–275.

CHAPTER XX.

POLE AND POLAR.

250. LET any point S be taken in the plane of a conic (Fig. 175), and through it let any number of transversals be drawn to cut the conic in pairs of points A and A', B and B', C and C',.... The tangents a and a', b and b', c and c' at these points will, by Arts. 203, 204, intersect in pairs on a fixed straight line s, on which lie also the points of contact of the tangents from S to the conic (when the position of S is such that tangents can be drawn). Further, the pairs of chords AB' and $A'B$, AC' and $A'C$,... BC' and $B'C$, ... AB and $A'B'$, AC and $A'C'$, ... BC and $B'C'$, ... will intersect on s. Another property of the straight line s may be noticed. In the complete quadrangle $AA'BB'$, each of the straight lines AA' and BB' is divided harmonically by the diagonal point S and the point where it is cut by the straight line s which joins the other diagonal points (Art. 57); consequently A and A' (and similarly B and B', C and C',...) are harmonic conjugates with regard to S and the point where AA' (or BB', CC',...) is cut by s.

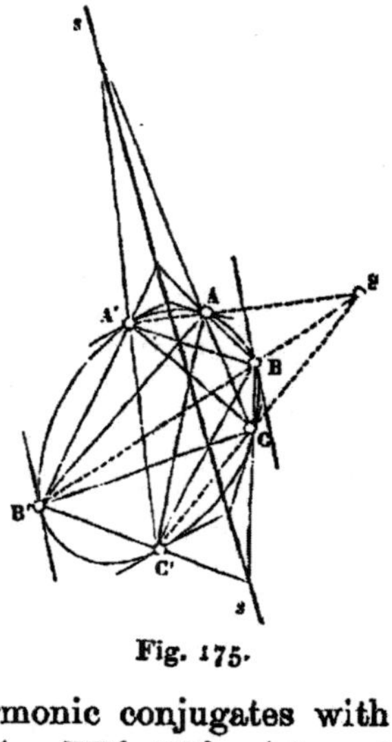

Fig. 175.

The straight line s determined in this manner by the point S is called the *polar* of S with respect to the conic; and, reciprocally, the point S is said to be the *pole* of the straight line s.

The polar of a given point S is therefore at the same time: (1) *the* *locus of the points of intersection of tangents to the conic at the pairs of points where it is cut by any transversal through S;* (2) *the locus of the points of intersection of pairs of opposite sides of quadrangles inscribed in the conic such that their diagonals meet in S;* (3) *the locus of points taken on any transversal through S such that they are harmonically conjugate to S with regard to the pair of points in which the transversal is cut by the conic;* (4) *the chord of contact of the tangents from S to the conic, when S has such a position that it is possible to draw these* * †.

251. Reciprocally, any given straight line *s* determines a point *S*, of which it is the *polar*. For let *A* and *B* (Fig. 176) be any two points on the conic; the tangents *a* and *b* at these points will cut *s* in two points from which can be drawn two other tangents *a′* and *b′* to the conic. Let *A′* and *B′* be the points of contact of these, and let *AA′*, *BB′* meet in *S*; then the polar of *S* will pass through the points *aa′* and *bb′*, and must therefore coincide with *s*.

If then from any point on s a pair of tangents can be drawn to the conic, their chord of contact will pass through S.

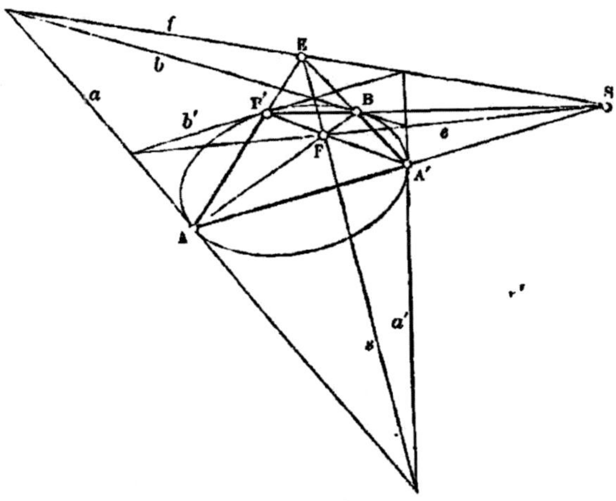

Fig. 176.

252. The complete quadrangle *AA′ BB′* and the complete quadrilateral *aa′ bb′* (Fig. 176) have the same diagonal

* Apollonius, *loc. cit.*, lib. vii. 37; Desargues, *loc. cit.*, pp. 164 sqq.; De la Hire, *loc. cit.*, books i. and ii.

† (4) follows from (3) by what has been proved in Art. 71.

triangle (Art. 169). The vertices of this triangle are S, the point of intersection F of AB and $A'B'$, and the point of intersection E of AB' and $A'B$; its sides are s, the straight line f joining the points ab and $a'b'$, and the straight line e joining the points ab' and $a'b$. Thus *if from any two points taken on the straight line s pairs of tangents a and a', b and b' be drawn to the conic, the diagonals of the quadrilateral aba'b' will pass through S.*

253. The straight lines a, a', b, b' (Fig. 177) form a quadrilateral circumscribed about the conic, one of whose diagonals is s, and whose other two diagonals meet in S. *Thus if from any point on s a pair of tangents be drawn to the conic, they will be harmonically conjugate with regard to s and the straight line joining the point to S* (Art. 56).

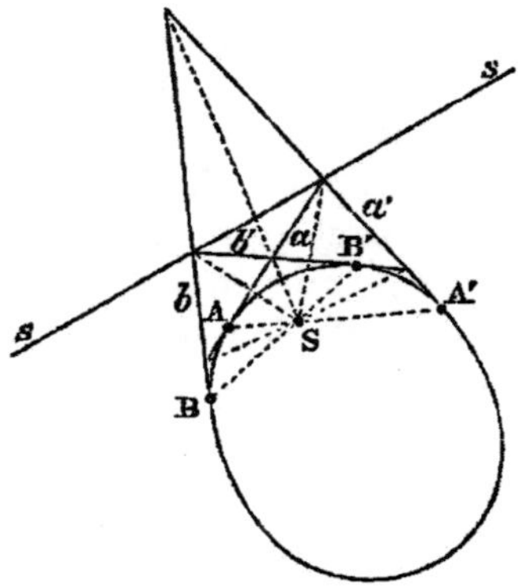

Fig. 177.

254. *If then a conic is given, every point in its plane has its polar and every straight line has its pole*.* The given conic, with reference to which the pole and polar are considered, may be called the *auxiliary* conic.

I. If a point in the plane of a conic is such that from it two tangents can be drawn to the curve, it is said to lie *outside* the conic, or to be an *external* point; if it is such that no tangent can be drawn, it is said to lie *inside* the conic, or to be an *internal* point. If then the pole lies outside the conic (Art. 203, VIII) the polar cuts the curve, and it cuts it at the points of contact of the tangents from the pole to the conic†.

If the pole lies inside the curve, the polar does not cut the conic.

II. If a point on the conic itself be taken as pole and a transversal be made to revolve round this point, one of its points of intersection with the conic will always coincide with the pole itself. Since then the polar is the locus of the points where the tangents at these points of intersection meet, and

* Desargues, *loc. cit.*, p. 190.

† See also Art. 250, (4).

in this case one of the tangents is fixed, it follows that the polar of a point on the conic is the tangent at this point; or that *if the pole is a point on the conic, the polar is the tangent at this point.*

III. Reciprocally, if every point of the polar lies outside the conic, the pole lies inside the conic; if the polar cuts the conic, the pole is the point where the tangents at the two points of intersection meet; and if the polar touches the conic, the pole is its point of contact.

255. *If two points are such that the first lies on the polar of the second, then will also the second lie on the polar of the first.*

Consider Fig. 176; let E be taken as pole and let F be a point lying on the polar of E. If the straight line EF cuts the conic, it will cut it in two points which are harmonically conjugate with regard to E and F (Art. 250 [3]); consequently one of the points E, F will lie inside and the other outside the conic, and by Art. 250 (3) again, if F be taken as pole, E will be a point on its polar.

If the straight line EF does not cut the conic, the chord of contact of the tangents from E will pass through F, since this chord is the polar of E; and therefore by Art. 250 (1) E will lie on the polar of F.

The above proposition may also be expressed in the following manner:

If a straight line f pass through the pole of another straight line e, then will also e pass through the pole of f.

For let E, F be the poles of e, f respectively; since by hypothesis E lies on the polar of F, therefore F will lie on the polar of E that is to say, e will pass through F, the pole of f.

Two points such as E and F, which possess the property that each lies on the polar of the other, are termed *conjugate* or *reciprocal* points with respect to the conic. And two straight lines such as e and f, each of which passes through the pole of the other, are termed *conjugate* or *reciprocal* lines with respect to the conic.

The foregoing proposition may then be enunciated as follows:

If two points are conjugate to one another with respect to a conic, their polars also are conjugate to one another, and conversely.

256. The same proposition can be put into yet another form, *viz.*

Every point on the polar of a given point E has for its polar a straight line passing through E.

*Every straight line passing through the pole of a given straight line e has for its pole a point lying on e**.

In other words, if a variable pole F be supposed to describe a given straight line e, the polar of F will always pass through a fixed point E, the pole of the given line; and conversely, if a straight line f revolve round a fixed point E, the pole of f will describe a straight line e, the polar of the given point E.

Or again: *the pole of a given straight line e is the centre of the pencil formed by the polars of all points on e; and the polar of a given point E is the locus of the poles of all straight lines passing through E* †.

257. PROBLEM. *Given a point S, to construct its polar with respect to a given conic.*

I. Let the conic be determined by five points A, B, C, D, E (Fig. 178).

Given a straight line s, to construct its pole with respect to a given conic.

I. Let the conic be determined by five tangents a, b, c, d, e (Fig. 179).

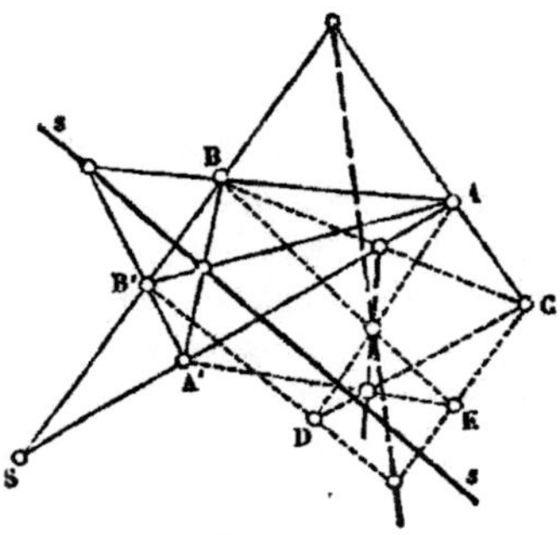

Fig. 178.

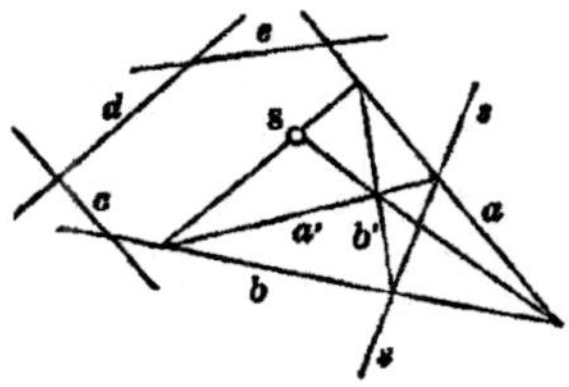

Fig. 179.

Join SA, SB, and find the points A', B' where these cut the conic again respectively (Art. 161, right). The straight line s which joins the point of intersection of AB' and $A'B$ to that of AB and

From the points sa, sb draw the second tangents a', b' respectively to the conic (Art. 161, left). The point S in which the diagonals of the quadrangle $aba'b'$ intersect one another will be

* DESARGUES, *loc. cit.*, p. 191.

† PONCELET, *loc. cit.*, Art. 195.

$A'B'$ will be the polar of the given point (Art. 250 [2]).

II. Let the conic be determined by five tangents a, b, c, d, e (Fig. 180).

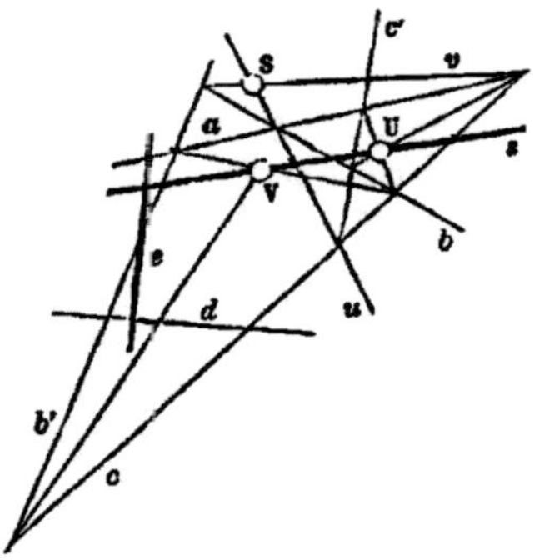

Fig. 180.

Through S draw two transversals u and v, and construct their poles U and V (as on the right-hand side above); UV will be the polar of S (Art. 256). To simplify matters the transversal u may be drawn through the point ab; if then the second tangent c' be drawn to the conic (Art. 161) from the point uc, U will be the point of intersection of the diagonals of the quadrilateral $acbc'$. So too if the transversal v be drawn through the point ac for example, and the second tangent b' be drawn to the conic from the point vb, then V will be the point of intersection of the diagonals of the quadrilateral $abcb'$.

the pole of the given straight line.

II. Let the conic be determined by five points A, B, C, D, E (Fig. 181).

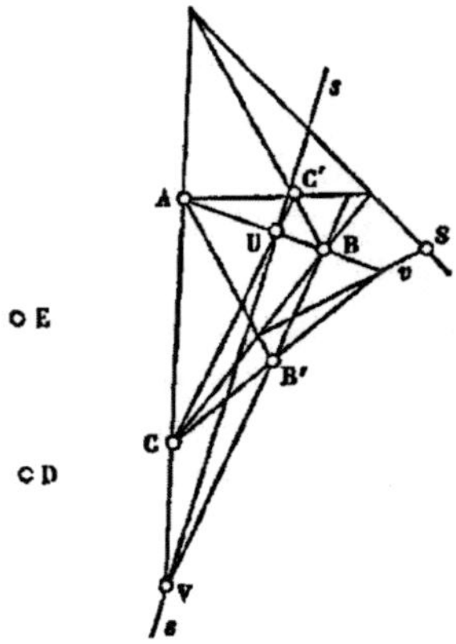

Fig. 181.

On s take two points U and V, and construct their polars u and v (as on the left-hand side above); the point uv will be the pole of s (Art. 256). To simplify matters the point U may be taken on the straight line AB; if then UC be joined, and the second point C' in which it meets the conic be constructed, u will be the straight line joining the points of intersection of the pairs of opposite sides of the quadrangle $ACBC'$. So too if V be taken on the straight line AC for example, and VB be joined, and its second point of intersection B' with the conic be constructed, then v will be the straight line joining the points of intersection of the pairs of opposite sides of the quadrangle $ABCB'$.

258. Let E and F (Fig. 182) be a pair of conjugate points and let G be the pole of EF; then G will be conjugate both to E and to F, so that the three points E, F, G are conjugate to one another two and two. Every side therefore of the triangle EFG is the pole of the opposite vertex, and the three sides are conjugate lines two and two.

A triangle such as EFG, in which each vertex is the pole of the opposite side with regard to a given conic is called a *self-conjugate* or *self-polar* triangle with regard to the conic.

259. *To construct a triangle self-conjugate with regard to a given conic.*

One vertex E (Fig. 182) may be taken arbitrarily; construct its

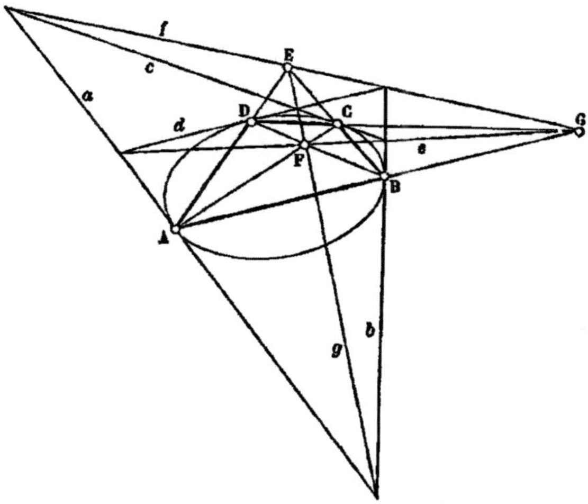

Fig. 182.

polar, take on this polar any point F, and construct the polar of F. This last will pass through E, since E and F are conjugate points; if G be the point where it cuts the polar of E, then E and G, F and G, will be pairs of conjugate points; and therefore EFG is a self-conjugate triangle.

In other words: take any point E and draw through it any two transversals to cut the conic in A and D, B and C respectively; join AC, BD, meeting in F, and AB, CD meeting in G; then EFG is a self-conjugate triangle.

Or again, one side e may be taken arbitrarily, and its pole E constructed; if through E any straight line f be drawn, and its pole (which will lie on e), be constructed and joined to the pole of e by the straight line g, then efg will be a triangle such as is required; for the straight lines e, f, g are conjugate two and two.

Thus, after having taken the side e arbitrarily, we may proceed as follows: take two points on e and from them draw pairs of tangents a and d, b and c, to the conic; join the points ac, bd by the straight line f, and the points ab, cd by the straight line g; then will efg be a self-conjugate triangle.

260. From what has been said above the following property is evident:

The diagonal points of the complete quadrangle formed by any four points on a conic are the vertices of a triangle which is self-conjugate with regard to the conic. And the diagonals of the complete quadrilateral formed by any four tangents to a conic are the sides of a triangle which is self-conjugate with regard to the conic.*

Or, in other words:

The triangle whose vertices are the diagonal points of a complete quadrangle is self-conjugate with regard to any conic circumscribing the quadrangle. And the triangle whose sides are the diagonals of a complete quadrilateral is self-conjugate with regard to any conic inscribed in the quadrilateral.

261. From the properties of the circumscribed quadrilateral and the inscribed quadrangle (Arts. 166 to 172) it follows moreover that:

If EFG (Fig. 182) is a triangle self-conjugate with regard to a given conic, and ABC is a triangle inscribed in the conic, such that two of its sides CA, AB pass through two of the vertices F, G respectively of the other triangle, then will the remaining side BC pass through the remaining vertex E, and every side of the inscribed triangle will be divided harmonically by the corresponding vertex of the self-conjugate triangle and the side which joins the other two vertices of it.

The three straight lines EA, FB, GC meet in one point D on the conic; the two triangles are therefore in perspective, and the three pairs of corresponding sides FG and BC, GE and CA, EF and AB, will meet in three collinear points.

Hence it follows that a self-conjugate triangle EFG and a point A of a conic determine an inscribed quadrangle $ABCD$, whose diagonal

* DESARGUES, *loc. cit.*, p. 186.

triangle is EFG. The points B, C, D are those in which the straight lines AG, AF, AE cut the conic again.

The enunciation of the correlative property is left to the student *.

262. Of the three vertices of the triangle EFG, one always lies inside the conic, and the two others outside it. For if E is an internal point, its polar does not cut the conic, and consequently F and G are both external to the conic. If, on the other hand, E is an external point, its polar cuts the conic, and F and G are harmonic conjugates with regard to the two points of intersection; of the two points F and G therefore, one must be internal and the other external to the conic.

From this property and that of Art. 254, I, we conclude that of the three sides of any self-conjugate triangle, two always cut the curve, and the third does not.

263. (1). *On every straight line there are an infinite number of pairs of points which are conjugate to one another with respect to a given conic, and these form an involution* †.

(2). *Through every point pass an infinite number of pairs of straight lines which are conjugate to one another with respect to a given conic, and these form an involution* †.

(3). *If a point describes a range, its polar with respect to a given conic will trace out a pencil which is projective with the given range. And, conversely, if a straight line describes a pencil, its pole with respect to a given conic will trace out a range which is projective with the given pencil* ‡.

To prove these theorems, consider Fig. 183, and suppose in it the conic and the three points A, B, G to be given. Let the point C be supposed to move along the conic. Then the rays AC, BC will trace out two pencils which are projective with one another (Art. 149 [1]); and therefore the ranges in which these pencils cut the polar of G will be projective also; that is to say, the conjugate points F and E will describe two collinear projective ranges. In these ranges the points F and E correspond to one another doubly, since the polar of E passes through F, and the polar of F passes through E; consequently the ranges in question are in involution.

From what has been said it follows also that the pairs of

* Poncelet, *loc. cit.*, p. 104. † Desargues, *loc. cit.*, pp. 192, 193.

‡ Möbius, *Baryc. Calc.*, § 290.

conjugate lines GF, GE in like manner form an involution, and that the range of poles E, F, ... is projective with the pencil of polars GF, GE, ...

264. If the straight line EF cuts the conic, the two points of

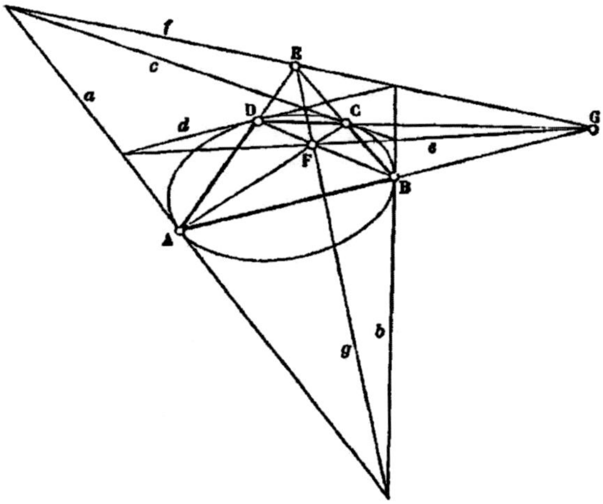

Fig. 183.

intersection are the double points of the involution formed by the pairs of conjugate poles. The centre of the involution lies on the diameter which passes through the pole G of the given straight line (Art. 290).

If the point G is external to the conic, the tangents from G to the conic are the double rays of the involution formed by the pairs of conjugate polars.

Consequently (Art. 125):

A chord of a conic is harmonically divided by any pair of points lying on it which are conjugate with respect to the conic; and

The pair of tangents drawn from any point to a conic are harmonic conjugates with respect to any pair of straight lines meeting in the given point which are conjugate with respect to the conic.

If the point G lies at infinity, the pairs of conjugate straight lines form an involution of parallel rays, the central ray of which is a diameter of the conic (Arts. 129, 276).

265. THEOREM. *If two complete quadrangles have the same diagonal points, their eight vertices lie either four and four on two straight lines or else they all lie on a conic.*

Let $ABCD$ and $A'B'C'D'$ (Fig. 184) be two quadrangles having the same diagonal points E, F, G; so that

$BC, AD, B'C', A'D'$ all meet in E,
$CA, BD, C'A', B'D'$ „ „ F,
$AB, CD, A'B', C'D'$ „ „ G.

(1). In the first place let the eight vertices be such that some three of them are collinear. Suppose for example that A' lies on AB. Since AB and $A'B'$ meet in G, therefore B' also must lie on AB; and since the straight lines GE, GF are harmonically conjugate with regard both to AB, CD and to $A'B', C'D'$, and AB coincides with $A'B'$, therefore also CD coincides with $C'D'$. Thus the four points C, D, C', D' are collinear, and the eight points $A, B, C, D, A', B', C', D'$ lie four and four on two straight lines.

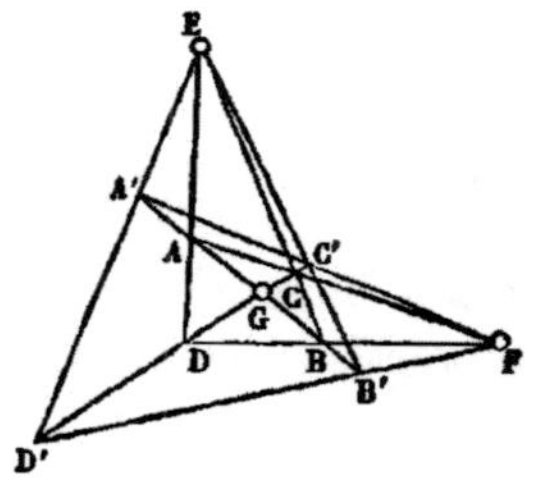

Fig. 184.

(2). But if this case be excluded, *i. e.* if no three of the eight vertices lie in a straight line, then a conic can be drawn through any five of them. Let a conic be drawn through A, B, C, D, A' (Fig. 185); then shall B', C', D' lie on the same conic. For since E, F, G are the diagonal points of the inscribed quadrangle $ABCD$, G is the pole of EF, and therefore G and the point where its polar EF meets the transversal $GB'A'$ are harmonically conjugate with regard to the points where this transversal cuts the conic. But one of these last points is A', therefore the other is B'; for since E, F, G are also the diagonal points of the quadrangle $A'B'C'D'$, the points A' and B' are harmonically conjugate with regard to G and the point where EF cuts $A'B'$. In a similar manner it can be shown that C' and D' also lie on the same conic. The eight vertices $A, B, C, D, A', B', C', D'$ therefore lie on a conic, and the proposition is proved.

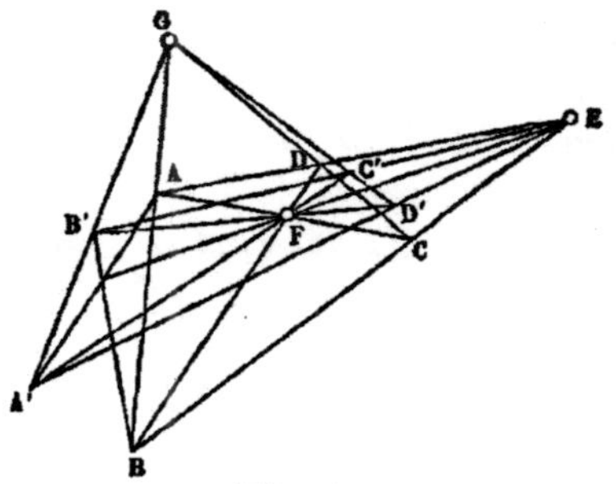

Fig. 185.

Since the straight lines AB and $A'B'$ meet in G, therefore AA' and BB', as also AB' and $A'B$, will meet on EF, the polar of G. This property gives the means of constructing the point B' when the points A, B, C, D, A' are given. The point C' will then be found as the point of intersection of $A'F$ and $B'E$, and the point D' as that of $B'F, A'E$, and $C'G$.

266. Suppose now that two conics are given which are inscribed in the same quadrilateral. Let the four common tangents which form this quadrilateral be a, b, c, d, and let their points of contact with the conics be A, B, C, D and A', B', C', D' respectively. By the theorem of Art. 169, the triangle formed by the diagonals of the circumscribed quadrilateral $abcd$ has for its vertices the diagonal points of the inscribed quadrangle $ABCD$ and also those of the inscribed quadrangle $A'B'C'D'$; thus $ABCD$ and $A'B'C'D'$ have the same diagonal points. Accordingly, by the theorem of Art. 265, *the eight points* $A, B, C, D, A', B', C', D'$ *lie either four and four on two straight lines, or they lie all on a conic.*

267. By writing, as usual, line for point, and point for line, the propositions correlative to those of Arts. 265 and 266 can be proved, *viz.*

If two complete quadrilaterals have the same three diagonals, their eight sides either pass four and four through two points, or else they all touch a conic.

If two conics intersect in four points, the eight tangents to them at these points either pass four and four through two points, or they all touch a conic.*

268. If there be given the diagonal points E, F, G and one vertex A of a quadrangle $ABCD$, the quadrangle is completely determined, and can be constructed. For D is that point on AE which is harmonically conjugate to A with respect to E and the point where FG cuts AE; so C is that point on AF which is harmonically conjugate to A with respect to F and the point where GE cuts AF; and B is that point on AG which is harmonically conjugate to A with respect to G and the point where EF cuts AG.

But if there be given the diagonal points E, F, G of a quadrangle $ABCD$ and the conic with respect to which EFG is a self-conjugate triangle, the quadrangle is not completely

* STAUDT, *loc. cit.*, p. 293.

determined. For we may take arbitrarily on the conic a point A as one vertex of the quadrangle $ABCD$; then the other vertices B, C, D are the second points of intersection of the conic with the straight lines AG, AF, AE respectively. Hence it follows that:

All conics with respect to which a given triangle EFG is self-conjugate, and which pass through a fixed point A, pass also through three other fixed points B, C, D.

269. Problem. *To construct a conic passing through two given points A and A', and with respect to which a given triangle EFG shall be self-conjugate.*

Solution. Construct, in the manner just shown, the three points B, C, D which form with A a complete quadrangle having E, F, and G for its diagonal points. Five points A, A', B, C, D on the conic are then known, and by means of Pascal's theorem any number of other points on it may be found. Or we may construct the three points B', C', D' which form with A' a complete quadrangle having E, F, and G for its diagonal points; the eight points $A, B, C, D, A', B', C', D'$ will then all lie on the conic required.

270. Consider again the problem (Art. 218) of describing a conic to touch four given straight lines a, b, c, d and to pass through a given point S (Fig. 186). The diagonals of the quadrilateral $abcd$ form a triangle EFG which is self-conjugate with regard to the conic; consequently, if the three points P, Q, R be constructed which together with S form a quadrangle having E, F, and G for its diagonal points, the three points so constructed will lie also on the required conic. Now it may happen that there is no conic which satisfies the problem, or again there may be two conics which satisfy it (Art. 218, right); in the second case, since the construction for the points P, Q, R is linear, the two conics will both pass through these points. Thus:

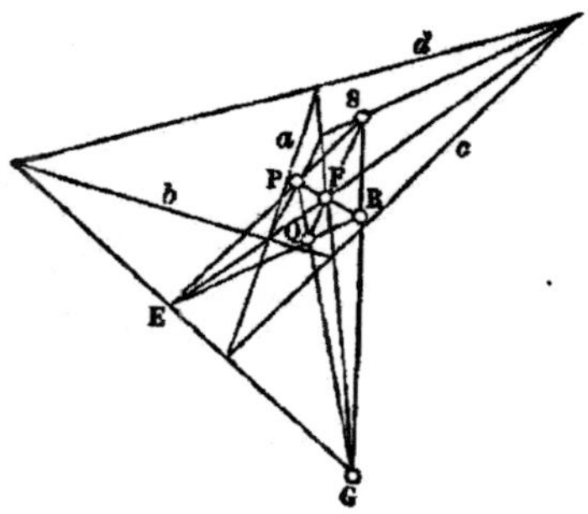

Fig. 186.

If two conics inscribed in the same quadrilateral abcd pass through the same point S, they will intersect in three other points P, Q, R; and the triangle formed by the diagonals of the circumscribed quadrilateral abcd will coincide with that formed by the diagonal points of the inscribed quadrangle $PQRS$.

In order to find a construction for the points P, Q, R, consider the point P for example which lies on ES (Fig. 186). It is seen that the segment SP must be divided harmonically by E and its polar FG (Art. 250); but the diagonal (ab) (cd) which passes through E is also divided harmonically, at E and F. We have therefore two harmonic ranges, which are of course projective (Art. 51) and which are in perspective since they have a self-corresponding point at E; therefore the straight lines P (ab), S (cd), and FG, which join the other pairs of corresponding points, will meet in a point (Art. 80). We must therefore join S to one extremity of one of the diagonals passing through E, for example to the point cd, and take the point where the joining line meets FG. This point, when joined to the other extremity ab of the diagonal, will give a straight line which will meet ES in the required point P *.

271. The propositions and constructions correlative to those of the last three Articles, and which will form useful exercises for the student, are the following:

All conics with respect to which a given triangle is self-conjugate, and which touch a fixed straight line, touch three other fixed straight lines.

To construct a conic to touch two given straight lines, and with respect to which a given triangle shall be self-conjugate.

If two conics circumscribing the same quadrangle have a common tangent, they have three other common tangents.

To construct the three remaining common tangents to two conics which pass through four given points and touch a given straight line (Art. 218, left).

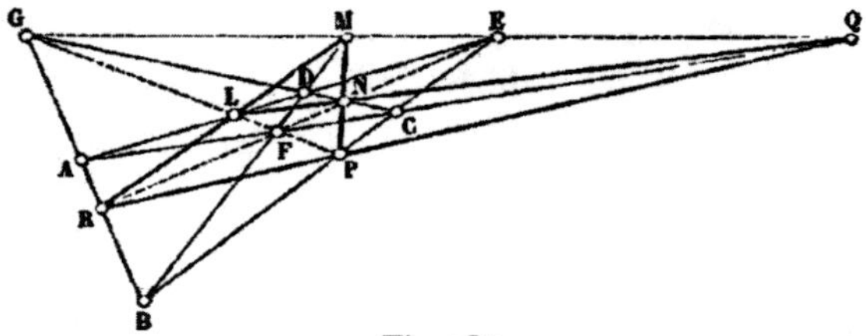

Fig. 187.

272. Let $ABCD$ (Fig. 187) be a complete quadrilateral whose diagonal points are E, F, and G. Let also

L and P be the points where FG meets AD and BC respectively.
M and Q „ „ GE „ BD and CA „
N and R „ „ EF „ CD and AB „

The six points so obtained are the vertices of a complete quadrilateral. For the triangle EFG is in perspective with each of the

* BRIANCHON, *loc. cit.*, p. 45; MACLAURIN, *De lin. Geom.*, § 43.

triangles ABC, DCB, CDA, BAD, the centres of perspective being D, A, B, C respectively; whence it follows that the four triads of points PQR, PMN, LQN, and LMR lie on four straight lines (the axes of perspective).

These four axes form a quadrilateral whose diagonals LP, MQ, NR form the triangle EFG. Accordingly, a conic inscribed in the quadrangle $ABCD$ and passing through L will pass also through N, P, and R (Art. 270); similarly a conic can be inscribed in the quadrangle $ABDC$ to pass through R, M, N, and Q; and a conic can be inscribed in the quadrangle $ACBD$ to pass through Q, P, M, and L.

It will be seen that for each of these conics the four tangents shown in the figure (the four sides of the complete quadrangle $ABCD$) are harmonic, and that the same will therefore be the case with regard to their points of contact (Arts. 148, 204). For take one of the sides of the quadrangle, for example AB; a consideration of the complete quadrangle $CDEF$ shows that this side is harmonically divided in R and G. Now the points A, B, G are the points of intersection of the tangent AB with the other three tangents, and R is the point of contact of AB; therefore the four tangents are cut by any other tangent to the conic in four harmonic points *.

273. If $ABCD$ is a parallelogram, the points E, G, M, Q pass off to infinity, and $LNPR$ also becomes a parallelogram. Of the three conics considered above the first will in this case be an ellipse which touches the sides of the parallelogram $ABCD$ at their middle points; the second a hyperbola which touches the sides AB and CD at their middle points and has AC and BD for asymptotes; and the third a hyperbola having the same asymptotes and touching the sides AD and BC at their middle points.

274. From that corollary to Brianchon's theorem which has reference to a quadrilateral circumscribed about a conic (Art. 172) we have already, in Art. 173, deduced a method for the construction of tangents to a conic when we are given three tangents a, b, c and the points of contact B, C of two of them (Fig. 183). We take any point E on BC and join it to the points ab, ac by the straight lines g, f, respectively; if the point in which g meets c be joined to that in which f meets b, the joining line d will be a tangent to the conic.

The four tangents a, b, c, d form a complete quadrilateral two of whose diagonals $g \equiv (ab)(cd)$ and $f \equiv (ac)(bd)$ intersect

* STEINER, *loc. cit.*, p. 160, § 43, 4; Collected Works, vol i. p. 347; STAUDT, *Beiträge zur Geometrie der Lage*, Art. 329.

in E; therefore also (Art. 172) the chords of contact AD and BC of the tangents a and d, b and c respectively will intersect in E. The straight lines joining E to the points ab and ac, being two of the diagonals of the quadrilateral $abcd$, are conjugate lines with respect to the conic; consequently:

If a triangle abc is circumscribed about a conic, the straight lines which join two of its vertices ab and ac to any point E on the polar of the third vertex bc are conjugate to one another with respect to the conic.

And conversely:

If two straight lines (c and b) touch a conic, any two conjugate straight lines (f and g) drawn from any point (E) on their chord of contact will cut the two given tangents in points such that the straight line (a) joining them touches the conic.

275. Let us now investigate the correlative property. Suppose three points A, B, C on a conic to be given, and the tangents b, c at two of these points (Fig. 183). If a straight line e drawn arbitrarily through the point bc cut AB in G and AC in F; then if GC and FB be joined they will intersect in a point D lying on the conic.

The four points A, B, D, C form a complete quadrangle two of whose diagonal points lie on e; therefore (Art. 166) the point bc and the point of intersection of the tangents at A and D will lie on e. The points G and F, being two of the diagonal points of the quadrilateral $ABCD$, are conjugate with respect to the conic; consequently

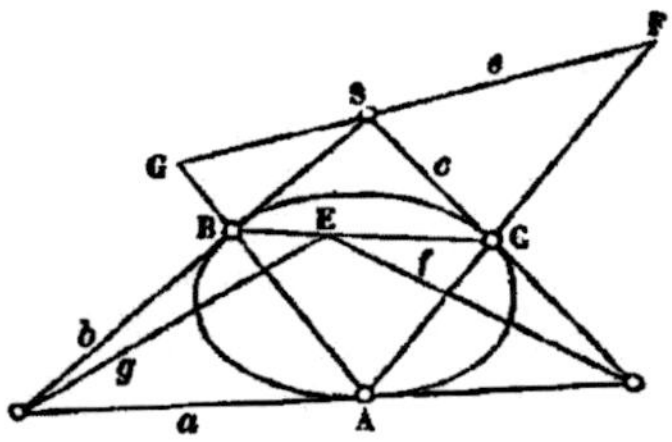

Fig. 188.

If a triangle ABC (Fig. 188) *is inscribed in a conic, the points F and G in which two of the sides are cut by any straight line drawn through the pole S of the third side are conjugate to one another with respect to the conic.*

And conversely:

If two given points (B, C) on a conic be joined to two conjugate points (G, F) which are collinear with the pole (S) of the chord (BC) joining the given points, then the joining lines will intersect in a point (A) lying on the conic.

CHAPTER XXI.

THE CENTRE AND DIAMETERS OF A CONIC.

276. LET an infinitely distant point be taken as pole, and through it let a transversal be drawn (Fig. 189) to cut the conic in two points A and A'. The segment AA' will be harmonically divided by the pole and the point where it is cut by the polar (Art. 250); this point will therefore be the middle point of AA' (Art. 59). That is to say:

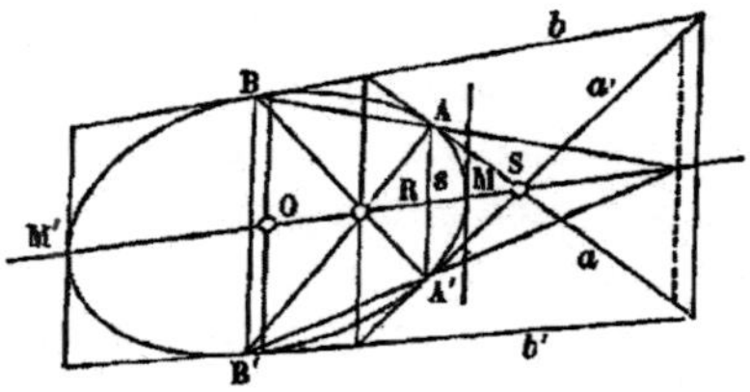

Fig. 189.

If any number of parallel chords of a conic be drawn, the locus of their middle points is a straight line; and this straight line is the polar of the point at infinity in which the chords intersect *.

277. This straight line is termed the *diameter* of the chords which it bisects. If the diameter meets the conic in two points, these will be the points of contact of the tangents drawn to the conic from the pole, *i.e.* of those tangents which are parallel to the bisected chords. If the tangents at the extremities A and A' of one of these chords be drawn, they will meet in a point on the diameter. If AA' and BB' are two of the bisected chords, the straight lines AB and $A'B'$, AB' and $A'B$ will intersect in pairs on the diameter (Art. 250).

If, conversely, from a point on the diameter can be drawn a pair of tangents a and a' to the conic, their chord of contact AA' will be bisected by the diameter; and if through the same point there be drawn the straight line which is harmonically conjugate to the diameter with respect to the two

* APOLLONIUS, *Conic.*, lib. i. 46, 47, 48; lib. ii. 5, 6, 7, 28–31, 34–37.

tangents, this straight line will be parallel to the bisected chords. If from two points on the diameter there be drawn two pairs of tangents a and a', b and b', the straight line joining the points ab and $a'b'$ and that joining the points ab' and $a'b$ will both be parallel to the bisected chords (Art. 252).

278. To each point at infinity, that is, to each pencil of parallel rays, corresponds a diameter. The diameters all pass through one point; for they are the polars of points lying on one straight line, *viz.* the straight line at infinity; the point in which the diameters intersect is the pole of the straight line at infinity (Art. 256).

279. Since every parabola is touched by the straight line at infinity, and the point of contact is the pole of this straight line (Art. 254, II), it follows (Art. 278) that all diameters of a parabola are parallel to one another (they all pass through the point at infinity on the curve); and conversely, every straight line which cuts a parabola at infinity is a diameter of the curve.

280. If S is any point from which a pair of tangents a and a' can be drawn to the conic (Fig. 189), the chord of contact AA', the polar of S, will be bisected at R by the diameter which passes through S; for S and the point at infinity on AA' are conjugate points with respect to the conic. If the diameter cuts the curve in M and M', the tangents at these points are parallel to AA', and MM' is divided harmonically by the pole S and the polar AA' (Art. 250).

If then the conic is a parabola (Fig. 190) the point M' moves off to infinity, and therefore M is the middle point of the segment SR; thus

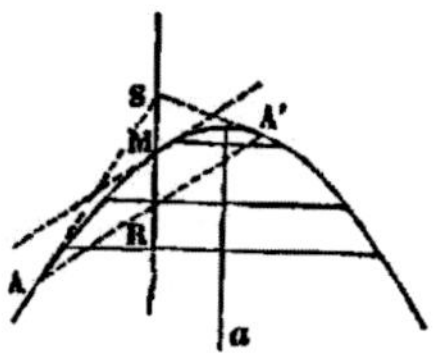

Fig. 190.

*The straight line which joins the middle point of a chord of a parabola to the pole of the chord is bisected by the curve**.

281. When the conic is not a parabola, the straight line at infinity is no longer a tangent to the curve, and consequently the pole of this straight line, or the point of intersection of the diameters, is a point lying at a finite distance. Since any two points on the conic which are collinear with the pole are separated harmonically by the pole and the polar (Art. 250), the pole will lie midway between the two points on the curve

* APOLLONIUS, *loc. cit.*, lib. i. 35.

when the polar lies at infinity. Every chord of the conic therefore which passes through the pole of the straight line at infinity is bisected at this point.

On account of this property the pole of the straight line at infinity or the point in which all the diameters intersect is called the *centre* of the conic.

282. Applying the properties of poles and polars in general (Arts. 250—253) to the case of the centre and the straight line at infinity, it is seen (Fig. 191) that:

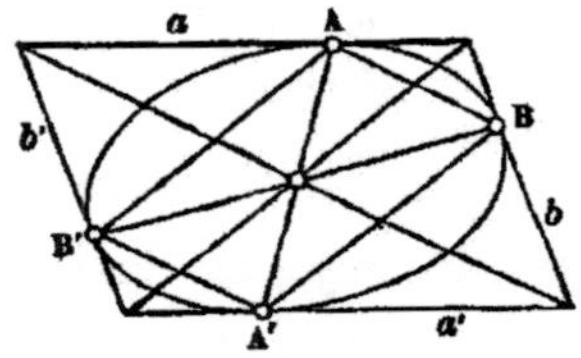

Fig. 191.

If A and A' are any pair of points on the conic collinear with the centre, the tangents at A and A' are parallel.

If A and A', B and B' are any two pairs of points on the conic which are collinear with the centre, the pairs of chords AB and $A'B'$, AB' and $A'B$ are parallel, so that the figure $ABA'B'$ is a parallelogram.

If a and a' are any pair of parallel tangents, their chord of contact passes through the centre, as also does the straight line lying midway between a and a' and parallel to both.

If a and a', b and b' are any two pairs of parallel tangents, the straight line joining the points ab and $a'b'$ and that joining the points ab' and $a'b$ both pass through the centre; in other words, if $aba'b'$ is a parallelogram circumscribed to the conic, its diagonals intersect in the centre.

283. If the conic is a hyperbola, the straight line at infinity cuts the curve; consequently the centre is a point exterior to the curve (Art. 254, I) in which intersect the tangents at the infinitely distant points, *i.e.* the asymptotes (Fig. 197).

If the conic is an ellipse, the straight line at infinity does not cut the curve; consequently the centre is a point inside the curve (Figs. 191, 192).

284. Two diameters of a central conic (ellipse or hyperbola *) are termed *conjugate* when they are conjugate straight

* In the case of the parabola there are no pairs of conjugate diameters; for since the centre lies at infinity, the diameter drawn parallel to the chords which are bisected by a given diameter must coincide always with the straight line at infinity.

lines with respect to the conic, *i.e.* when each passes through the pole of the other (Art. 255).

Since the pole of a diameter is the point at infinity on any of the chords which the diameter bisects, it follows that the diameter b' conjugate to a given diameter b is parallel to the chords bisected by b; conversely, b' bisects the chords which are parallel to b*.

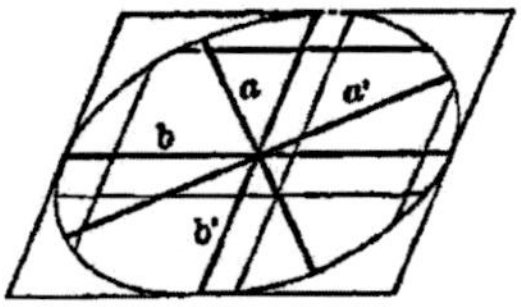

Fig. 192.

Any two conjugate diameters form with the straight line at infinity a self-conjugate triangle (Art. 258), of which one vertex is the centre of the conic and the other two are at infinity.

Since in a self-conjugate triangle two of the sides cut the conic and the third side does not (Art. 262), and since the straight line at infinity cuts a hyperbola but does not cut an ellipse, it follows that of every two conjugate diameters of a hyperbola one only cuts the curve, while an ellipse is cut by all its diameters.

285. PROBLEM. *Given five points A, B, C, D, E on a conic, to determine its centre.*

Solution. We have only to repeat the construction given in Art. 257, II (right), assuming the straight line s to lie in this case at infinity. Draw through C a parallel to AB, and determine the point C' in which this parallel meets the conic again; draw also through B a parallel to AC, and determine the point B' in which this parallel meets the conic again. The straight line u which joins the points of intersection of the pairs of opposite sides of the quadrangle $ACBC'$, and the straight line v which joins the points of intersection of the pairs of opposite sides of the quadrangle $ABCB'$, will meet in the required point O, which is the pole of the straight line at infinity and therefore the centre of the conic †.

The straight lines u and v are the diameters conjugate respectively to AB and AC; if through O there be drawn the straight lines u', v' parallel to AB, AC respectively, then u and u', v and v' will be two pairs of conjugate diameters.

If the conic is determined by five tangents, its centre may be found by a method which will be explained further on (Art. 319).

* APOLLONIUS, *loc. cit.*, lib. ii. 20.

† If u and v should be parallel, the conic is a parabola, whose diameters are parallel to u and v.

286. Four tangents to a conic form a complete quadrilateral whose diagonals are the sides of a self-conjugate triangle (Art. 260). Suppose the four tangents to be parallel in pairs (Fig. 191); then one diagonal will pass to infinity, and consequently the other two will be conjugate diameters (Art. 284); thus:

The diagonals of any parallelogram circumscribed to a conic are conjugate diameters.

The points of contact of the four tangents form a complete quadrangle whose diagonal points are the vertices of the self-conjugate triangle (Arts. 169, 260). In the case where the four tangents are parallel in pairs one of these diagonal points is the centre of the conic, and the other two lie at infinity. That is to say, the six sides of the quadrangle are the sides and diagonals of an inscribed parallelogram; its sides are parallel in pairs to the diagonals of the circumscribed parallelogram, and its diagonals intersect in the centre of the conic.

287. Conversely, let $ABA'B'$ (Fig. 191) be any inscribed parallelogram, and consider it as a complete quadrangle. Since its three diagonal points must be the vertices of a self-conjugate triangle, one of them will be the centre of the conic, and the other two will be the points at infinity on two conjugate diameters; thus:

In any parallelogram inscribed in a conic, the sides are parallel to two conjugate diameters and the diagonals intersect in the centre.

Or again:

The chords which join a variable point A on a conic to the extremities B and B' of a fixed diameter are always parallel to two conjugate diameters.

288. The following conclusions can be drawn at once from Art. 286.

Any two parallel tangents (a and a') are cut by any pair of conjugate diameters in two pairs of points, the straight lines connecting which give two other parallel tangents (b and b').

If from the extremities (A and A') of any diameter straight lines be drawn parallel to any two conjugate diameters, they will meet in two points on the curve, and the chord joining these will be a diameter.

Given any two parallel tangents a and a' whose points of contact are A and A' respectively, and any third tangent b; if from A a parallel be drawn to the diameter passing through $a'b$ this parallel will meet the tangent b at its point of contact B.

Given any two parallel tangents a and a' whose points of contact are A and A' respectively, and another point B on the conic; the tangent at B will meet the tangent a in a point lying on that diameter which is parallel to $A'B$, and it will meet the tangent a' in a point lying on that diameter which is parallel to AB.

289. Suppose now that the conic is a circle (Fig. 193), *i. e.* the locus of the vertex of a right angle AMB whose arms AM and BM turn round fixed points A and B respectively. These arms in moving generate two equal and consequently projective pencils; therefore the tangent at A will be the ray of the first pencil which corresponds to the ray BA of the second (Art. 143). The tangent at A must therefore make a right angle with BA; and similarly the tangent at B will be perpendicular to AB. The tangents at A and B are therefore parallel, and consequently AB is a diameter, and the middle point O of AB is the centre of the circle (Art. 282).

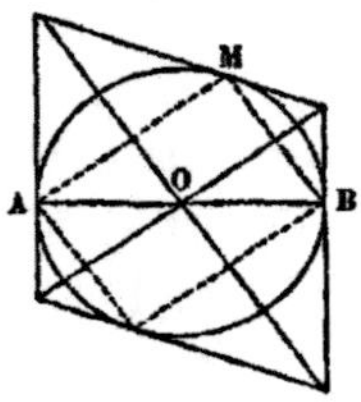

Fig. 193.

I. Since AB is a diameter, the straight lines AM and BM will be parallel to a pair of conjugate diameters, whatever be the position of M (Art. 287); therefore:

Every pair of conjugate diameters of a circle are at right angles to one another.

II. Since the diagonals of any parallelogram circumscribed about the circle are conjugate diameters, they will intersect at right angles; thus *any parallelogram which circumscribes a circle must be a rhombus.*

III. In a rhombus, the distance between one pair of opposite sides is equal to the distance between the other pair; thus by allowing one pair of opposite sides of the circumscribed rhombus to vary while the other pair remain fixed, we see that the distance between two parallel tangents is constant. This distance is the length of the straight line joining the points of contact of the tangents, for this straight line, which is a diameter, cuts at right angles the conjugate diameter and the tangents parallel to it; therefore *all diameters of a circle are equal in length.*

IV. The diagonals of any inscribed parallelogram are diameters; but all diameters are equal in length; therefore *any parallelogram inscribed in a circle must be a rectangle.*

290. Returning to the general case where the conic is any whatever (Fig. 189), let s be any straight line and S its pole. All chords parallel to s will be bisected by the diameter passing through S; for since S and the point at infinity on s are conjugate points with respect to the conic, the polar of the second point will pass through the first. We may also say that:

If a diameter pass through a fixed point, the conjugate diameter will be parallel to the polar of this point.

I. If the diameter passing through S cuts the conic in two points M and M', then MM' is divided harmonically by the pole S and the polar s*; thus if O is the middle point of MM', that is, the centre of the conic, and R the point where MM' is cut by the polar s, we have (Art. 69)

$$OS.OR = OM^2.$$

II. From this follows a *construction for the semi-diameter conjugate to a chord AA' of a conic, having given the extremities A and A' of the chord and three other points on the conic.* We determine (Art. 285) the centre O, and join it to the middle point R of AA'; we then construct the tangent at A and take its point of intersection S with OR. If now a point M be taken on OR such that OM is the mean proportional between OR and OS, then OM will be the required semi-diameter.

If O lie between R and S, so that OR and OS have opposite signs, the diameter OR will not cut the conic; but in this case also the length OM, the mean proportional between OR and OS, is called the *magnitude of the semi-diameter* conjugate to the chord AA'.

An analogous definition can be given for the case of any straight line (Art. 294).

III. If the conic is a circle, the perpendicularity of the conjugate diameters in this case gives the theorem:

* APOLLONIUS, *loc. cit.*, i. 34, 36; ii. 29, 30.

The polar of any point with respect to a circle is perpendicular to the diameter which passes through the pole.

291. From this last property can be derived a second demonstration of the very important theorem of Art. 263 (3), *viz.*

The range formed by any number of collinear points, and the pencil formed by their polars with respect to any given conic, are two projective forms.

Consider as poles the points $A, B, C, \ldots$ lying on a straight line s (Fig. 194); the diameters $O(A, B, C, \ldots)$ obtained by joining them to the centre O of the conic will form a pencil which is in perspective with the range $A, B, C, \ldots$. Another pencil will be formed by the polars $a, b, c, \ldots$ of the points $A, B, C, \ldots$ since these polars all pass through a point S (Art. 256), the pole of s. If now the conic is a circle, then by the property proved in Art. 290, III, the straight lines $O(A, B, C, \ldots)$ are perpendicular respectively to $a, b, c, \ldots$; and the two pencils are in this case equal. The range of poles $A, B, C, \ldots$ is therefore projective with the pencil of polars $a, b, c, \ldots$ with regard to a circle.

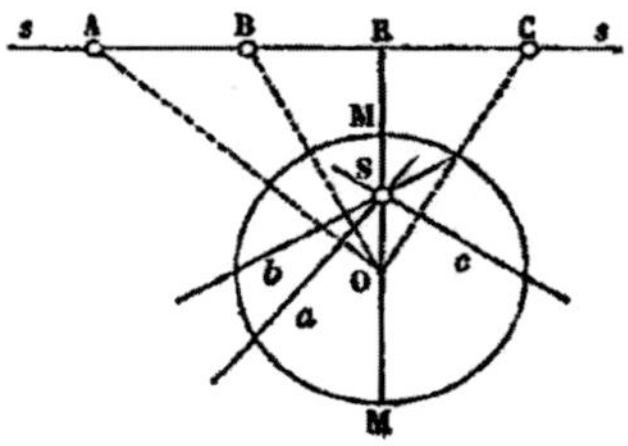

Fig. 194.

This result may now be extended and shown to hold not only for a circle but for any conic. For any given conic may be regarded as the projection of a circle (Arts. 149, 150). In the projection, to harmonic forms correspond harmonic forms (Art. 51); consequently to a point and its polar with regard to the conic will correspond a point and its polar with regard to the circle, and to a range of poles and the pencil formed by their polars with regard to the conic will correspond a range of poles and the pencil formed by their polars with regard to the circle. But it has been seen that this range and pencil are projective in the case of the circle; therefore the same is true with regard to the range and pencil in the case of the conic, and the theorem is proved.

292. Theorem. *A quadrangle is inscribed in a conic, and a point is taken on the straight line which joins the points of intersection of the pairs of opposite sides. If from this point be drawn the straight lines connecting it with the two pairs of opposite vertices, and also a pair of* *tangents to the conic, these straight lines will be three conjugate pairs of an involution.*

Let $ABCD$ be a simple quadrangle inscribed in a conic (Fig. 195); let the diagonals AC, BD meet in F, and the pairs of opposite sides BC, AD and AB, CD in E and G respectively; the points E, F, G will then be conjugate two and two with respect to the conic (Art. 259). Take any point I on EG and join it to the vertices of the quadrangle, and draw also the tangents IP, IQ to the conic. The two tangents are harmonically separated by IE, IF (Art. 264), since these are conjugate straight lines, F being the pole of IE. But the rays IE, IF are harmonically conjugate also with regard to IA, IC; for the diagonal AC of the complete quadrilateral formed by AB, BC, CD, and DA is divided harmonically by the other two diagonals BD and EG, and the two pairs of rays in question are formed by joining I to the four harmonic points on AC. For a similar reason the rays IE, IF are harmonically conjugate with regard to IB, ID. The pair of tangents, the rays IA, IC, and the rays IB, ID are therefore three conjugate pairs of an involution, of which IE, IF are the double rays (Art. 125).

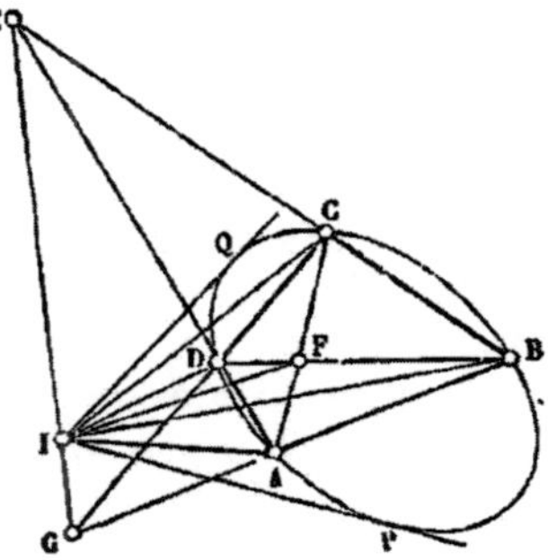

Fig. 195.

I. By virtue of the theorem correlative to that of Desargues (Art. 183, right), a conic can be inscribed in the quadrilateral $ABCD$ so as to touch the straight lines IP and IQ.

II. The theorem correlative to the one proved above may be thus enunciated:

If a simple quadrilateral $ABCD$ (Fig. 196) *is circumscribed about a conic, and if through the point of intersection of its diagonals any transversal be drawn, this will cut the conic and the pairs of opposite sides* AB *and* CD, BC *and* AD, *in three pairs of conjugate points of an involution.*

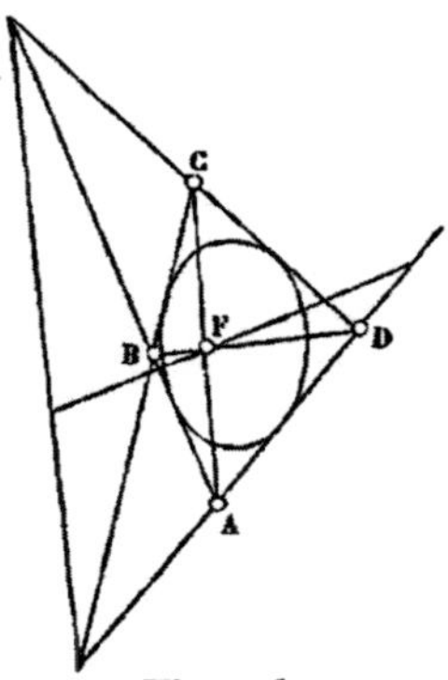

Fig. 196.

III. By virtue of Desargues' theorem (Art. 183, left), a conic can be described to pass through the four vertices of the quadrilateral and through the two points where the conic is cut by the transversal *.

* CHASLES, *Sections coniques*, Arts. 122, 126.

293. The theory of conjugate points with regard to a conic gives a solution of the problem :

To construct the points of intersection of a given straight line s with a conic which is determined by five points or by five tangents.

Take on s any two points U and V, construct their polars u and v (Art. 257), and let U' and V' be the points where these meet s. If the involution determined by the two pairs of reciprocal points U and U', V and V', has two double points M and N, these will be the required points of intersection of the conic with s. If U' and V' should coincide, the conic touches s at the point in which they coincide. If the involution has no double points, the conic does not cut s *.

By a correlative method may be solved the problem: *to draw from a given point S a pair of tangents to a conic which is determined by five tangents or by five points.*

294. Let A and A' be a pair of points lying on a straight line s which are conjugate with respect to the conic, and let O be the point where s meets the diameter passing through its pole S (the diameter bisecting chords parallel to s). Then O will be the centre of the involution formed on s by the pairs of conjugate points such as A and A', and therefore (Art. 125)

$$OA \,.\, OA' = \text{constant}.$$

If s cuts the conic in two points M and N, these will be the double points of the involution, and

$$OA \,.\, OA' = OM^2 = ON^2.$$

If s does not cut the conic, the constant value of $OA \,.\, OA'$ will be negative (Art. 125); in this case there exists a pair H and H' of conjugate points of the involution, or of conjugate points with regard to the conic, such that O lies midway between them, and

$$OA \,.\, OA' = OH \,.\, OH' = -OH^2 = -OH'^2.$$

The segment HH' has been called an *ideal chord* † of the conic, just as MN in the first case is a real chord. Accepting this definition we may say that a diameter contains the middle points of all chords, real and ideal, which are parallel to the conjugate diameter.

When two conics are said to have a real common chord MN, it is meant that they both pass through the points M and N. When two conics are said to have an ideal common chord HH', this signifies that H and H' are conjugate points with regard to both conics, and that the diameters of the two conics which pass through the respective poles of HH' both pass through the middle point of HH'.

* STAUDT, *Geometrie der Lage*, Art. 305.

† PONCELET, *loc. cit.*, p. 29.

295. A pencil of rays in involution has in general (Art. 207) one pair of conjugate rays which include a right angle. Therefore

Through a given point can always be drawn one pair of straight lines which are conjugate with respect to a given conic and which include a right angle; and these are the internal and external bisectors of the angle made with one another by the tangents drawn from the given point, when this is exterior to the conic.

296. In Art. 263 (Fig. 183) let the point G be taken to coincide with the centre O of the conic (hyperbola or ellipse); two conjugate lines such as GF, GE will then become conjugate diameters, and we see that *the pairs of conjugate diameters of a conic form an involution.* If the conic is a hyperbola, the asymptotes are the double rays of the involution (Arts. 264, 283); thus *any two conjugate diameters of a hyperbola are harmonically conjugate with regard to the asymptotes* *. If the conic is an ellipse, the involution has no double rays.

Consider two pairs of conjugate elements of an involution; the one pair either overlaps or does not overlap the other, and according as the first or the second is the case, the involution has not, or it has, double points (Art. 128); thus:

Of any two pairs of conjugate diameters of an ellipse, the one aa' is always separated by the other bb' (Fig. 192);

Of any two pairs of conjugate diameters of a hyperbola, the one aa' is never separated by the other bb' (Fig. 197).

297. The involution of conjugate diameters will have one pair of conjugate diameters including a right angle (Art. 295). If there were a second such pair, every diameter would be perpendicular to its conjugate (Art. 207), and in that case the angle subtended at any point on the curve by a fixed diameter would be a right angle (Art. 287), and consequently the conic would be a circle. Every conic therefore which is not a parabola or a circle has a single pair of conjugate diameters which are at right angles to one another. These two diameters a and a' are called the *axes* of the conic (Figs. 192, 197). In the

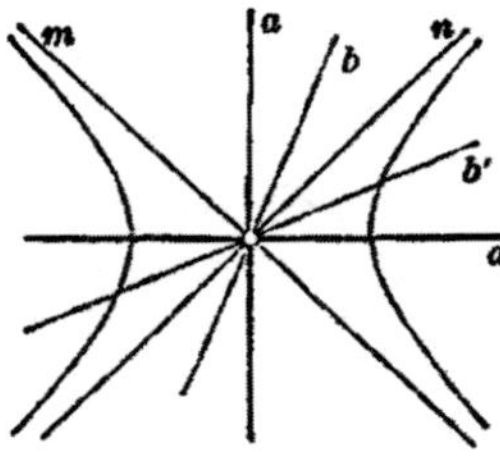

Fig. 197.

* De la Hire, *loc. cit.*, book ii. prop. 13, Cor. 4.

hyperbola (Fig. 197) the axes are the bisectors of the angle between the asymptotes m and n (Arts. 296, 60).

In the ellipse both axes cut the curve (Art. 284); the greater (a') is called the *major*, the smaller (a) the *minor* axis. In the hyperbola only one of the axes cuts the curve; this one (a') is called the *transverse* axis, the other (a) the *conjugate* axis. The points in which the conic is cut by the axis a' in either case are called the *vertices*.

Regarding an axis as a diameter which bisects all chords perpendicular to itself, it is seen that the parabola also has an axis. For since all chords at right angles to the common direction of the diameters are parallel to one another, their middle points lie on one straight line, which is the axis a of the parabola (Fig. 190). The parabola has one vertex at infinity; the other, the finite point in which the axis a cuts the curve, is generally called *the* vertex of the parabola.

298. Since each of the orthogonal conjugate diameters of a central conic (ellipse or hyperbola) bisects all chords perpendicular to itself, it follows that the conic is symmetrical with respect to each of the diameters in question (Art. 76). The ellipse and the hyperbola have therefore each two axes of symmetry; the parabola, on the other hand, has only one such axis.

The ellipse and hyperbola are also symmetrical with respect to a point; the centre of symmetry being in each case the pole of the straight line at infinity.

In general, given a conic, a point S, and s the polar of S with respect to the conic; if S be taken as centre and s as axis of harmonic homology (Art. 76), the conic is homological with itself (Art. 250)*.

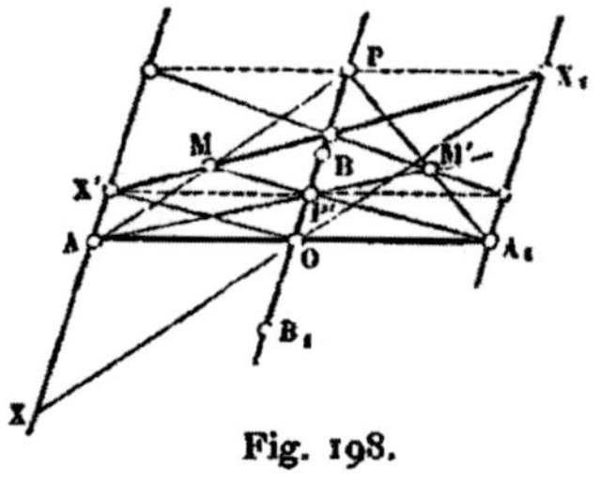

Fig. 198.

299. In the theorem of Art. 275 suppose the inscribed triangle to be AA_1M (Fig. 198); that is, let two of its vertices A and A_1 be collinear with the centre O of the conic, which is taken to be an ellipse or hyperbola. The pole of the side AA_1 will be the point at infinity common to the chords bisected by the diameter AA_1, and the theorem will become the following:

* See also Art. 396, below.

The straight lines which join two conjugate points P and P' to the extremities A and A_1 of that diameter whose conjugate is parallel to PP' intersect on the conic.

300. The pairs of conjugate points taken, similarly to P and P', on the diameter conjugate to AA_1 form an involution (Art. 263) whose centre is the centre O of the conic. If this involution has two double points B and B_1, these lie on the curve, which is therefore an ellipse. If the involution has no double points, the conic is a hyperbola (Art. 284); in this case two points B and B_1 can be found which are conjugate in the involution and consequently conjugate with respect to the conic, and which lie at equal distances on opposite sides of O (Art. 125). In both cases the *length* of the diameter conjugate to AA_1 is interpreted as being the segment BB_1 (Arts. 290, 294).

In the ellipse we have (Art. 294)

$$OP \,.\, OP' = \text{constant} = OB^2 = OB_1^2,$$

and in the hyperbola

$$OP.OP' = \text{constant} = OB.OB_1 = -OB^2 = -OB_1^2.$$

301. The foregoing theorem enables us to solve the problem:

To construct by points a conic, having given a pair of conjugate diameters AA_1 and BB_1 in magnitude and position.

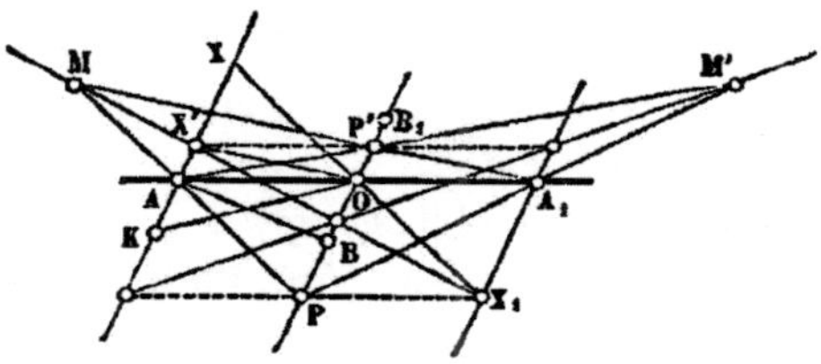

Fig. 199.

In the case of the ellipse (Fig. 198) the four points A, A_1, B, B_1 all lie on the curve; in the case of the hyperbola (Fig. 199) let AA_1 be that one of the two given diameters which meets the conic.

Construct on the diameter BB_1 several pairs of conjugate points P and P' of the involution determined by having O as centre and B and B_1 in the first case as double points, in the second case as conjugate points. The straight lines AP and A_1P' (as also A_1P and AP') will intersect on the curve.

302. The straight lines OX, OX' drawn parallel to AP, A_1P' respectively are a pair of conjugate diameters (Art. 287). The pairs of conjugate diameters form an involution (Art. 296); consequently the pairs of points analogous to X, X' (in which the diameters cut the tangent at A) also form an involution, the centre of which is A, since OA and the diameter OB parallel to AX are a pair of conjugate diameters. If the conic is a hyperbola, the involution of conjugate diameters has two double rays, which are the asymptotes; therefore the points K and K_1, in which AX meets the asymptotes, are the double points of the involution $XX', \ldots$*.

303. Since $OPAX$ is a parallelogram, $AX = -OP$; and from the similar and equal triangles $OP'A_1$ and $AX'O$, $AX' = OP'$ †. But $OP . OP' = \pm OB^2$ (Art. 125); therefore $AX . AX' = \mp OB^2$; or

The rectangle contained by the segments intercepted on a fixed tangent to a conic between its point of contact and the points where it is cut by any two conjugate diameters is equal to the square ($\mp OB^2$) *on the semi-diameter drawn parallel to the tangent.*

304. We have seen (Art. 302) that in the case of the hyperbola K and K_1 are the double points of the involution of which A is the centre and X, X' a pair of conjugate points; thus

$$AX . AX' = AK^2 = OB^2.$$

Therefore $AK = OB$, and $OAKB$ is a parallelogram. Accordingly:

If a parallelogram be described so as to have a pair of conjugate semi-diameters of a hyperbola as adjacent sides, one of its diagonals will coincide with an asymptote ‡.

Further, *the other diagonal AB is parallel to the second asymptote.* For consider the harmonic pencil (Art. 296) formed by the two asymptotes and the two conjugate diameters OA, OB. The four points in which this pencil cuts AB will be harmonic; but one of the asymptotes OK meets AB in its middle point, therefore the other will meet it at infinity (Art. 59).

305. Let X_1 be the point where the diameter OX meets the tangent at A_1. Since OX' and OX_1 are a pair of conjugate lines which meet in a point on the chord of contact AA_1 of

* In Fig. 199 only one of the points K, K_1 is shown.

† In order to account for the signs, it need only be observed that in the case of the ellipse OP and OP' are similar, but AX and AX' opposite to one another in direction; while in the case of the hyperbola OP and OP' are opposite, but AX and AX' similar as regards direction.

‡ APOLLONIUS, *loc. cit.*, book ii. 1.

the tangents AX and A_1X_1, the straight line $X'X_1$ (Art. 274) will be a tangent to the conic.

The point of contact of this tangent is M, the point of intersection of AP and A_1P' (Art. 299).

306. It is seen moreover that $X'X_1$ is one diagonal of the parallelogram formed by the tangents at A and A_1 and the parallels to AA_1 drawn through P and P'; this may also be proved in the following manner. All points of a diameter have for their polars straight lines which are parallel to the conjugate diameter (Art. 284); if then through the conjugate points P and P' parallels be drawn to AA_1, the first will be the polar of P' and the second the polar of P; consequently these parallels are conjugate lines. If now the theorem of Art. 274 be applied to these conjugate lines and the two tangents at A and A_1, we obtain the following proposition:

If a parallelogram is such that one pair of its opposite sides are tangents to a conic, and the other pair are straight lines, conjugate with regard to the conic and drawn parallel to the chord of contact of the two tangents, then its diagonals also will be tangents to the conic.

307. This gives the following solution of the problem:

To construct a conic by tangents, having given a pair of conjugate diameters AA_1 and BB_1 in magnitude and direction.

Suppose BB_1 to be that diameter which meets the conic in the case where the latter is a hyperbola. On BB_1 determine a pair of conjugate points P and P' of the involution which has the centre O of the conic as centre and the points B, B_1 either as double points or as conjugate points, according as the conic to be drawn is an ellipse or a hyperbola. Draw through A and A_1 parallels to BB_1, and through P and P' parallels to AA_1; the diagonals of the parallelogram so obtained will be tangents to the required conic.

308. The segments AX and A_1X_1 are equal in magnitude and opposite in sign; and it has been seen that $AX.AX' = \mp OB^2$; therefore $AX'.A_1X_1 = \pm OB^2$; or

The rectangle contained by the segments intercepted upon two parallel fixed tangents between their points of contact and the points where they are cut by a variable tangent ($X'X_1$) is equal to the square ($\pm OB^2$) on the semi-diameter parallel to the fixed tangents.*

309. Since the straight line OB is parallel to AX and A_1X_1 and half-way between them, the segments determined by AM

* See Art. 160.

and A_1M respectively on A_1X_1 and AX (measured from A_1 and A respectively) are double of OP and OP'; but by the theorem of Art. 300 the rectangle $OP.OP'$ is constant; thus

*The straight lines connecting the extremities of a given diameter with any point on the conic meet the tangents at these extremities in two points such that the rectangle contained by the segments of the tangents intercepted between these points and the points of contact is constant**.

310. Since X is (Art. 288) the point of intersection of the tangent at A and the tangent parallel to $X'X_1$, the proposition of Art. 303 may also be expressed as follows:

The rectangle contained by the segments (AX, AX') determined by two variable parallel tangents upon any fixed tangent is equal to the square ($\mp OB^2$) on the semi-diameter parallel to the fixed tangent.

311. From the theorems of Arts. 299, 300 is derived the solution of the following problem:

Given the two extremities A and A_1 of a diameter of a conic, a third point M on the conic, and the direction of the diameter conjugate to AA_1, to determine the length of the latter diameter (Fig. 199).

Through O, the middle point of AA_1, draw the diameter whose direction is given; let it be cut by AM and A_1M in P and P' respectively, and take OB the mean proportional between OP and OP'; then OB will be the half of the length required.

312. The proposition of Art. 303 gives a *construction for pairs of conjugate diameters, and in particular for the axes, of an ellipse of which two conjugate semi-diameters OA and OB are given in magnitude and direction* (Fig. 200).

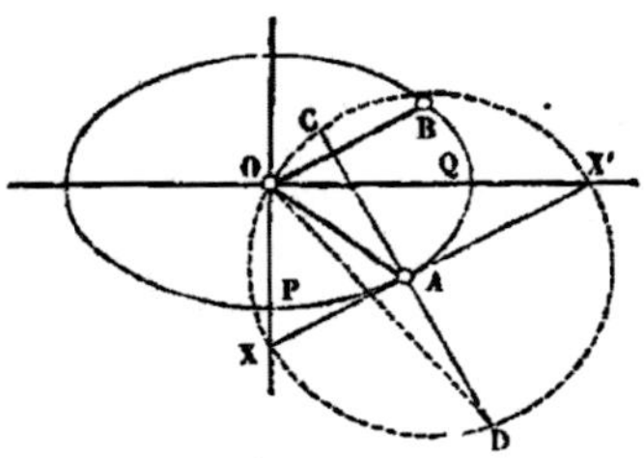

Fig. 200.

Through A draw a parallel to OB; this will be the tangent at A, and will be cut by any two conjugate diameters in two points X and X' such that

$$AX \,.\, AX' = -OB^2.$$

If now there be taken on the normal at A two segments AC and AD each equal to OB, every circle passing through C and D will cut this tangent in two points X and X' which possess the property expressed by the above equation; these points are therefore such that the straight lines joining them to the centre O will give the directions of a pair of conjugate diameters. If the circle be drawn

* APOLLONIUS, *loc. cit.*, lib. iii. 53.

through O the angle XOX' becomes a right angle, and consequently OX, OX' will be the directions of the axes.

Since the circular arcs CX', $X'D$ are equal, the angles COX', $X'OD$ are equal; consequently OX', OX are the internal and external bisectors of the angle which OC, OD make with one another. In order then to construct the semi-axes OP, OQ in magnitude, let fall perpendiculars AX_1, AX_1' on OX, OX' respectively. Then X and X_1, X' and X_1' are pairs of conjugate points; therefore OP will be the geometric mean between OX and OX_1, and OQ the geometric mean between OX' and OX_1' *.

313. Through the extremities A and A' (Fig. 201) of two conjugate semi-diameters OA and OA' of a conic draw any two parallel chords AB and $A'B'$. To find the points B and B' we have only to join the poles of these chords; this will give the diameter OX' which passes through their middle points.

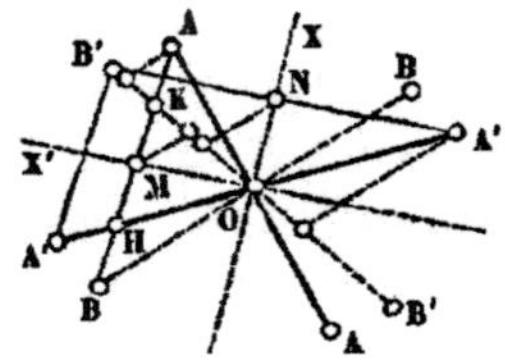

Fig. 201.

Let OX be the diameter conjugate to OX', *i.e.* that diameter which is parallel to the chords AB, $A'B'$. The pencils $O(XX'AB)$ and $O(X'XA'B')$ are each harmonic (Art. 59), and are therefore projective with one another; consequently the pairs of rays $O(XX', AA', BB')$ are in involution (Art. 123). But the two pairs $O(XX', AA')$ determine the involution of conjugate diameters (Arts. 127, 296); therefore also OB and OB' are conjugate diameters. Thus

If through the extremities A and A' of two conjugate semi-diameters parallel chords AB, $A'B'$ be drawn, the points B and B' will be the extremities of two other conjugate semi-diameters.

Two diameters AA and BB determine four chords AB which form a parallelogram (Arts. 260, 287). The diameters conjugate respectively to them form in the same way another parallelogram, which has its sides parallel to those of the first; that is, every chord AB is parallel to two chords $A'B'$, and not parallel to two other chords $A'B'$.

314. Let H, K be the points where AB is cut by OA', OB' respectively. The diameter OX' which bisects $A'B'$ will also bisect HK; therefore AB and HK have the same middle point; thus $AH = KB$ and $AK = HB$. The triangles OAK and OBH

* CHASLES, *Aperçu historique*, pp. 45, 362; *Sections coniques*, Art. 205.

are therefore equal in area (Euc. I. 37), as also AKB' and BHA', and therefore also OAB' and $OA'B$ are equal. Accordingly:

The parallelogram described on two semi-diameters (OA, OB') as adjacent sides is equal in area to the parallelogram described similarly on the two conjugate semi-diameters.

In the same way the triangles OAB and $OA'B'$ can be proved equal.

The triangles AHA', BKB' are equal for the same reason; and OAH, OBK are equal, and therefore also OAA' and OBB'. Therefore

The parallelogram described on a pair of conjugate semi-diameters as adjacent sides is of constant area.*

315. Let M and N be the middle points of the non-parallel chords AB and $A'B'$. Since AB and $A'B'$ are parallel to a pair of conjugate diameters (Art. 287) and since ON is the diameter conjugate to the chord $A'B'$, therefore ON will be parallel to AB; so also OM will be parallel to $A'B'$. The angles OMA and ONA' are therefore equal or supplementary; and since the triangles OMA and ONA' are equal in area (being halves of the equal triangles OAB and $OA'B'$), we have (Euc. VI. 15),

$$OM \cdot AM = \pm ON \cdot NA' \dagger .$$

Now project (Fig. 202) the points A, M, B, A', N, B' from the point at infinity on OB as centre upon the straight line $B'B'$. The ratio of the parallel segments AM and ON, OM and NA' is equal to that of their projections; we conclude therefore from the equality just proved that the rectangle contained by the projections of OM and AM is equal to that contained by the projections of ON and NA'. As the projecting rays are parallel to OB, the projections of OM and MA are

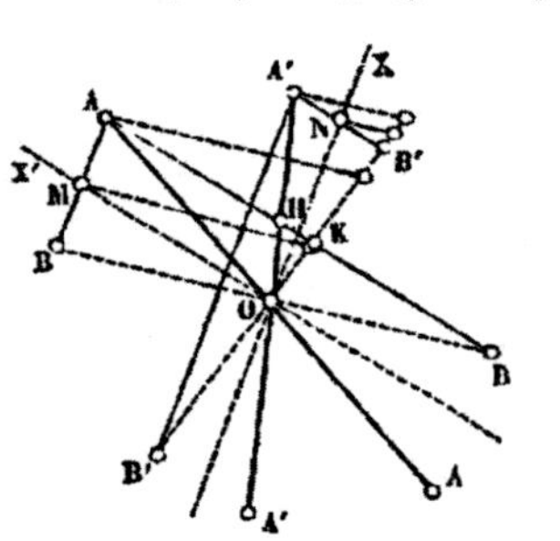

Fig. 202.

* APOLLONIUS, *loc. cit.*, lib. vii. 31, 32.

† The signs + and − caused by the relative direction of the segments OM, NA' and ON, AM correspond respectively to the case of the ellipse (Fig. 201) and to that of the hyperbola (Fig. 202).

each equal to half the projection of BA or of OA. Since N is the middle point of $A'B'$, the projection of ON will be equal to half the sum of the projections of OA' and OB', and the projection of NA' will be equal to half the projection of $B'A'$, that is, to half the difference between the projections of OA' and OB'. We have therefore

$$(\text{proj. } OA)^2 = \pm \text{proj. } (OA' + OB') \times \text{proj. } (OB' - OA'),$$

or

$$(\text{proj. } OA')^2 \pm (\text{proj. } OA)^2 = (\text{proj. } OB')^2.$$

In the same manner, by projecting the same points on OB by means of rays parallel to OB' (Fig. 203), we should obtain

$$(\text{proj. } OA)^2 \pm (\text{proj. } OA')^2 = (\text{proj. } OB)^2.$$

This proves the following proposition:

If any pair of conjugate diameters are projected upon a fixed diameter by means of parallels to the diameter conjugate to this last, then the sum (in the ellipse) or difference (in the hyperbola) of the squares on the projections is equal to the square on the fixed diameter.

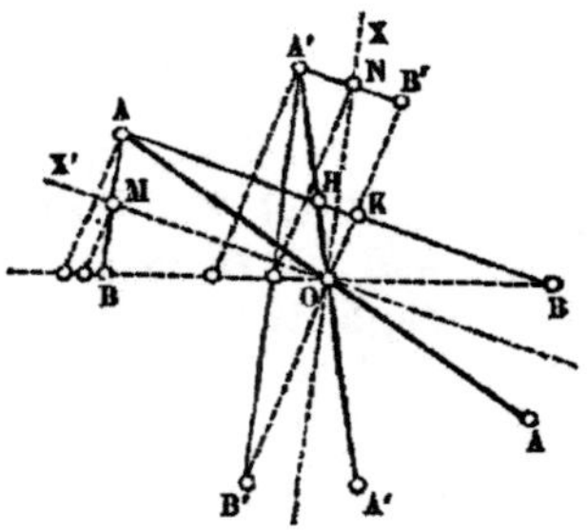

Fig. 203.

By the Pythagorean theorem (Euc. I. 47) the sum of the squares on the orthogonal projections of a segment on two straight lines at right angles to one another is equal to the square on the segment itself. If then a pair of conjugate diameters are projected orthogonally on one of the axes of a conic and the squares on the projections of each diameter on the two axes are added together, the following proposition will be obtained:

The sum (for the ellipse) or difference (for the hyperbola) of the squares on any pair of conjugate diameters is constant, and is equal to the sum or the difference of the squares on the axes.*

316. If five points on a conic are given, then by the method explained in Art. 285 the centre O and two pairs of conjugate diameters u and u', v and v' can be constructed. If these pairs overlap one another, the conic is an ellipse; in the contrary case it

* APOLLONIUS, *loc. cit.*, lib. vii. 12, 13, 22, 25.

is a hyperbola (Art. 296). If in this second case the double rays of the involution determined by the two pairs u and u', v and v' be constructed, they will be the asymptotes of the hyperbola.

If in either case the orthogonal pair of conjugate rays of the involution be constructed, they will be the axes of the conic.

The direction of the axes can be found without first constructing the centre and two pairs of conjugate diameters *. Let A, B, C, F, G be the five given points (Fig. 168); describe a circle round three of them ABC, and construct (Art. 227, I) the fourth point of intersection C' of this circle with the conic determined by the five given points. Any transversal will cut the two curves and the two pairs of opposite sides of the common inscribed quadrangle $ABCC'$ in pairs of points forming an involution (Art. 183). The double points P and Q (if such exist) of this involution will be conjugate with regard to each of the curves (Arts. 125, 263); *i.e.* they will be the pair common (Art. 208) to the two involutions which are formed on the transversal by the pairs of points conjugate with regard to the circle and by the pairs of points conjugate with regard to the conic (Art. 263). Suppose that the straight line at infinity is taken as the transversal. As this straight line does not meet the circle, one at least of these two involutions will have no double points, and consequently (Art. 208) the points P and Q do really exist. Since these points are infinitely distant and are conjugate with regard to both curves they will be (Arts. 276, 284) the poles of two conjugate diameters of the circle and also of two conjugate diameters of the conic; but conjugate diameters of the circle are perpendicular to one another (Art. 289); therefore P and Q are the poles of the axes of the conic. Further, the segment PQ is harmonically divided by either pair of opposite sides of the quadrangle $ABCC'$; consequently P and Q are the points at infinity on the bisectors of the angles included by the pairs of opposite sides (Art. 60). In order then to find the required directions of the axes, we have only to draw the bisectors † of the angle included by a pair of opposite sides of the quadrangle $ABCC'$, for example by AB and CC' (Fig. 168).

317. Let $qrst$ (Fig. 161) be a complete quadrilateral, and S any point. It has already been seen (Art. 185, right) that the pairs of rays a and a', b and b', which join S to two pairs of opposite vertices, belong to an involution of which the tangents drawn from S to any conic inscribed in the quadrilateral are a pair of conjugate rays. Suppose the involution to have two double rays m and n; they will be harmonically conjugate

* PONCELET, *loc. cit.*, Art. 394.

† See also the note to Art. 387.

with regard to such a pair of tangents (Art. 125), and will consequently be conjugate lines with respect to the conic. But (Art. 218, right) m and n are the tangents at S to the two conics which can be inscribed in the quadrilateral $qrst$ so as to pass through S. Therefore

If two conics which are inscribed in a given quadrilateral pass through a given point, their tangents at this point are conjugate lines with respect to any conic inscribed in the quadrilateral.

Instead of taking an arbitrary point S, let m be supposed given. If this straight line does not pass through any of the vertices of the quadrilateral, there will be one conic, and only one, which touches the five straight lines m, q, r, s, t (Art. 152). Let S be the point where this conic touches m; there will be a second conic which is inscribed in the quadrilateral and which passes through S; let the tangent to this at S be n. The straight lines m and n will then be conjugate to one another with respect to all conics inscribed in the quadrilateral; and therefore (Art. 255),

The poles of any straight line m with respect to all conics inscribed in the same quadrilateral lie on another straight line n.

Moreover, since the straight lines m and n are the double rays of the involution of which the rays drawn from S to two opposite vertices are a conjugate pair, therefore *m and n divide harmonically each diagonal of the quadrilateral.*

318. I. The correlative propositions to those of Art. 317 are the following:

If a straight line touches two conics which circumscribe the same quadrangle, the two points of contact are conjugate to one another with respect to all conics circumscribing the quadrangle.

The polars of any given point M with respect to all the conics circumscribing the same quadrangle meet in a fixed point N. The segment MN is divided harmonically at the two points where it is cut by any pair of opposite sides of the complete quadrangle.

II. Suppose in the second theorem of Art. 317 that the straight line m lies at infinity; then the poles of m will be the centres of the conics (Art. 281), and n will bisect each of the diagonals of the quadrilateral (Art. 59); therefore:

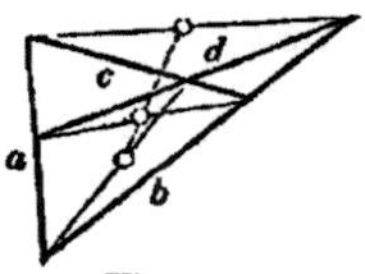

Fig. 204.

The centres of all conics inscribed in the same quadrilateral lie *on the straight line* (Fig. 204) *which passes through the middle points of the diagonals of the quadrilateral**.

III. Suppose similarly in theorem I of the present Article that the point M lies at infinity; the polars of M will become the diameters conjugate to those which have M as their common point at infinity; thus:

In any conic circumscribing a given quadrangle, the diameter which is conjugate to one drawn in a given fixed direction will pass through a fixed point.

319. Newton's theorem (Art. 318, II) gives a simple *method for finding the centre of a conic determined by five tangents a, b, c, d, e* (Fig. 205). The four tangents a, b, c, d form a quadrilateral; join the middle points of its diagonals. Let the same be done with regard to the quadrilateral *abce*; the two straight lines thus obtained will meet in the required centre O.

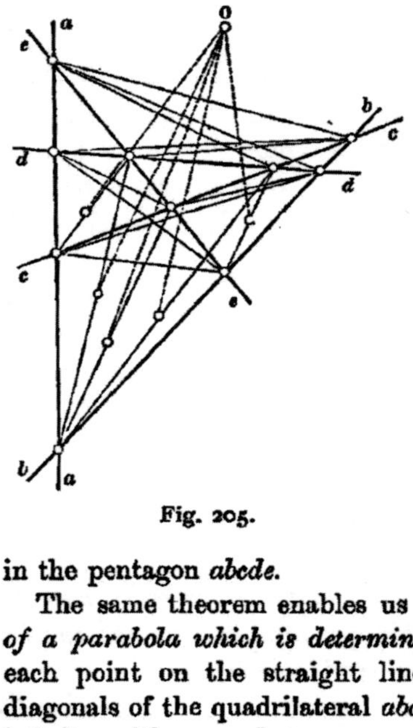

Fig. 205.

The five tangents, taken four and four together, form five quadrilaterals; the five straight lines which join the middle points of the diagonals of each of the quadrilaterals will therefore all meet in the centre O of the conic inscribed in the pentagon *abcde*.

The same theorem enables us *to find the direction of the diameters of a parabola which is determined by four tangents a, b, c, d.* For each point on the straight line joining the middle points of the diagonals of the quadrilateral *abcd* is the pole of the straight line at infinity with regard to some conic inscribed in the quadrilateral (Art. 318, II); therefore the point at infinity on the line will be the pole with regard to the inscribed parabola (Arts. 254 III, and 23). The straight line therefore which joins the middle points of the diagonals is itself a diameter of the parabola (Fig. 204).

* NEWTON, *Principia*, book i. lemma 25. Cor. 3.

CHAPTER XXII.

POLAR RECIPROCAL FIGURES.

320. An auxiliary conic K being given, it has been seen (Art. 256) that if a variable pole describes a fixed straight line its polar turns round a fixed point, and reciprocally, that if a straight line considered as polar turns round a fixed point, its pole describes a fixed straight line.

Consider now as polars all the tangents of a given curve C, or in other words suppose the polar to move, and to envelope the given curve. Its pole will describe another curve, which may be denoted by C′. Thus the points of C′ are the poles of the tangents of C.

But it is also true that, reciprocally, the points of C are the poles of the tangents of C′. For let M' and N' be two points on C′ (Fig. 206); their polars m and n will be two tangents to C and the point mn where they meet will be the pole of the chord $M'N'$ (Art. 256). Now suppose the point N' to approach M' indefinitely; the chord $M'N'$ will approach more and more nearly to the position of the tangent at M' to the curve C′; the straight line n will at the same time approach more and more nearly to coincidence with m, and the point mn will tend more and more to the point where m touches C. In the limit, when the distance $M'N'$ becomes indefinitely small, the tangent to C′ at M' will become the polar of the point of contact of m with C. Just then as the tangents of C are the polars of the points of C′, so also are the tangents of C′ the polars of the points of C; if a straight line m touches the curve C at M, the pole M' of m

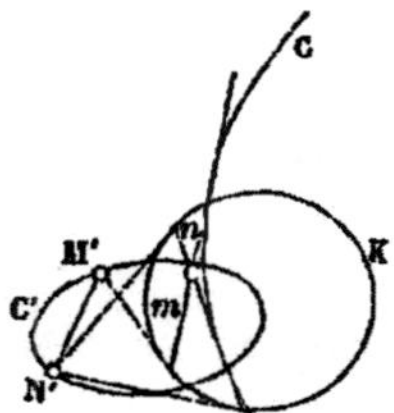

Fig. 206.

is a point of the curve **C**$'$ and the polar m' of M is a tangent to the curve **C**$'$ at M'.

Two curves **C** and **C**$'$ such that each is the locus of the poles of the tangents of the other, and at the same time also the envelope of the polars of the points of the other, are said to be *polar reciprocals* * one of the other with respect to the *auxiliary conic* **K**.

321. An arbitrary straight line r meets one of the reciprocal curves in n points say; the polars of these points are n tangents to the other curve all passing through the pole R' of r. To the second curve therefore can be drawn from any given point R' the same number of tangents as the first curve has points of intersection with the straight line r, the polar of R'; and *vice versa*. In other words, *the degree and class of a curve are equal to the class and degree respectively of its polar reciprocal with respect to a conic.*

322. Now suppose the curve **C** to be a conic, and a, b two tangents to it; they will be cut by all the other tangents $c, d, e, \ldots$ in corresponding points of two projective ranges (Art. 149). In other words, **C** may be regarded as the curve enveloped by the straight lines $c, d, e, \ldots$ which connect the pairs of corresponding points of two projective ranges lying on a and b respectively (Art. 150).

The curve **C**$'$ will pass through the poles $A', B', C', D', E', \ldots$ of the tangents $a, b, c, d, e, \ldots$ of **C**. The straight lines $A'(C', D', E', \ldots)$ will be the polars of the points $a\,(c, d, e, \ldots)$ and will form a pencil projective with the range of poles lying on the straight line a (Art. 291); so too the straight lines $B'(C', D', E', \ldots)$ will be the polars of the points $b\,(c, d, e, \ldots)$ and will form a pencil projective with the range of poles lying on b. But the ranges $a\,(c, d, e, \ldots)$ and $b\,(c, d, e, \ldots)$ are projective; therefore also the pencils $A'(C', D', E', \ldots)$ and $B'(C', D', E', \ldots)$ are projective. Consequently **C**$'$ is the locus of the points of intersection of corresponding rays of two projective pencils; that is (Art. 150) a conic. Accordingly:

The polar reciprocal of a conic with respect to another conic is a conic †.

323. When an auxiliary conic **K** is given and another conic

* PONCELET, *loc. cit.*, Art. 232.

† Ibid., Art. 231.

C whose polar reciprocal C′ is to be determined, the question arises whether C′ is an ellipse, a hyperbola, or a parabola. The straight line at infinity is the polar of the centre O of K; therefore the points at infinity on C′ correspond to the tangents of C which pass through O. It follows that *the conic* C′ *will be an ellipse or a hyperbola according as the point O is interior or exterior to the conic* C, *and* C′ *will be a parabola when O lies upon* C.

If A is the pole of a straight line a with respect to C, and a' the polar of A and A' the pole of a with respect to K, then will A' be the pole of a' with respect to C′, since to four poles forming a harmonic range correspond four polars forming a harmonic pencil (Art. 291) and *vice versa*. Therefore the centre M' of C′ will be the pole with respect to K of the straight line m which is the polar of O with respect to C. To two conjugate diameters of C′ will correspond two points of m which are conjugate with respect to C, &c.

324. Let there be given in the plane of the auxiliary conic a *figure* (Art. 1) or complex of any kind composed of points, straight lines, and curves; and let the polar of every point, the pole of every line, and the polar reciprocal of every curve, be constructed. In this way a new figure will be obtained; the two figures are said to be *polar reciprocals* one of the other, since each of them contains the poles of the straight lines of the other, the polars of its points, and the curves which are the polar reciprocals of its curves. To the method whereby the second figure has been derived from the first the name of *polar reciprocation* is given.

Two figures which are polar reciprocals one of the other are *correlative figures* in accordance with the law of duality in plane Geometry (Art. 33); for to every point of the one corresponds a straight line of the other, and to every range in the one corresponds a pencil in the other. They lie moreover in the same plane; their positions in this plane are determinate, but may be interchanged, since every point in the one figure and the corresponding straight line in the other are connected by the relation that they are pole and polar with respect to a fixed conic. Thus two polar reciprocal figures are correlative figures which are coplanar, and which have a special relation to one another with respect to their positions in the plane in which they lie. On the other hand, if two figures are merely correlative in accordance with the law of duality, there is no relation of any kind between them as regards their position *.

325. If one of the reciprocal figures contains a range (of poles) the other contains a pencil (of polars), and these two corresponding forms are projective (Art. 291). If then the points of the range are in involution, the rays of the corresponding pencil will also be in involution, and to the double points of the first involution will correspond the double rays of the second (Art. 124). If there is a conic in one of the figures there will also be one in the other figure (Art. 322); to the points of the first conic will correspond the tangents of the second, and to the tangents of the first will correspond the points of the second; to an inscribed polygon in the first figure will correspond a circumscribed polygon in the second (Art. 320). If the first figure exhibits the proof of a theorem or the solution of a problem, the second will show the proof of the correlative theorem or the solution of the correlative problem; that namely which is obtained by interchanging the elements 'point' and 'line.'

326. THEOREM. *If two triangles are both self-conjugate with regard to a given conic, their six vertices lie on a conic, and their six sides touch another conic* †.

Let ABC and DEF be two triangles (Fig. 207) each of which is self-conjugate (Art. 258) with regard to a given conic K. Let DE and DF cut BC in B_1 and C_1 respectively, and let AB and AC cut EF in E_1 and F_1 respectively. The point B is the pole of CA, and C is the pole of AB; B_1 is the pole of the straight line joining the poles of BC and DE, *i.e.* of AF; and C_1 is the pole of the straight line joining the poles of BC and DF, *i.e.* of AE. The range of poles BCB_1C_1 is therefore (Art. 291) projective with the pencil of polars $A(CBFE)$, and therefore with the range of points F_1E_1FE in which this pencil is cut by the transversal EF. Thus

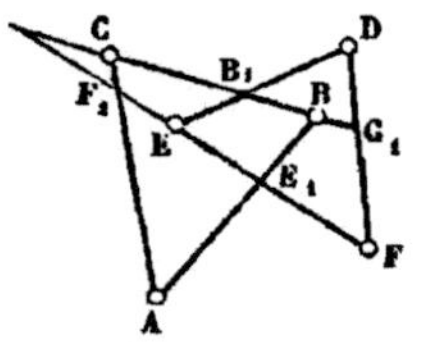

Fig. 207.

$$(BCB_1C_1) = (F_1E_1FE)$$
$$= (E_1F_1EF) \text{ by Art. 45,}$$

which shows that the two ranges in which the straight lines BC and EF respectively are cut by AB, CA, DE, FD are projectively related.

* STEINER, *loc. cit.*, p. vii of the preface; Collected Works, vol. i. p. 234.

† STEINER, *loc. cit.*, p. 308, § 60, Ex. 46; Collected Works, vol. i. p. 448; CHASLES, *Sections coniques*, Art. 215.

These six straight lines therefore, the six sides of the given triangles, all touch a conic **C** (Art. 150, II).

The poles of these six sides are the six vertices of the triangles; these vertices therefore all lie on another conic **C′** which is the polar reciprocal of **C** with regard to the conic **K** *.

327. The proposition of the preceding Article may also be expressed as follows: Given two triangles which are self-conjugate with regard to the same conic **K**; if a conic **C** touch five of the six sides it will touch the sixth side also, and if a conic pass through five of the six vertices it will pass through the sixth vertex also.

It follows that *if a conic* **C** *touch the sides of a triangle abc which is self-conjugate with regard to another conic* **K**, *there are an infinite number of other triangles which are self-conjugate with regard to the second conic and which circumscribe the first.*

For let d be any tangent to **C**; from D, its pole with regard to **K**, draw a tangent e to **C**, and let f be the polar with regard to **K** of the point de; then the triangle def will be self-conjugate with regard to **K** (Art. 259). But **C** touches five sides a, b, c, d, e of two triangles which are both self-conjugate with respect to **K**; therefore it must also touch the sixth side f; which proves the proposition.

328. If the point D is such that from it a pair of tangents e' and f' can be drawn to **K**, the four straight lines e, f, e', f' will form a harmonic pencil (Art. 264), since e and f are conjugate straight lines with respect to the conic **K**; consequently the straight lines e' and f' are conjugate to one another with respect to **C**.

The locus of D is the conic **C′** which is the polar reciprocal of **C** with regard to **K**; therefore:

If a conic **C** *is inscribed in a triangle which is self-conjugate with respect to another conic* **K**, *the locus of a point such that the pairs of tangents drawn from it to the conics* **C** *and* **K** *form a harmonic pencil is a third conic* **C′** *which is the polar reciprocal of* **C** *with respect to* **K**.

329. Correlatively: *If a conic* **C′** *circumscribes a triangle which is self-conjugate with respect to another conic* **K**, *there are an infinite number of other triangles which are inscribed in* **C′** *and are self-conjugate with respect to* **K**; *and the straight lines which cut* **C′** *and* **K** *in two pairs of points which are harmonically conjugate to one another all touch a third conic* **C** *which is the polar reciprocal of* **C′** *with regard to* **K**.

* We may show independently that the six vertices lie on a conic as follows. It has been seen that the pencil of polars $A(CBFE)$ is projective with the range of poles BCB_1C_1; it is therefore projective with the pencil $D(BCB_1C_1)$ formed by joining these to the point D. Therefore

$$\begin{aligned} A(CBFE) &= D(BCB_1C_1) = D(BCEF) \\ &= D(CBFE) \text{ by Art. 45,} \end{aligned}$$

which shows (Art. 150, I) that A, B, C, D, E, F lie on a conic.

330. THEOREM. *If two triangles circumscribe the same conic, their six vertices lie on another conic.*

Let $OQ'R'$ and $O'PS$ be two triangles each circumscribing a given conic C (Fig. 208). The two tangents PS and $Q'R'$ are cut by the four other tangents $O'P, OQ', OR', O'S$ in two groups of corresponding points $PQRS$ and $P'Q'R'S'$ of two projective ranges u and u' (Art. 149); consequently the pencils $O(PQRS)$ and $O'(P'Q'R'S')$ formed by connecting these points with O and O' respectively are projective. Therefore the points P, Q', R', S, in which their pairs of corresponding rays intersect, lie on a conic C′ (Art. 150, I) passing through the centres O and O'; which proves the theorem.

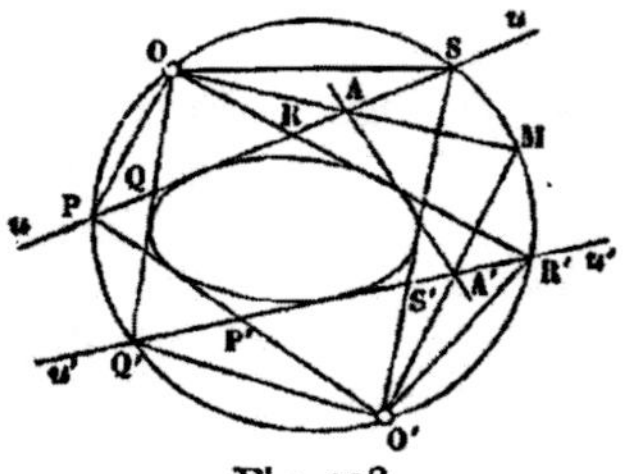

Fig. 208.

331. The theorem correlative and converse to the foregoing one is the following:

If two triangles are inscribed in the same conic, their six sides touch another conic.*

This may be proved by considering the triangles $OQ'R'$ and $O'PS$ as both inscribed in the conic C′, and by reasoning in a manner exactly analogous, but correlative, to that above.

332. It follows at once that:

If two triangles circumscribe the same conic, the conic which passes through five of their vertices passes through the sixth vertex also.	If two triangles are inscribed in the same conic, the conic which touches five of their sides touches the sixth side also.

Or:

If two conics are such that a triangle can be inscribed in the one so as to circumscribe the other, then there exist an infinite number of other triangles which possess the same property†.

333. There are in the figure (Fig. 208) four projective forms: the two ranges u and u', which determine the tangents to the conic C, and the two pencils O and O', which determine the points of C′; the pencil O is in perspective with the range u

* BRIANCHON, *loc. cit.*, p. 35; STEINER, *loc. cit.*, p. 173, § 46, II; Collected Works, vol. i. p. 356.

† PONCELET, *loc. cit.*, Art. 565.

and the pencil O' is in perspective with the range u'. If then any tangent to **C** cut the bases u and u' of the two ranges in A and A' respectively, the rays OA and $O'A'$ will meet in a point M lying on **C**$'$; and, conversely, if any point M on **C**$'$ be joined to the centres O and O', the joining lines will cut u and u' respectively in two points A and A' such that the straight line joining them is a tangent to **C**. Therefore:

If a variable triangle $AA'M$ is such that two of its sides pass respectively through two fixed points O' and O lying on a given conic, and the vertices opposite to them lie respectively on two fixed straight lines u and u', while the third vertex lies always on the given conic, then the third side will touch a fixed conic which touches the straight lines u and u'.

If a variable triangle $AA'M$ is such that two of its vertices lie respectively on two fixed tangents u and u' to a given conic, and the sides opposite to them pass respectively through two fixed points O' and O, while the third side always touches the given conic, then the third vertex will lie on a fixed conic which passes through the points O and O'.

334. THEOREM. *If the extremities of each of two diagonals of a complete quadrilateral are conjugate points with respect to a given conic, the extremities of the third diagonal also will be conjugate points with respect to the same conic*.*

Let $ABXY$ (Fig. 209) be a complete quadrilateral such that A is conjugate to X, and B to Y, with respect to a given conic **K** (not shown in the figure). Let the sides AB, XY meet in C, and the sides AY, BX in Z; then shall C and Z be conjugate points with respect to the conic **K**.

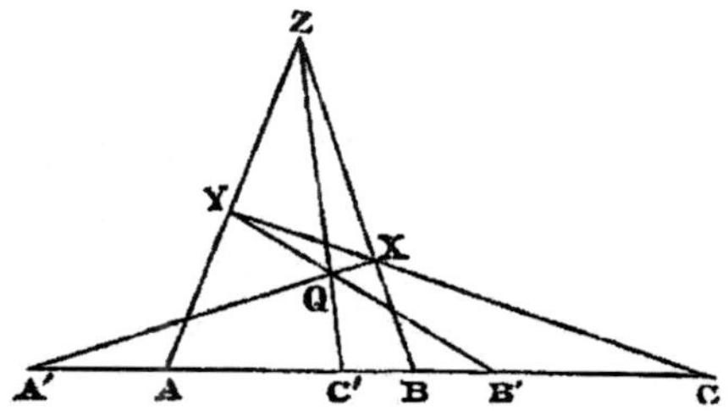

Fig. 209.

Suppose the polars of the points A, B, C (with respect to **K**) to cut the straight line ABC in A', B', C' respectively. The three pairs of conjugate points A and A', B and B', C and C' are in involution; consequently, considering XYZ as a triangle cut by a transversal $A'B'C'$, it follows by Art. 135 that the

* HESSE, *De octo punctis intersectionis trium superficierum secundi ordinis* (Dissertatio pro venia legendi, Regiomonti, 1840), p. 17.

straight lines XA', YB', ZC' meet in one point Q. Since evidently XA' is the polar of A and YB' the polar of B with respect to K, their point of intersection Q is the pole of AB. Since then C is a point on AB and is conjugate to C', its polar will be QC'; but QC' passes through Z; therefore C and Z are conjugate points, which was to be proved.

335. The proof of the following, the correlative theorem, is left as an exercise to the student:

If two pairs of opposite sides of a complete quadrangle are conjugate lines with respect to a conic, the two remaining sides also are conjugate lines with respect to the same conic.

In order to obtain such a complete quadrangle, it is only necessary to take the polar reciprocal of the quadrilateral considered in Hesse's theorem, *i.e.* the figure which is formed by the polars of the six points A and X, B and Y, C and Z.

336. The following proposition is a corollary to that of Art. 334:

Two triangles which are reciprocal with respect to a conic are in homology.*

Let ABC (Fig. 210) be any triangle; the polars of its vertices with respect to a given conic form another triangle $A'B'C'$ reciprocal to the first, that is, such that the sides of the first triangle are also the polars of the vertices of the second. Let the sides CA and $C'A'$ meet in E, and the sides AB and $A'B'$ in F.

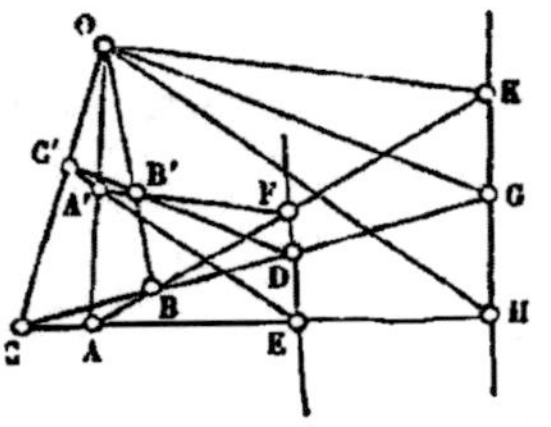

Fig. 210.

The points B and E are conjugate with respect to the conic, since E lies on $C'A'$, the polar of B; similarly C and F are conjugate points. Thus in the quadrilateral formed by BC, CA, AB, and EF, two pairs of opposite vertices B and E, C and F are conjugate; therefore the third pair are conjugate also, viz. A and the point D where BC meets EF. The polar $B'C'$ of A therefore passes through D; thus BC and $B'C'$ meet in a point D lying on EF. Since then the pairs of opposite sides of the two triangles meet one another in three collinear points, the triangles are in homology, and the straight lines AA', BB', CC' which join

* CHASLES, *loc. cit.*, Art. 135.

the pairs of vertices meet (Art. 17) in a point O, the pole of the straight line DEF.

337. By combining this theorem with that of Art. 155 the following property may be enunciated:

If two triangles are reciprocals with respect to a given conic **K**, *the six points in which the sides of the one intersect the non-corresponding* sides of the other lie on a conic* **C**, *and the six straight lines which connect the vertices of the one with the non-corresponding vertices of the other touch another conic* **C′**, *the polar reciprocal of* **C** *with respect to* **K** (Art. 322); these straight lines are in fact the polars with regard to **K** of the six points just mentioned.

If one of the triangles $A'B'C'$ is inscribed in the other ABC, the three conics **C**, **C′**, and **K** coincide in one which is circumscribed about the former triangle and inscribed in the latter (Arts. 174, 176).

338. PROBLEM. *Given two triangles ABC, $A'B'C'$ which are in homology; to construct (when it exists) the conic with regard to which they are reciprocal.*

Take one of the sides, BC for example; the points in which it is cut by $C'A'$ and $A'B'$ are conjugate to the points B and C respectively, and these two pairs of conjugate points determine an involution (Art. 263), the double points of which (if they exist) are the points where BC is cut by the conic in question. In order then to find the points in which this conic cuts BC, it is only necessary to construct these double points. In this way the points in which the sides of the triangles meet the conic can be found, and the latter is determined. Since A' and B are the poles of BC and $C'A'$, these points and that in which $C'A'$ meets BC will be the vertices of a self-conjugate triangle (Art. 258). If then, in finding the points of intersection of the conic and the straight lines BC and $C'A'$ in the manner just explained, it should happen that the two involutions found have neither of them double points, the conclusion is that no conic exists such as is required; for if it did exist, it must be cut by two of the sides of the self-conjugate triangle (Art. 262).

339. The centre of homology O of the given triangles (Fig. 210) is the pole of the axis of homology DEF; and the projective correspondence (Art. 291) between the points (poles) lying on the axis and the straight lines (polars) radiating from the centre of homology is determined by the three pairs of corresponding elements D and

* Two sides BC and $B'C'$ of the triangles may be termed *corresponding*, when each lies opposite to the pole of the other. And two vertices A and A' may be termed *corresponding*, when each lies opposite to the polar of the other.

AA', E and BB', F and CC'. Consequently it is possible to construct with the ruler only (Art. 84) the polar of any other point on the axis, and the pole of any other ray passing through the centre O.

What has just been said with regard to the point O and the axis of homology may also be said with regard to any vertex of one of the triangles and its polar (the corresponding side of the other triangle). For if *e.g.* the vertex A' and the side BC be considered, the projective correspondence between the straight lines radiating from A' and the points lying on BC is determined by the three pairs of corresponding elements $A'B'$ and C, $A'C'$ and B, $A'O$ and D.

This being premised, it will be seen that the polar of any point P and the pole of any straight line p can be constructed with the help of the ruler only. For suppose P to be given; it has been shown that the poles of the straight lines PO, PA, PB, PC, PA', ... can be constructed, and these all lie on a straight line X which is the required polar of P. So again if the straight line p is given, the polars of the points in which it meets BC, CA, ... can be constructed, and will meet in a point which is the pole of p.

It will be noticed that all these determinations of poles and polars are linear (*i.e.* of the first degree) and independent of the construction (Art. 338) of the auxiliary conic, which is of the second degree, since it depends on finding the double elements of an involution. The construction of the poles and polars is therefore always possible, even when the auxiliary conic does not exist. In other words: the two given triangles in homology determine between the points and the straight lines of the plane a reciprocal correspondence such that to every point corresponds a straight line and to every straight line a point, to the rays of a pencil the points of a range projective with the pencil, and *vice versa*. Any point and the straight line corresponding to it may be called *pole* and *polar*, and this assemblage of poles and polars, which possesses all the properties of that determined by an auxiliary conic (Art. 254), may be called a *polar system*.

Two triangles in homology accordingly determine a polar system. If an auxiliary conic exists, it is the locus of the points which lie on the polars respectively corresponding to them, and it is at the same time the envelope of the straight lines which pass through the poles respectively corresponding to them. If no auxiliary conic exists, there is no point which lies on its own polar *.

* STAUDT, *loc. cit.*, Art. 241.

CHAPTER XXIII.

FOCI *.

340. It has been seen (Art. 263) that the pairs of straight lines passing through a given point S and conjugate to one another with respect to a given conic form an involution. Let a plane figure be given, containing a conic **C**; and let the figure homological with it be constructed, taking S as centre of homology; let **C′** be the conic corresponding to **C** in the new figure. Since in two homological figures a harmonic pencil corresponds to a harmonic pencil, any pair of straight lines through S which are conjugate with respect to **C** will be conjugate also with respect to **C′**. The polars of S with respect to the two conics will be corresponding straight lines; if then the polar of S with respect to **C** be taken as the vanishing line in the first figure, the polar of S with respect to **C′** will lie at infinity; *i.e.* the point S will be the centre of the conic **C′**.

In this case therefore any two straight lines through S which are conjugate with respect to **C** will be a pair of conjugate diameters of **C′**. If S is external to **C**, the double rays of the involution formed by the conjugate lines through S are the tangents from S to **C**, and therefore the asymptotes of **C′**, which is in this case a hyperbola. If S is internal to **C**, the involution has no double rays, and therefore **C′** is an ellipse.

We conclude then that *to every point S in the plane of a given conic* **C** *corresponds a conic* **C′** *homological with* **C** *and having its centre at S; which conic* **C′** *is a hyperbola or an ellipse according as S is external or internal to the given conic* **C**.

* Steiner, *Vorlesungen über synthetische Geometrie* (ed. Schröter), II[ter] Abschnitt, § 35; Zech, *Höhere Geometrie* (Stuttgart, 1857), § 7; Reye, *Geometrie der Lage* (2nd ed., Hannover, 1877), *Vortrag* 13.

341. For certain positions of the point S the conic C' will be a circle. When S has one of these positions it is called a *focus** of the conic C. Since all pairs of conjugate diameters of a circle cut one another orthogonally the involution at S of conjugate lines with respect to C will in this case consist entirely of orthogonal pairs.

If C *is a circle, its centre O is a focus;* for every pair of conjugate lines which meet in O, *i.e.* every pair of conjugate diameters of C, cut orthogonally. And *a circle* C *has no other focus but its centre O.* For let any point S be taken (Fig. 211) distinct from O and a straight line SQ be drawn not passing through O; and let P be the pole of SQ. Then since PO must be perpendicular to SQ, the conjugate lines SP, SQ cannot be orthogonal, and therefore S cannot be a focus of C.

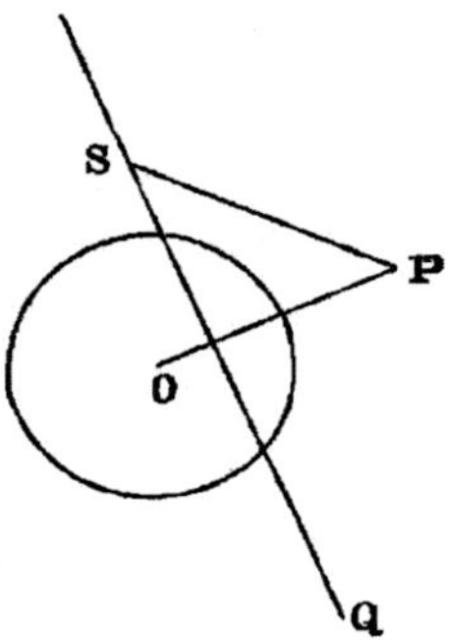

Fig. 211.

The foci of a conic C which is not a circle *are of necessity internal points;* this follows from what has been said above (Art. 340). Further, *they lie on the axes.* For if F is a focus and O the centre of the conic, the pole of the diameter FO will lie on the perpendicular drawn through F to FO; therefore FO is perpendicular to its conjugate diameter, *i.e.* FO is an axis of the conic.

Again, *the straight line connecting two foci F and F_1 is an axis.* For if straight lines perpendicular to FF_1 be drawn through F and F_1 these will both be conjugate to FF_1, and their point of intersection will therefore be the pole of FF_1; but this point lies at infinity; therefore FF_1 is an axis.

342. Let a point P be taken arbitrarily on an axis a of a conic; through P draw a straight line r, and from R, the pole of r, draw the straight line r' perpendicular to r; let P' be the point where r' meets the axis. The straight lines passing through P and those passing through P' and conjugate to them respectively form two projective pencils; for the second pencil is composed of rays which project from P' the range

* DE LA HIRE, *Sectiones conicae* (Parisiis, 1685), lib. viii. prop. 23; PONCELET, *Propriétés projectives*, Art. 457 et seqq.

formed by the poles of the rays of the first pencil, which range is (Art. 291) projective with the first pencil itself. The two pencils in question have three pairs of corresponding rays which are mutually perpendicular; for if A be the point at infinity which is the pole of the axis a, the rays PA, PP', r of the first pencil correspond to the rays $P'P$, $P'A$, r' of the second, and the three former rays are severally perpendicular to the three latter. The two pencils therefore by the intersection of corresponding rays generate a circle of which PP' is a diameter; and therefore every pair of corresponding rays of the two pencils P and P' intersect at right angles. Thus:

To every point P lying on an axis of the conic corresponds a point P' on the same axis such that any two conjugate straight lines which pass one through P and the other through P' are perpendicular to one another.

The pairs of points analogous to P, P' form an involution. For let the ray r move parallel to itself; the corresponding rays r' (which are all perpendicular to r) will all be parallel to each other. The pencil of parallels r is projective (Art. 291) with the range which the poles R of the rays r determine upon the diameter conjugate to that drawn parallel to r; and the pencil of parallels r' is in perspective with this same range. Therefore the pencils r, r' are projective, and consequently the points P, P' in which a pair of corresponding rays r, r' of the pencils cut the axis a trace out two projective ranges. To the straight line at infinity regarded as a ray r corresponds in the second pencil the diameter parallel to the rays r'; and similarly, to the line at infinity regarded as a ray r' corresponds in the first pencil the diameter parallel to the rays r. Therefore the point at infinity on the axis has the same correspondent whether it be regarded as a point P or as a point P': *viz.* the centre O of the conic. We conclude that *the pairs of points P, P' constitute an involution of which the centre is the centre O of the conic.*

343. If the involution formed by the points P, P' on the axis a has double points, each of them will be a focus of the conic, since every straight line through such a double point will be conjugate to the perpendicular drawn to it through the point itself.

If the involution has no double points, each of the two points (Art. 128) at which the pairs PP' subtend a right angle will be a focus of the conic. For every pair of mutually perpendicular straight lines which meet in such a point will pass through two points P, P', and will therefore be conjugate lines with respect to the conic.

From this it follows that one at least of the two axes of a conic contains two foci. Further, a conic has only two foci; for every straight line which joins two foci is an axis (Art. 341), and no conic (except it be a circle) has more than two axes.

Consequently *a central conic (ellipse or hyperbola) has two foci, which are the double points of the involution PP' on an axis and are also the points at which the pairs of points PP' of the involution on the other axis subtend a right angle.*

The axis which contains the foci may on this account be called the *focal axis*. Since the foci are internal points, it is seen that in the hyperbola the focal axis is that one which cuts the curve (the transverse axis).

Since the centre O of the conic is the centre of the involution PP', it bisects the distance between the two foci.

From what has been said it follows that *two perpendicular straight lines which are conjugate with respect to a conic meet the focal axis in two points which are harmonically conjugate with respect to the foci; and they determine upon the other axis a segment which subtends a right angle at either focus.*

344. The *normal* at any point on a curve is the perpendicular at this point to the tangent. Since the tangent and normal at any point on a conic are conjugate lines at right angles, they meet the focal axis in a pair of points harmonically conjugate with respect to the foci; and they determine on the other axis a segment which subtends a right angle at either focus (Art. 343). Accordingly:

If a circle be drawn to pass through the two foci and through any point on the conic, it will have the two points in which the non-focal axis is cut by the tangent and normal at that point as extremities of a diameter.

And again (Art. 60):

The tangent and normal at any point on a conic are the bisectors *of the angle made with one another by the rays which join that point to the foci**.

These rays are called the *focal radii* of the given point.

345. A pair of conjugate lines which intersect at right angles in a point S external to the conic are harmonically conjugate with respect to the tangents from S to the conic (Art. 264) as well as with respect to the rays joining S to the foci (Art. 343); therefore:

The angle between two tangents and that included by the straight lines which join the point of intersection of the tangents to the foci have the same bisectors†.

346. In the parabola, the point at infinity on the axis, regarded as a point P, coincides with its correspondent P'; for the straight line at infinity, being a tangent to the conic at the said point P, passes through its own pole.

Accordingly one of the double points of the involution determined on the axis by the pairs of conjugate orthogonal rays, *i.e.* one of the foci, is at infinity. The other double point lies at a finite distance, and is generally spoken of as *the* focus of the parabola.

Since in the case of the parabola one focus is at infinity, the theorems proved above (Arts. 343–345) become the following:

Two conjugate orthogonal rays, and in particular the tangent and normal at any point on the parabola, meet the axis in two points which are equidistant from the focus.

The tangent and normal at a point on a parabola are the bisectors of the angle which the focal radius of the point makes with the diameter passing through the point‡.

The straight line which connects the focus with the point of intersection of two tangents to a parabola makes with either of the tangents the same angle that the axis makes with the other tangent.

347. From the last of these may be immediately deduced the following theorem:

The circle circumscribing a triangle formed by three tangents to a parabola passes through the focus.

Let PQR (Fig. 212) be a triangle formed by three

* APOLLONIUS, *loc. cit.*, iii. 48.

† Ibid., iii. 46.

‡ DE LA HIRE, *loc. cit.*, lib. viii. prop. 2.

tangents to a parabola, and let F be the focus. Considering the tangents which meet in P, the angle FPQ is equal to that made by PR with the axis; and considering the tangents which meet in R, the angle FRQ is equal to that made by RP with the axis. Hence the angles FPQ, FRQ are equal, and therefore P, Q, R, F lie on the same circle.

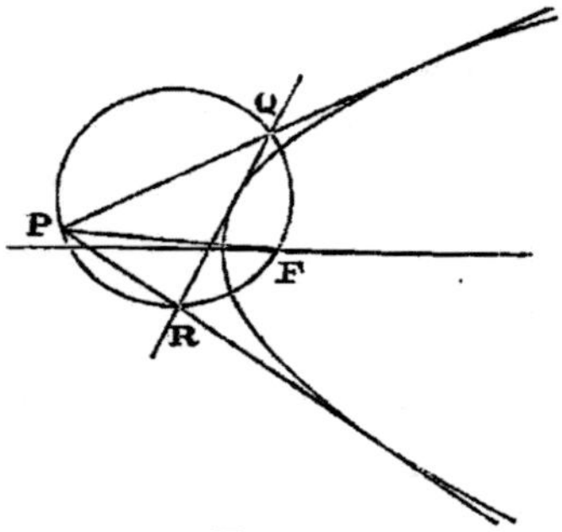

Fig. 212.

COROLLARY. *The locus of the foci of all parabolas which touch the three sides of a given triangle is the circumscribing circle of the triangle.*

This corollary gives the construction for the focus of a parabola which touches four given straight lines. And since only one such parabola can be drawn (Art. 157), we conclude that:

Given four straight lines, the circles circumscribing the four triangles which can be formed by taking the lines three and three together all pass through the same point.

348. The polar of a focus is called a *directrix*.

The two directrices are straight lines perpendicular to the transverse axis and external to the conic, since the foci lie on the transverse axis and are internal to the conic (Art. 343).

In the case of the parabola, the straight line at infinity is one directrix; the other lies at a finite distance, and is generally spoken of as *the* directrix of the parabola.

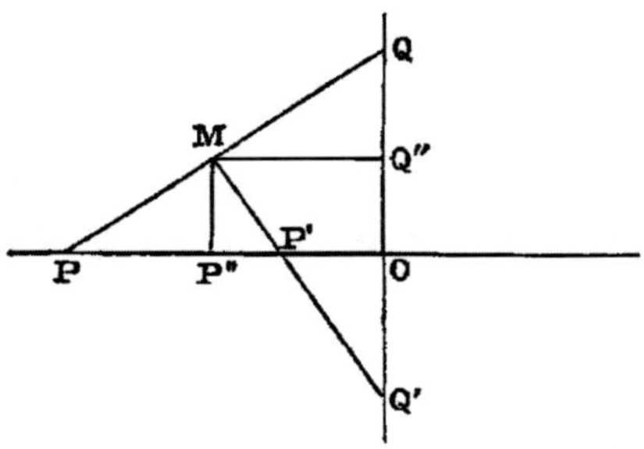

Fig. 213.

If F be a focus, and if the tangent at any point X on a conic cut the corresponding directrix in Y, this point Y will be the pole of the focal radius FX. Therefore FX, FY are conjugate lines with respect to the conic, and since they meet in a focus, they will be at right angles: consequently:

The part of a tangent to a conic intercepted between its point of contact and a directrix subtends a right angle at the corresponding focus.

349. Let the tangent and normal at any point M on a conic meet the focal axis in P, P' respectively, and let them meet the other axis in Q, Q' respectively (Fig. 213). From M let perpendiculars MP'', MQ'' be drawn to the axes.

From the similar triangles OPQ, $Q''MQ$

$$OP : OQ = Q''M : Q''Q,$$

and from the right-angled triangle $Q'MQ$

$$Q''M : Q''Q = Q'Q'' : Q''M;$$

$$\therefore \quad OP : OQ = Q'Q'' : Q''M$$

$$= Q'Q'' : OP'',$$

or

$$OP \,.\, OP'' = OQ \,.\, Q'Q''$$

$$= OQ\,(Q'O + OQ''),$$

so that

$$OP \,.\, OP'' - OQ \,.\, OQ'' = OQ \,.\, Q'O. \quad . \quad . \quad . \quad (1)$$

But P and P'' are a pair of conjugate points, since MP'' is the polar of P; similarly Q and Q'' are conjugate points. Therefore (Art. 294)

$$OP \,.\, OP'' = OA^2 \text{ and } OQ \,.\, OQ'' = \pm OB^2,$$

where OA, OB are the lengths of the semiaxes, and the double sign refers to the two cases of the ellipse and the hyperbola. Again, the points Q, Q' subtend a right angle at either of the two foci F, F' (Art. 343) so that

$$OQ \,.\, Q'O = OF^2.$$

Substituting, (1) becomes

$$OF^2 = OA^2 \mp OB^2.$$

This shows that in the ellipse $OA > OB$; so that the focal axis is the axis major.

Referring now to Figs. 214 and 215,

$$FA = FO + OA,$$

$$FA' = FO + OA' = FO - OA;$$

$$\therefore \quad FA \,.\, FA' = FO^2 - OA^2$$

$$= \mp OB^2.$$

If D be the point in which a directrix cuts the focal axis, the vertices A and A' of the conic will be harmonically conjugate with respect to F and the point D where the polar of F cuts AA' (Art. 264); therefore, since O bisects AA',

$$OA^2 = OF \,.\, OD.$$

The parabola has one vertex at infinity; consequently the other lies midway between the focus and the directrix (Fig. 218).

350. If a focus F of a conic $\mathbf{C}$ be taken as centre of homology, and a conic $\mathbf{C}'$ be constructed homological with $\mathbf{C}$ and

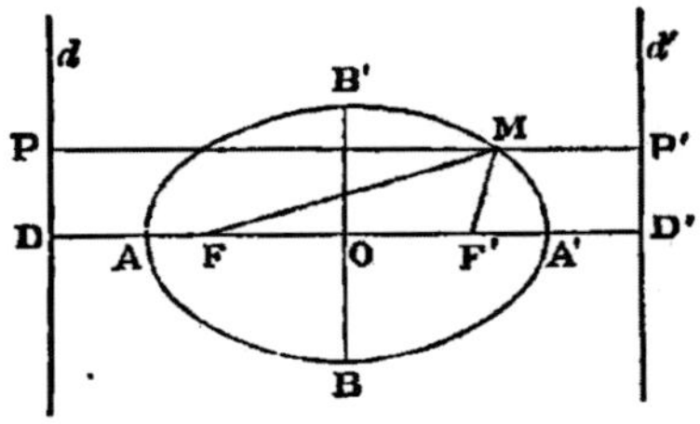

Fig. 214.

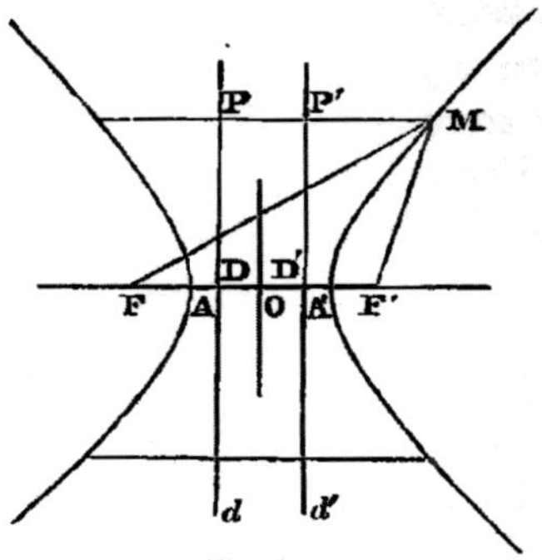

Fig. 215.

having its centre at F, it has been seen (Arts. 340, 341) that $\mathbf{C}'$ is a circle. But by what has been proved in Art. 77, if M and M' are a pair of corresponding points of $\mathbf{C}$ and $\mathbf{C}'$,

$$\frac{FM}{FM'} : MP = \text{constant},$$

or

$$\frac{FM}{MP} = FM' \times \text{constant},$$

where MP (Figs. 214, 215) is the distance of M from the vanishing line, that is from the polar of F, *i.e.* the corresponding directrix. Now FM' is constant, because $\mathbf{C}'$ is a circle; therefore

The distance of any point on a conic from a focus bears a constant ratio to its distance from the corresponding directrix.

Moreover, *this ratio is the same for the two foci.* For let O (Figs. 214, 215) be the centre of the conic, F, F' the foci, A, A' the vertices lying on the focal axis, D, D' the points in which this axis is cut by the directrices; then (Art. 294)

$$OA^2 = OA'^2 = OF \,.\, OD = OF' \,.\, OD'.$$

But $OF' = -OF$, so that $A'D' = -AD$ and $F'A' = -FA$,

and therefore $FA : AD = F'A' : A'D'$,

which shows that the ratio is the same for F and for F'.

In the case of the parabola the ratio in question is unity,

because (Art. 349) the vertex of a parabola is equally distant from the focus and the directrix. Therefore

The distance of any point on a parabola from the focus is equal to its distance from the directrix.

351. Conversely, *the locus of a point M which is such that its distance from a fixed point F bears a constant ratio ϵ to its distance from a fixed straight line d is a conic of which F is a focus and d the corresponding directrix**.

For let MP (Figs. 214, 215) be drawn perpendicular to d; then by hypothesis

$$\frac{FM}{MP} = \epsilon.$$

Let now the figure be constructed which is homological with the locus of M; F being taken as centre of homology, and d as vanishing line. If M' be the point corresponding to M, then (Art. 77)

$$\frac{FM}{FM'} : MP = \text{constant}.$$

These two equations show that FM' is constant; thus the locus of M' is a circle, centre F. The locus of M is therefore a conic (Art. 23) having one focus at F (Art. 341). And since the straight line at infinity is the polar of F with respect to the circle, the straight line d is the polar of F with respect to the conic; *i.e.* it is the directrix corresponding to F.

352. The length of a chord of a conic drawn through a focus perpendicular to the focal axis is called the *latus rectum* or the *parameter* of the conic.

Let MFM' (Fig. 216) be a chord of a conic drawn through a focus F, and let N be the point where it cuts the corresponding directrix. Let LFL' be the latus rectum drawn through F. Then since the directrix is the polar of the focus, N and F are harmonic conjugates with regard to M and M'. Therefore

$$\frac{2}{NF} = \frac{1}{NM} + \frac{1}{NM'},$$

and if perpendiculars MK, FD, $M'K'$ be let fall on the directrix,

$$\frac{2}{FD} = \frac{1}{M'K'} + \frac{1}{MK}.$$

* Pappus, *Math. Collect.*, lib. vii. prop. 238.

But by Art. 350

$$M'K' : FD : MK = M'F : FL : FM;$$

$$\therefore \quad \frac{2}{FL} = \frac{1}{M'F} + \frac{1}{FM},$$

that is to say:

In any conic, half the latus rectum is a harmonic mean between the segments of any focal chord.

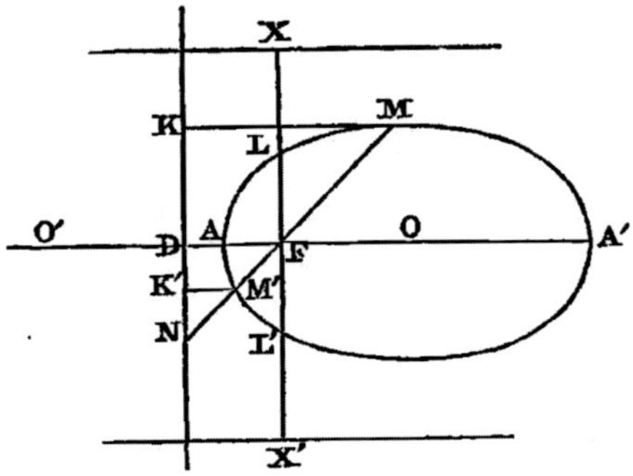

Fig. 216.

COROLLARY. If M, M' be taken at A', A respectively,

$$\frac{1}{FL} = \frac{1}{2}\left(\frac{1}{AF} + \frac{1}{FA'}\right)$$

$$= \frac{1}{2}\frac{AA'}{AF.FA'}$$

$$= \frac{OA}{\pm OB^2} \text{ (by Art. 349)},$$

so that $$FL = \pm\frac{OB^2}{OA},$$

which gives the length of the semi-latus rectum in terms of the semi-axes.

In the parabola $\frac{1}{FA'} = 0$, so that $FL = 2FA$.

353. THEOREM. *In the ellipse the sum, and in the hyperbola the difference, of the focal radii of any point on the curve is constant*.*

Let M be any point on a central conic (Figs. 214, 215) whose

* APOLLONIUS, *loc. cit.*, iii. 51, 52.

foci are F, F' and directrices d, d'; and let (M, d) &c. denote as usual the distance of M from d, &c. By Art. 351

$$\frac{FM}{(M, d)} = \frac{F'M}{(M, d')} = \epsilon;$$

$$\therefore \quad \frac{FM \pm F'M}{(M, d) \pm (M, d')} = \epsilon.$$

But (Fig. 214) in the ellipse $(M, d) + (M, d')$, and (Fig. 215) in the hyperbola $(M, d) - (M, d')$ is equal to the distance DD' between the two directrices; therefore

$$FM \pm F'M = \epsilon . DD',$$

which proves the proposition.

Conversely: *The locus of a point the sum (difference) of whose distances from two fixed points is constant is an ellipse (a hyperbola) of which the given points are the foci.*

354. If in the proposition of the last Article the point M be taken at a vertex A,

$$\begin{aligned} \epsilon . DD' &= FA \pm F'A \\ &= 2OA \\ &= AA', \end{aligned}$$

so that *the length of the focal axis is the constant value of the sum or difference of the focal radii.* It is seen also that the constant ϵ is equal to the ratio of the length of the focal axis to the distance between the directrices.

355. Since by Art. 294

$$OA^2 = OF . OD,$$

or

$$AA'^2 = FF' . DD',$$

$$\therefore \quad \epsilon = \frac{AA'}{DD'} = \frac{FF'}{AA'};$$

so that the constant ϵ is equal to the ratio of the distance between the foci to the length of the focal axis. Now in the ellipse $FF' < AA'$, in the hyperbola $FF' > AA'$, in the parabola $FF' = AA' = \infty$, in the circle $FF' = 0$. Therefore the conic is an ellipse, a hyperbola, a parabola, or a circle, according as $\epsilon < 1$, $\epsilon > 1$, $\epsilon = 1$, or $\epsilon = 0$. This constant ϵ is called the *eccentricity* of the conic.

356. THEOREM. *The locus of the feet of perpendiculars let fall from a focus upon the tangents to an ellipse or hyperbola is the circle described on the focal axis as diameter*.*

* APOLLONIUS, *loc. cit.*, iii. 49, 50.

Take the case of the ellipse (Fig. 217). If F, F' are the foci, and M is any point on the curve, join $F'M$ and produce it to G making MG equal to MF. Then $F'G$ will (Art. 354) be equal to AA' whatever be the position of M; thus the locus of G is a circle, centre F' and radius equal to AA'.

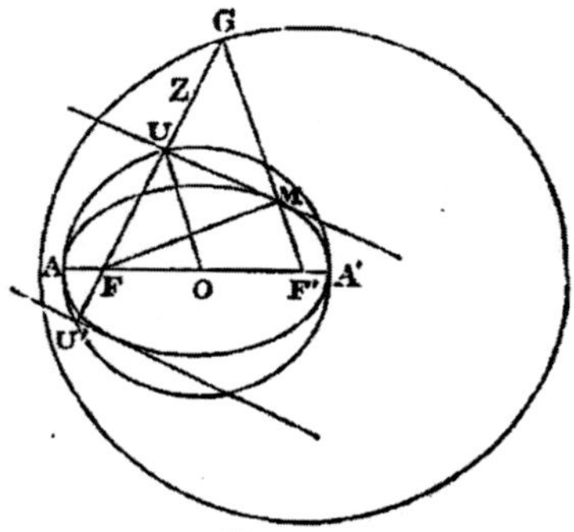

Fig. 217.

If FG be joined, it will cut the tangent at M perpendicularly, since this tangent (Art. 344) bisects the angle FMG; and the point U where the two lines intersect will be the middle point of FG because FMG is an isosceles triangle. Therefore OU is parallel to $F'G$ and equal to $\frac{1}{2}F'G$, that is, to OA; *i.e.* the locus of U is the circle on AA' as diameter.

A similar proof holds good for the hyperbola, except that from the greater of the two MF, MF' must be *cut off* a part MG equal to the less.

357. If FU, FU' (Fig. 217) are the perpendiculars let fall from a focus F on a pair of parallel tangents, U, F, U' will evidently be collinear. And since U and U' both lie on the circle described on AA' as diameter,

$$FU \, . \, FU' = FA \, . \, FA'$$
$$= \mp OB^2 \text{ (Art. 349)},$$

according as the conic is an ellipse or a hyperbola.

Thus *the product of the distances of a pair of parallel tangents from a focus is constant.*

Since the perpendicular let fall from the other focus F' on the tangent at M is equal to FU', it follows that

The product of the distances of any tangent to an ellipse (hyperbola) from the two foci is constant, and equal to the square of half the minor (conjugate) axis.

Conversely: *The envelope of a straight line which moves in such a way that the product of its distances from two fixed points is constant is a conic; an ellipse if the value of the constant is positive, a hyperbola if it is negative.*

358. Let F (Fig. 218) be the focus of a parabola, A the vertex, M any point on the curve, N the point of intersection of the tangents at M and A. If NF' be drawn to the infinitely distant focus F' (*i. e.* if NF' be drawn parallel to the axis), the angles ANF', FNM will be equal (Art. 346). But ANF' is a right angle; therefore FNM is a right angle also. Thus

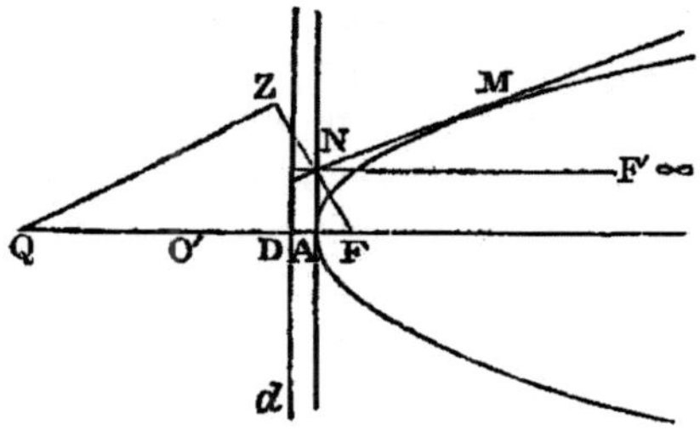

Fig. 218.

The foot of the perpendicular let fall from the focus of a parabola on any tangent lies on the tangent at the vertex.

COROLLARY. Since any point on the circumscribing circle of a triangle may be regarded (Art. 347) as the focus of a parabola inscribed in the triangle, it follows at once from the theorem just proved that *if from any point on the circumscribing circle of a triangle perpendiculars be let fall on the three sides, their feet will be collinear* *.

359. The theorem of Art. 356 may be put into the following form:

If a right angle move in its plane in such a way that its vertex describes a fixed circle, while one of its arms passes always through a fixed point, the envelope of its other arm will be a conic concentric with the given circle, and having one focus at the fixed point. The conic is an ellipse or a hyperbola according as the given point lies within or without the given circle †.

So too the corresponding theorem (Art. 358) for the parabola may be expressed in a similar form as follows:

If a right angle move in its plane in such a way that its vertex describes a fixed straight line, while one of its arms passes always through a fixed point, the other arm will envelope a parabola having the fixed point for focus and the fixed straight line for tangent at its vertex.

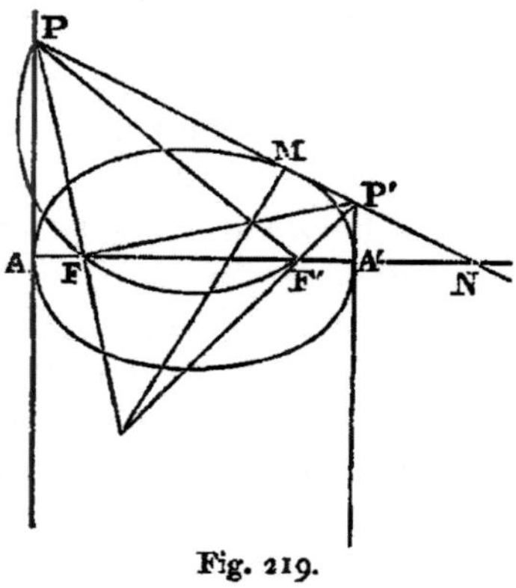

Fig. 219.

360. I. Let the tangents at the vertices of a central conic be cut in P, P' by the tangent at any point M (Fig. 219). The three tangents form a triangle circumscribed about the conic, two of the vertices of which are P and P', the third (at infinity) being the pole of the

* For other proofs of this see Art. 416.

† MACLAURIN, *Geometria Organica*, pars II[a]. prop. xi.

axis AA'. Therefore (Art. 274) the straight lines drawn from P and P' to any point on the axis will be conjugate to one another with respect to the conic. Thus, in particular, the straight lines joining P and P' to a focus will be conjugate to one another; but conjugate lines which meet in a focus are mutually perpendicular (Art. 343); consequently the circle on PP' as diameter will cut the axis AA' at the foci *.

II. Let the tangent PMP' cut the axis AA' at N; then N is the harmonic conjugate of M with respect to P, P' (Art. 194).

Consider now the complete quadrilateral formed by the lines $FP, F'P, FP', F'P'$. Two of its diagonals are FF' and PP'; the third diagonal must then cut FF' and PP' in points which are harmonically conjugate to N with regard to F, F' and P, P' respectively. It must therefore be the normal at M to the conic †.

361. Let TM, TN (Fig. 220) be a pair of tangents to a conic, M and N their points of contact, F a focus, d the corresponding directrix. If the chord MN cut d in P, this point is the pole of TF; therefore TFP is a right angle (Art. 343) ‡.

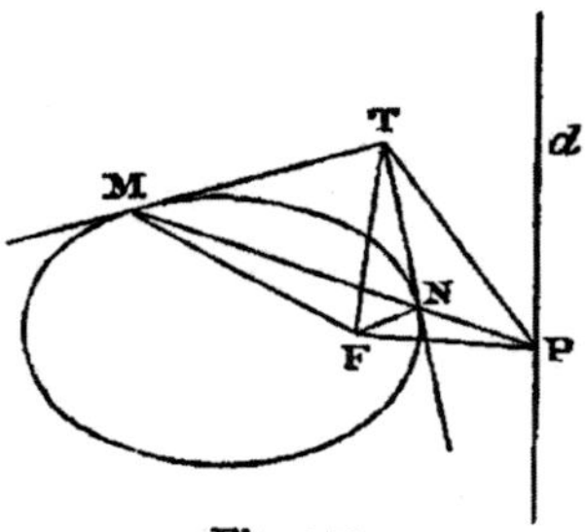

Fig. 220.

But MN is divided harmonically by FT and its pole P; thus $F(MNTP)$ is a harmonic pencil, and consequently FT, FP are the bisectors of the angle MFN. Accordingly:

One of the bisectors of the angle which a chord of a conic subtends at a focus passes through the pole of the chord. The other bisector meets the chord at its point of intersection with the directrix corresponding to the focus.

Or the same thing may be stated in a different manner, thus:

The straight line which joins a focus to the point of intersection of a pair of tangents to a conic makes equal (or supplementary) angles with the focal radii of their points of contact §.

* APOLLONIUS, *loc. cit.*, iii. 45; DESARGUES, *Œuvres*, i. pp. 209, 210.

† APOLLONIUS, *loc. cit.*, iii. 47.

‡ If the points M and N are taken indefinitely near to one another, this reduces to the theorem already proved in Art. 348.

§ DE LA HIRE, *loc. cit.*, lib. viii. prop. 24.

362. Let the tangents TM, TN be cut by any third tangent in M', N' respectively (Figs. 221, 222); let L be the point of contact of this third tangent. The following relations will hold among the angles of the figures:

$$N'FL = NFN' = \tfrac{1}{2} NFL,$$
$$LFM' = M'FM = \tfrac{1}{2} LFM,$$

whence by addition

$$N'FL + LFM' = \tfrac{1}{2}(NFL + LFM),$$

or $$N'FM' = \tfrac{1}{2} NFM = NFT = TFM.*$$

Let now the tangents TM, TN be fixed, while the tangent $M'N'$ is supposed to vary. By what has just been proved, the angle subtended at the focus by the part $M'N'$ of the

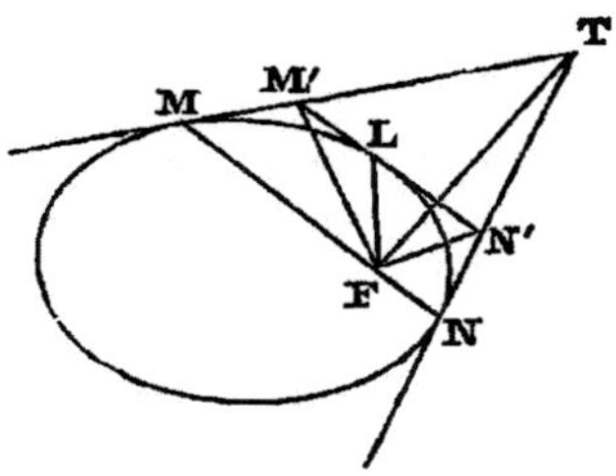

Fig. 221.

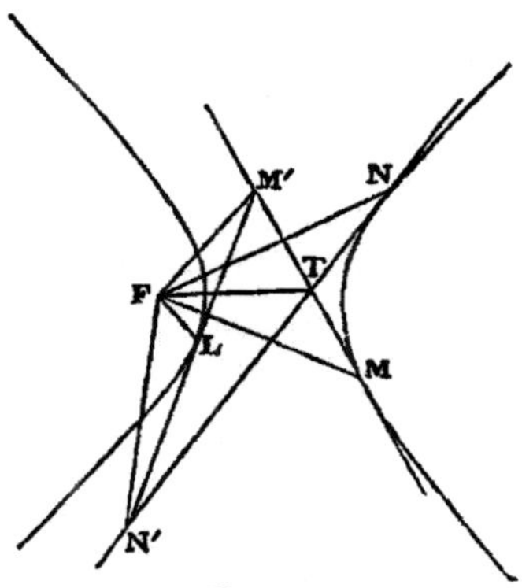

Fig. 222.

variable tangent intercepted between the two fixed ones is constant. As the variable tangent moves, the points M', N' describe two projective ranges (Art. 149), and the arms FM', FN' of the constant angle $M'FN'$ trace out two concentric and directly equal pencils (Art. 108). Accordingly:

* In this reasoning it is supposed that FM', FN', FT are all internal bisectors; *i.e.* that either the conic is an ellipse or a parabola, or that if it is a hyperbola, the three tangents all touch the same branch (Fig. 221). If on the contrary two of the tangents, for example TM and TN, touch one branch and the third $M'N'$ the other branch (Fig. 222), then FM' and FN' will be external bisectors. In that case,

$$N'FL = \tfrac{1}{2} NFL - \frac{\pi}{2}$$
$$LFM' = \tfrac{1}{2} LFM + \frac{\pi}{2}$$

(the angles being measured all in the same direction);

$\therefore$ $N'FM' = \tfrac{1}{2} NFM$, just as in the case above.

The ranges which a variable tangent to a conic determines on two fixed tangents are projected from either focus by means of two directly equal pencils.

This theorem clearly holds good for the cases of the parabola and its infinitely distant focus, and the circle and its centre. For the parabola it becomes the following:

Two fixed tangents to a parabola intercept on any variable tangent to the same a segment whose projection on a line perpendicular to the axis is of constant length.

The general theorem may also be put into the following form:

One vertex F of a variable triangle $M'FN'$ is fixed, and the angle $M'FN'$ is constant, while the other vertices M', N' move respectively on fixed straight lines TM, TN. The envelope of the side $M'N'$ is a conic of which F is a focus, and which touches the given lines TM, TN.

363. The problem, *Given the two foci F, F' of a conic and a tangent t, to construct the conic*, is determinate, and admits of a single solution, as follows.

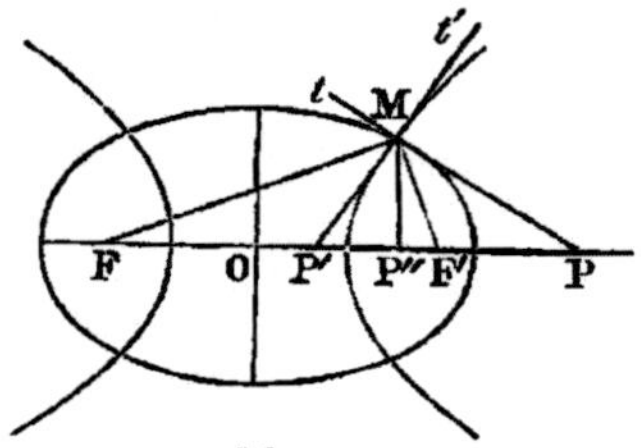

Fig. 223.

Join FF' (Figs. 223, 224) and let it cut t in P; take P' the harmonic conjugate of P with respect to F and F'. If a straight line $P'M$ be drawn perpendicular to t, it will be the normal corresponding to the tangent t (Art. 344), *i.e.* M will be the point of contact of t. Draw MP'' perpendicular to FF'; it will be the polar of P, and P, P'' will be conjugate points with respect to the

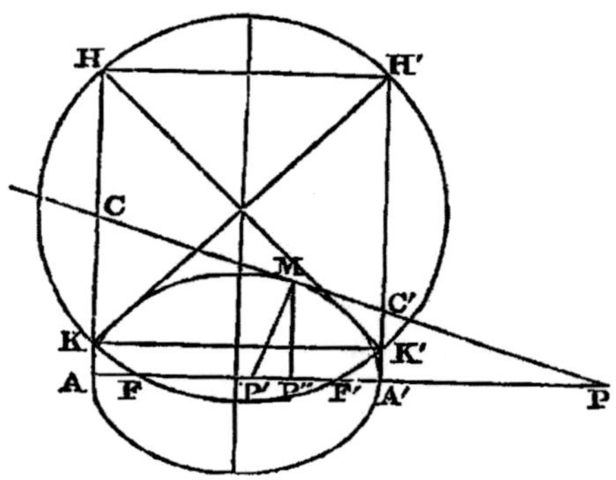

Fig. 224.

conic. If then FF' be bisected at O, and on FF' there be taken two points A, A' such that $OA^2 = OA'^2 = OP \, . \, OP''$, A and A' will be the vertices of the conic. The conic is therefore completely determined; for three points on it are known (M, A, A') and the tangents at these three points (t and the straight lines AC, $A'C'$ drawn through A, A' at right angles to AA').

An easy method of constructing the conic by tangents is to describe any circle through F and F', cutting AC, $A'C'$ in H and K, H' and K' respectively (Fig. 224). Then if the chords HK', $H'K$ be drawn which intersect crosswise in the centre of the circle (which lies on the non-focal axis), these will be tangents to the conic (Art. 360). Every circle through F and F' which cuts AC and $A'C'$ thus determines two tangents to the conic.

The conic is an ellipse or a hyperbola according as t cuts the segment FF' externally or internally.

The conic is a parabola when F' is at infinity (Fig. 225). In this case produce the axis PF to P' making FP' equal to PF, and draw $P'M$ perpendicular to t; then M will be the point of contact of the given tangent t. Draw MP'' perpendicular to the axis; then P and P'' will be conjugate points with respect to the parabola. And since the involution of conjugate points on the axis has one double point at infinity, the middle point A of PP'' will be the other double point, *i.e.* the vertex of the parabola. The parabola is therefore completely determined, since two points on it are known (M and A), and the tangents at these points (t and the straight line drawn through A at right angles to the axis).

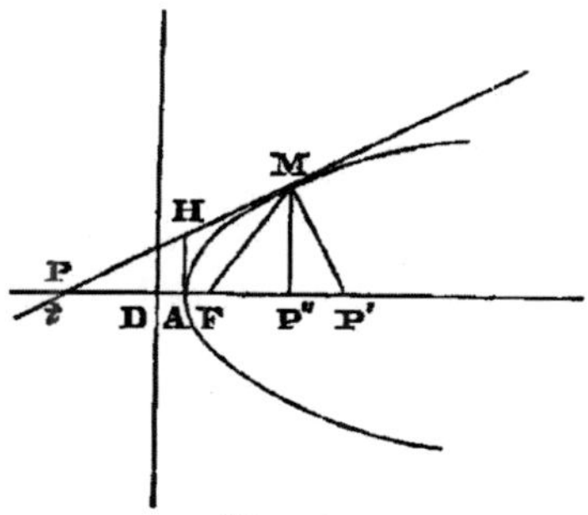

Fig. 225.

364. On the other hand, the problem, *To construct the conic which has its foci at two given points F, F' and which passes through a given point M*, which is also a determinate one, admits of two solutions. For if the locus of a point be sought the *sum* of whose distances from F and F' is equal to the constant value $FM + F'M$, an *ellipse* is arrived at; but if the locus of a point be sought the *difference* of whose distances from F and F' is equal to $FM \backsim FM'$, a *hyperbola* is found.

This may also be seen from the theorem of Art. 344, which shows that if the straight lines t, t' be drawn bisecting the angle FMF' (Fig. 223) each of these lines will be a tangent at M to a conic which satisfies the problem, the other line being the corresponding normal to this conic. The finite segment FF' is cut or not by the tangents according as the conic is a hyperbola or an ellipse. There will consequently be two conics which have F, F' for foci and which pass through M; a hyperbola having for tangent at M that bisector t' which cuts the segment FF', and for normal the other bisector t; and an ellipse having t for tangent at M and t' for normal.

These two conics, having the same foci, are concentric and have their axes parallel. They will cut one another in three other points besides M; and their four points of intersection will form a rectangle inscribed in the circle of centre O and radius OM; in other words, the three other points will be symmetrical to M with respect to the two axes and the centre. This is evident from the fact that a conic is symmetrical with respect to each of its axes.

365. Through every point M in the plane then pass two conics, an ellipse and a hyperbola, having their foci at F and F'. In other words, the system of *confocal conics* having their foci at F and F' is composed of an infinity of ellipses and an infinity of hyperbolas; and through every point in the plane pass one ellipse and one hyperbola, which cut one another there orthogonally and intersect in three other points.

Two conics of the system which are of the same kind (both ellipses or both hyperbolas) clearly do not intersect at all.

Two conics of the system however which are of opposite kinds (one an ellipse, the other a hyperbola) always intersect in four points, and cut one another orthogonally at each of them. This may be seen by observing that the vertices of the hyperbola are points lying within the segment FF', and therefore within the ellipse. On the other hand, there must be points on the hyperbola which lie outside the ellipse; for the latter is a closed curve which has all its points at a finite distance, while the former extends in two directions to infinity. The hyperbola therefore, in passing from the inside to the outside of the ellipse, must necessarily cut it.

No two conics of the system can have a common tangent; because (Art. 363) only one conic can be drawn to have its foci at given points and to touch a given straight line.

Any straight line in the plane will touch a determinate conic of the system, and will be normal, at the same point, to another conic of the system, belonging to the opposite kind. The first of these conics is a hyperbola or an ellipse according as the given straight line does or does not cut the finite segment FF'.

366. If first point F' lies at infinity, the problem of Art. 364 becomes the following: *Given the axis of a parabola, the focus F, and a point M on the curve, to construct the parabola.*

Just as in Art. 364, there are two solutions (Fig. 226). The tangents at M to the two parabolas which satisfy the problem are the bisectors of the angle made by MF with the diameter passing through M; therefore the parabolas cut orthogonally at M and consequently intersect at another point, symmetrical to M with respect to the axis. The parabolas cannot intersect in any other finite point, since they touch one another at infinity *.

The tangents to the two parabolas at M cut the axis in two points P, P' which lie at equal distances on opposite sides of F; and if P'' is the foot of the perpendicular let fall from M on the axis, the vertices A, A' of the parabolas are the middle points of the segments PP'', $P'P''$ respectively.

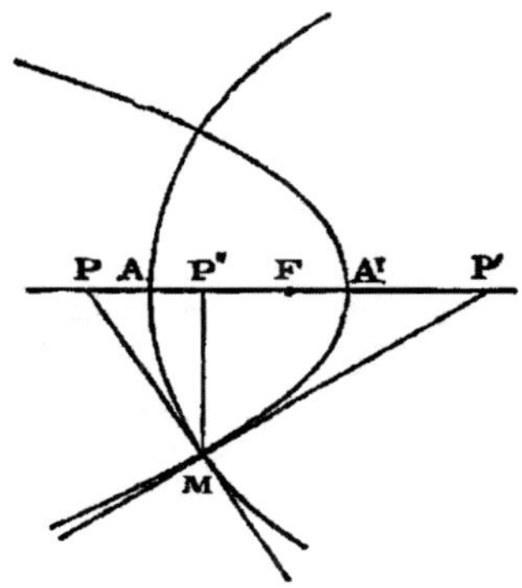

Fig. 226.

Suppose A and P'' to fall on the same side of F. Then since $P'P'' < P'P$, and $P'A'$ is the half of $P'P''$, and $P'F$ the half of $P'P$, therefore $P'A' < P'F$; *i.e.* A and A' fall on opposite sides of F. It follows that in the system composed of the infinity of parabolas which have a common axis and focus, two parabolas intersect (orthogonally and in two points) or do not intersect, according as their vertices lie on opposite sides or on the same side of the common focus.

Since F, A, A' are the middle points of PP', PP'', $P'P''$ respectively, we have the relations

$$FP + FP' = 0,$$
$$2FA = FP + FP'',$$
$$2FA' = FP' + FP'',$$

whence the following are easily deduced:

$$FP'' = FA + FA', \dagger$$
$$FP = FA - FA' = A'A,$$
$$FP' = FA' - FA = AA'.$$

These last relations enable us at once to find the points P, P', P'' when A and A' are known. The point M (and the symmetrical point in which the parabolas intersect again) can then be constructed by observing that FM is equal to FP or FP'.

367. It has been seen that a conic is determined when the two foci and a tangent are given. It can also be shown that *a conic is determined when one focus and three tangents are given*; this follows

* That is to say, if the figure be constructed which is homological with that formed by the two parabolas, it will consist of two conics touching one another at a point situated on the vanishing line of the new figure, and intersecting in two other points.

† Hence the middle point of AA' is also the middle point of FP''.

at once from the proposition at the end of Art. 362. For let LMN (Fig. 227) be the triangle formed by the three given tangents, and F the given focus. Then the conic is seen to be the envelope of the base $M'N'$ of a variable triangle $M'FN'$, which is such that the vertex F is fixed, the angle $M'FN'$ is always equal to the constant angle MFN, and the vertices M', N' move on the fixed straight lines LM, LN respectively.

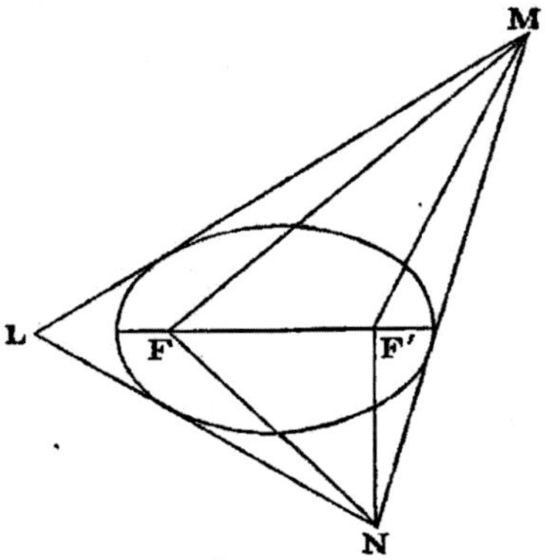

Fig. 227.

In order to determine the other focus F', we make use of the theorem of Art. 345. At the point M make the angle LMF' equal to FMN; and at the point N make the angle LNF' equal to FNM (all these angles being measured in the same direction); then the point of intersection of MF', NF' will be the second focus F'.

The investigation of the circumstances under which the conic is an ellipse, a hyperbola, or a parabola, is left as an exercise to the student. The following are the results:

(1) The conic is an ellipse if F lies within the triangle LMN; or if F lies without the circle circumscribing LMN and within one of the (infinite) spaces bounded by one of the sides of the triangle and the other two produced:

(2) a hyperbola if F lies inside the circle but outside the triangle; or if it lies within one of the (infinite) V-shaped spaces which have one of the angular points of the triangle LMN for vertex and are bounded by the sides meeting in that angular point, both produced backwards:

(3) a parabola if F lies on the circle circumscribing the triangle LMN, as we have seen already (Art. 347) *.

368. Let TM, TN (Fig. 228) be a pair of tangents to an ellipse or hyperbola which intersect at right angles. If perpendiculars FU, $F'U'$ and FV, $F'V'$ be let fall upon them respectively from the foci F and F', then evidently $TU = VF$ and $TU' = V'F'$. But by Art. 357 we have $VF \,.\, V'F' = \pm OB^2$; therefore $TU \,.\, TU' = \pm OB^2$. But since U and U' both lie on

* STEINER, *Développement d'une série de théorèmes relatifs aux sections coniques* (Annales de Gergonne, t. xix. 1828, p. 47); Collected Works, vol. i. p. 198.

the circle described upon the focal axis AA' as diameter (Art. 356), the rectangle $TU.TU'$ is the *power* of the point T with respect to this circle, and is equal to $OT^2 - OA^2$. Thus

$$OT^2 = OA^2 \pm OB^2 = \text{constant},$$

so that we have the following theorem *:

The locus of the point of intersection of two tangents to an ellipse or a hyperbola which cut at right angles is a concentric circle.

This circle is called the *director circle* of the conic †.

In the ellipse $OT^2 = OA^2 + OB^2$, so that the director circle circumscribes the rectangle formed by the tangents at the extremities of the major and minor axes. In the hyperbola $OT^2 = OA^2 - OB^2$, so that pairs of mutually perpendicular tangents exist only if $OA > OB$. If $OA = OB$, *i.e.* if the hyperbola is equilateral (Art. 395), the director circle reduces simply to the centre O; that is, the asymptotes are the only pair of tangents which cut at right angles. If $OA < OB$, the director circle has no real existence; the hyperbola has no pair of mutually perpendicular tangents.

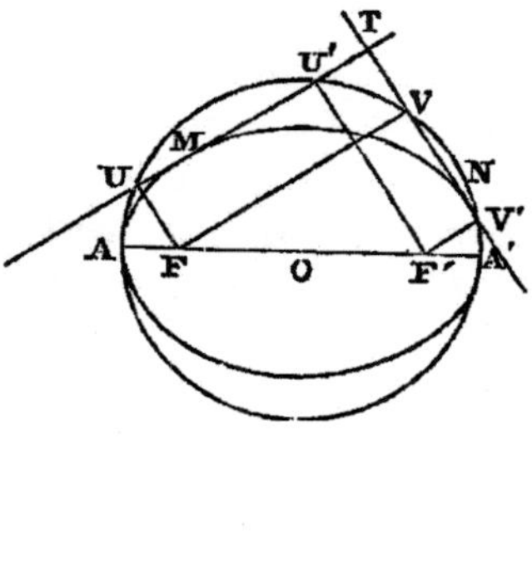

Fig. 228.

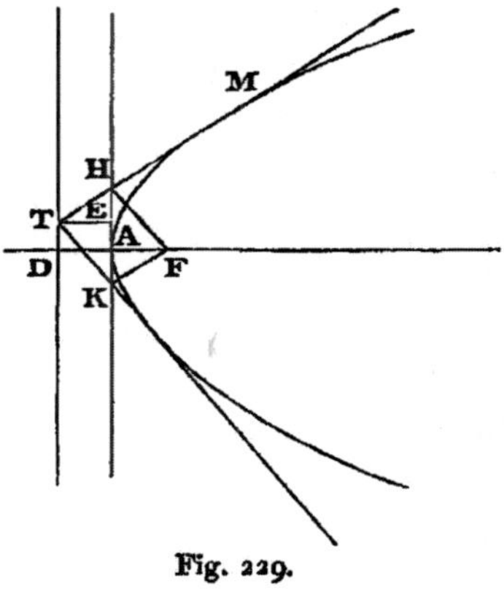

Fig. 229.

369. Consider now the case of the parabola (Fig. 229). Let F be the focus, A the vertex, TH and TK a pair of mutually perpendicular tangents. If these meet the tangent at the vertex in H and K respectively, the angles FHT, FKT will be right angles (Art. 358), so that the figure $THFK$ is a rectangle. Therefore $TH = KF$; and since the triangles TEH, FAK are evidently similar, $TE = AF$. The locus of the point T is

* De La Hire, *loc. cit.*, lib. viii. props. 27, 28.

† Gaskin, *The geometrical construction of a conic section*, . . . (Cambridge, 1852), chap. iii. prop. 10 et seqq.

therefore a straight line parallel to HK, and lying at the same distance from HK (on the opposite side) that F does. That is to say:

The locus of the point of intersection of two tangents to a parabola which cut at right angles is the directrix.*

Since the director circle of a conic is concentric with the latter, it must in the case of the parabola have an infinitely great radius. In other words, it must break up into the line at infinity and a finite straight line. And we have just seen that this finite straight line is the directrix.

370. The director circle possesses a property in relation to the self-conjugate triangles of the conic which we will now proceed to investigate. Let XYZ (Fig. 230) be a triangle which is self-conjugate with respect to a conic whose centre is O. Join OX and let it cut YZ in X' and the conic in A'. Draw OB' parallel to YZ; let it cut XY in L and the conic in B'; and draw ZL' parallel to OX to meet OB' in L'.

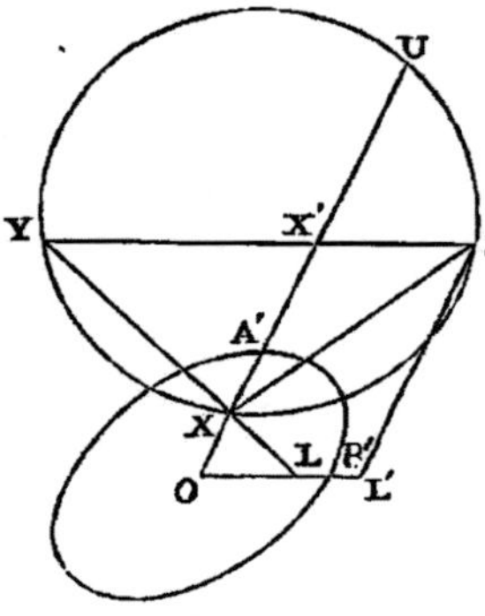

Fig. 230.

Then OA' and OB' are evidently conjugate semi-diameters; also X and X', L and L' are pairs of conjugate points with respect to the conic. Therefore

$$OX.OX' = \pm OA'^2, \text{ and } OL.OL' = \pm OB'^2,$$

where the positive or the negative signs are to be taken according as the semidiameters OA', OB' are real or ideal (Art. 294).

Thus for the ellipse

$$\begin{aligned} OX.OX' + OL.OL' &= OA'^2 + OB'^2 \\ &= OA^2 + OB^2, \end{aligned}$$

and for the hyperbola

$$\begin{aligned} OX.OX' + OL.OL' &= \pm(OA'^2 - OB'^2) \\ &= \pm(OA^2 - OB^2), \end{aligned}$$

so that in both cases (Art. 368)

$$OX.OX' + OL.OL' = OT^2, \quad . \quad . \quad . \quad . \quad . \quad (1)$$

where OT is the radius of the director circle.

* De la Hire, *loc. cit.*, lib. viii. prop. 26.

Now let a circle be described round the triangle XYZ, and let U be the point where it cuts OX again; then

$$X'Y.X'Z = X'X.X'U;$$

$$\therefore \quad X'U = \frac{X'Y}{X'X}\cdot X'Z$$

$$= \frac{OL}{OX}\cdot X'Z,$$

(from the similar triangles OLX, $X'YX$)

$$= \frac{OL}{OX}\cdot OL'.$$

Therefore equation (1) gives

$$OT^2 = OX.OX' + OX.X'U$$
$$= OX.OU,$$

that is to say: *The centre of a conic has with respect to the circumscribing circle of any triangle self-conjugate to the conic a constant power, which is equal to the square of the radius of the director circle.*

Or in other words:

The circle circumscribing any triangle which is self-conjugate with regard to a conic is cut orthogonally by the director circle.*

The following particular cases of this theorem are of interest:

I. *The centre of the circle circumscribing any triangle which is self-conjugate with respect to a parabola lies on the directrix.*

II. *The circle circumscribing any triangle which is self-conjugate with respect to an equilateral hyperbola passes through the centre of the conic.*

371. Consider a quadrilateral circumscribed about a conic. Since each of its diagonals is cut harmonically by the other two, the circle described on any one of the diagonals as diameter is cut orthogonally by the circle which circumscribes the diagonal triangle (Art. 69). But the diagonal triangle is self-conjugate with respect to the conic (Art. 260), and therefore its circumscribing circle cuts orthogonally the director circle (Art. 370). Consequently the director circle and the three circles described on the diagonals as diameters all cut orthogonally the circle circumscribing the diagonal triangle. Now by Newton's theorem (Art. 318) the centres of the four first-named circles are collinear; and circles whose centres are collinear and which all cut the same circle orthogonally have a common radical axis. Therefore:

The director circle of a conic, and the three circles described on

* GASKIN, *loc. cit.*, p. 33.

the diagonals of any circumscribed quadrilateral as diameters, are coaxial.

In the parabola the director circle reduces to the directrix and the straight line at infinity; in this case then the above theorem becomes the following:

If a quadrilateral is circumscribed about a parabola, the three circles described on the diagonals of the quadrilateral as diameters have the directrix for their common radical axis.

372. If in the theorem of Art. 371 the quadrilateral be supposed to be given, and the conic to vary, we arrive at the following theorem:

The director circles of all the conics inscribed in a given quadrilateral form a coaxial system, to which belong the three circles having as diameters the diagonals of the quadrilateral.

There is one circle of such a system which breaks up into two straight lines: that namely which degenerates into the radical axis and the straight line at infinity. Now the director circle breaks up into two straight lines—viz. the directrix and the line at infinity—in the case of a parabola (Art. 369). Therefore the common radical axis of the system of coaxial director circles is the directrix of the parabola which can be inscribed in the quadrilateral.

If the circles of the system do not intersect, there are two of them which degenerate into point-circles (the limiting points). Now the director circle degenerates into a point in the case of the equilateral hyperbola (Art. 368). Therefore when the circles do not cut one another, the two limiting points of the system are the centres of the two equilateral hyperbolas which can in this case be inscribed in the quadrilateral. If the circles do intersect, the system has no real limiting points; and in this case no equilateral hyperbola can be inscribed in the quadrilateral.

The circles which cut orthogonally the circles of a coaxial system form another coaxial system; if the first system has real limiting points, the second system has not, and *vice versa.* In order then *to inscribe an equilateral hyperbola in a given quadrilateral,* it is only necessary to describe circles on two of the diagonals of the quadrilateral as diameters, and then to draw two circles cutting the former two orthogonally. When the problem is possible, these two orthogonal circles will intersect; and their two points of intersection are the centres of the two equilateral hyperbolas which satisfy the conditions of the problem.

373. If five points are taken on a conic, five quadrangles may be formed by taking these points four and four together; and the diagonal triangles of these five quadrangles are each of them self-conjugate with respect to the conic. If the circumscribing circles of these five diagonal triangles be drawn, they will give, when taken together in pairs, ten radical axes. These ten radical axes will all meet in the same point, viz. the centre of the conic.

374. Consider again a quadrilateral circumscribing a conic; let P and P', Q and Q', R and R' be its three pairs of opposite vertices. If these be joined to any arbitrary point S, and if moreover from this point S the tangents t, t' are drawn to the conic, it is known by the theorem correlative to that of Desargues (Art. 183, right) that t and t', SP and SP', SQ and SQ', SR and SR' are in involution. Now let one of the sides of the quadrilateral (say $P'Q'R'$) be taken to be the straight line at infinity, so that the inscribed conic is a parabola; and let S be taken at the orthocentre (centre of perpendiculars) of the triangle PQR formed by the other three sides of the quadrilateral. Then each of the three pairs of rays SP and SP', SQ and SQ', SR and SR' cut orthogonally; therefore the same will be the case with the fourth pair t and t'. But tangents to a parabola which cut orthogonally intersect on the directrix (Art. 369); therefore:

The orthocentre of any triangle circumscribing a parabola lies on the directrix.

375. If in the theorem of the last Article the triangle be supposed to be fixed, and the parabola to vary, we obtain the theorem:

The directrices of all parabolas inscribed in a given triangle meet in the same point, viz. the orthocentre of the triangle.

Given a quadrilateral, one parabola (and only one) can always be inscribed in it. By taking the sides of the quadrilateral three and three together, four triangles are obtained; and the four orthocentres of these triangles must all lie on the directrix of the parabola. It follows that

Given four straight lines, the orthocentres of the four triangles formed by taking them three and three together are collinear.

376. Let C be any given conic, and let C′ be its polar reciprocal with respect to an auxiliary conic K. The particular case in which K is a circle whose centre coincides with a focus F of the conic C is of great interest; we shall now proceed to consider it.

If r, r' be any two straight lines which are conjugate with respect to C, and if R, R' be their poles with respect to K, it is known (Art. 323) that R, R' will be conjugate points with respect to C′. Consider now two such lines r, r' which pass through F; they will be at right angles since every pair of conjugate lines through a focus cut one another orthogonally.

They will therefore be perpendicular diameters of the circle K, and their poles R, R' with respect to K will be the points at infinity on r', r respectively. These points are conjugate with respect to C′, and the straight lines joining them to the centre of this conic are therefore a pair of conjugate diameters of C′; consequently two conjugate diameters of C′ are always mutually perpendicular. This proves that C′ is a circle; *i. e. the polar reciprocal of a conic, with respect to a circle which has its centre at one of the foci, is a circle.*

By taking the steps of the above reasoning in the opposite order, the converse proposition may be proved, viz.

The polar reciprocal of a circle with respect to an auxiliary circle is a conic having one focus at the centre of the auxiliary circle.

As in Art. 323, it is seen that the conic is an ellipse, a hyperbola, or a parabola, according as the centre of the auxiliary circle lies within, without, or upon the other circle.

377. If d be the directrix of the conic C corresponding to the focus F, and if its pole be taken with respect to the circle K, this point will evidently be the centre of the circle C′ (Art. 323).

The radius of the circle C′ may also easily be found. For in Fig. 216 let two points X, X' be taken in the latus rectum LFL' such that

$$FX.FL = FX'.FL' = k^2,$$

where k denotes the radius of the circle K; and let straight lines be drawn through X and X' perpendicular to XFX'. These straight lines are evidently parallel tangents of the circle C′, and the distance XX' between them is therefore equal in length to the diameter of C′. But

$$\tfrac{1}{2}X'X = FX = \frac{k^2}{FL},$$

so that the radius of the circle C′ is equal to $\dfrac{k^2}{FL}$.

The eccentricity e of the conic C may be expressed in a simple manner in terms of quantities depending upon the two circles K and C′. For if O' be the centre and ρ the radius of the latter circle, it has been seen that the directrix is the polar of O' with respect to **K**; therefore (Fig. 216)

$$FD \,.\, FO' = k^2.$$

But it has just been proved that

$$FL \,.\, \rho = k^2;$$

therefore (Art. 351), $\quad e = \dfrac{FL}{FD} = \dfrac{FO'}{\rho}.$

378. The proposition of Art. 376 may be proved in a different manner, so as to lead at once to the position and size of the circle **C**$'$.

Take any point M on the (central) conic **C** (Fig. 217); from the focus F draw FU perpendicular to the tangent at M, and on FU take a point Z such that $FZ \,.\, FU = k^2$, k being as before the radius of the circle **K**. Then the locus of Z is the polar reciprocal of **C** with respect to **K**.

Now it is known (Arts. 356, 357) that U lies on the circle on AA' as diameter, and that if UF cut this circle again at U'

$$FU \,.\, FU' = \mp OB^2.$$

Therefore $\qquad FZ : FU' = k^2 : \mp OB^2;$

which proves (Art. 23 [6]) that the locus of Z is a circle whose centre O' lies on FO and divides it so that $FO' : FO = k^2 : OB^2$, and whose radius ρ is equal to $k^2 \,.\, \dfrac{OA}{OB^2}$, that is, (Art. 352 Cor.) to $\dfrac{k^2}{FL}$. And again, since $OF \,.\, OD = OA^2$ and $FD = FO + OD$, (Figs. 214, 215),

$$\therefore\ FD \,.\, FO = OF^2 - OA^2 = \mp OB^2 = k^2 \frac{FO}{FO'}$$

by what has just been proved.

$$\therefore\ FO' \,.\, FD = k^2;$$

i.e. O' is the pole of the directrix d with respect to **K**.

In the particular case where $k = OB$, $\rho = OA$; that is to say:

The polar reciprocal of an ellipse (hyperbola) with respect to a circle having its centre at a focus and its radius equal to half the minor (conjugate) axis is the circle described on the major (transverse) axis as diameter.

379. In the case where **C** is a parabola, let M be any point on the curve (Fig. 218); let fall FN perpendicular to the tangent at M, and take on FN a point Z such that $FZ \,.\, FN = k^2$. Then, as before, the locus of Z will be the polar reciprocal of C with respect to K. Draw ZQ perpendicular to ZF to cut the axis of the parabola in Q.

Then a circle will evidently go round $QANZ$, so that

$$FA \,.\, FQ = FN \,.\, FZ = k^2;$$

therefore Q is a fixed point, and the locus of Z is the circle on QF as diameter. If O' be the centre, ρ the radius of this circle,

$$FO' = \rho = \tfrac{1}{2}\frac{k^2}{FA}.$$

In the particular case where k is equal to half the latus rectum, that is, to $2FA$, we have $\rho = k$; that is to say:

The polar reciprocal of a parabola with respect to a circle having its centre at the focus and its radius equal to half the latus rectum is a circle of the same radius, having its centre at the point of intersection of the axis with the directrix.

CHAPTER XXIV.

COROLLARIES AND CONSTRUCTIONS.

380. In the theorem of Art. 275 suppose the vertices B and C of the inscribed triangle ABC (Fig. 188) to be the points at infinity on a hyperbola; then S will be the centre of the curve, and the theorem will become the following:

If from any point A on a hyperbola parallels be drawn to the asymptotes, they will meet any given diameter in two points F and G which are conjugate to one another with regard to the curve. Or:

If through two points lying on a diameter of a hyperbola, which are conjugate to one another with regard to the curve, parallels be drawn to the asymptotes, they will intersect on the curve.

From this follows a method for *the construction of a hyperbola by points, having given the asymptotes and a point M on the curve.*

On the straight line SM, which joins M to the point of intersection S of the asymptotes, take two conjugate points of the involution determined by having S for centre and M for a double point. These points will be conjugate to one another with respect to the conic (Art. 263); if then parallels to the asymptotes be drawn through them, the two vertices of the parallelogram so formed will be points on the hyperbola which is to be constructed.

381. Let similarly the theorem of Art. 274 be applied to the hyperbola, taking the sides b and c of the circumscribed triangle abc to be the asymptotes; it will then become the following:

If through the points where the asymptotes are cut by any tangent to a hyperbola any two parallel straight lines be drawn, these will be conjugate to one another with respect to the conic. Or:

Two parallel straight lines which are conjugate to one another with respect to a hyperbola cut the asymptotes in points, the straight lines joining which are tangents to the curve.

From this we deduce a method for *the construction, by means of its tangents, of a hyperbola, having given the asymptotes b and c and one tangent m.*

Draw parallel to m two conjugate rays of the involution (Art. 129) determined by having m for a double ray and the parallel diameter for central ray. The two straight lines so drawn will be conjugate to one another with respect to the conic; if then the points where they cut the asymptotes be joined to one another, we shall have two tangents to the curve.

382. Let B and C be any two points on a parabola, and A the point where the curve is cut by the diameter which bisects the chord BC. Let F and G be two points lying on this diameter which are conjugate with respect to the parabola, *i.e.* two points equidistant from A (Art. 142); by the theorem of Art. 275, BF and CG, and likewise BG and CF, will meet on the curve.

This enables us *to construct by points a parabola which circumscribes a given triangle ABC and has the straight line joining A to the middle point of BC as a diameter.*

Or we may proceed according to the following method:

On BC take two points H and H' which shall be conjugate to one another with regard to the parabola, *i.e.* any two points dividing BC harmonically. Since H and H' are collinear with the pole of the diameter passing through A, therefore by the theorem of Art. 275, a point on the parabola will be found by constructing the point of intersection of AH with the diameter passing through H', and another will be found as the point where AH' meets the diameter passing through H.

383. In the theorem of Art. 274 suppose the tangent c to lie at infinity; then we see that

If a and b are two tangents to a parabola, and if from any point on the diameter passing through the point of contact of a there be drawn two straight lines, one passing through the point ab and the other parallel to b, these will be conjugate to one another with regard to the parabola.

This enables us *to construct by tangents a parabola, having given two tangents a and t, the point of contact A of one of them a, and the direction of the diameters.*

Draw the diameter through A and let it meet t in O; the second tangent t' from O will be the straight line which is harmonically conjugate to t with respect to the diameter OA (the polar of the point at infinity on a) and the parallel through O to a. If now two straight lines h and h' be drawn through O which shall be conjugate to one another with regard to the parabola, *i.e.* two straight lines which are harmonic conjugates with regard to t and t', the parallel to h' drawn from the point ha and the parallel to h drawn from the point $h'a$ will both be tangents to the required parabola.

384. If in the theorem of Art. 274 the straight line a be supposed to lie at infinity, and b and c to be two tangents to a parabola, we obtain the following:

The parallels drawn to two tangents to a parabola, from any point on their chord of contact, are conjugate lines with regard to the conic.

By another application of the same theorem we deduce a result already proved in Art. 178, viz. that

If, from a point on the chord of contact of a pair of tangents b and c to a parabola, two straight lines h and h′ be drawn parallel to b and c respectively, the straight line joining the points hc and h′b will be a tangent to the curve.*

From this may be deduced a *construction for the tangents to a parabola determined by two tangents and their points of contact.*

385. THEOREM. *If a conic cut the sides BC, CA, AB of a triangle ABC in the points D and D', E and E', F and F' respectively, then will*

$$\frac{BD.BD'}{CD.CD'}\cdot\frac{CE.CE'}{AE.AE'}\cdot\frac{AF.AF'}{BF.BF'}=1. \quad . \quad . \quad . \quad . \quad (1)$$

This celebrated theorem is due to CARNOT †.

Consider the sides of the triangle ABC (Fig. 231) as

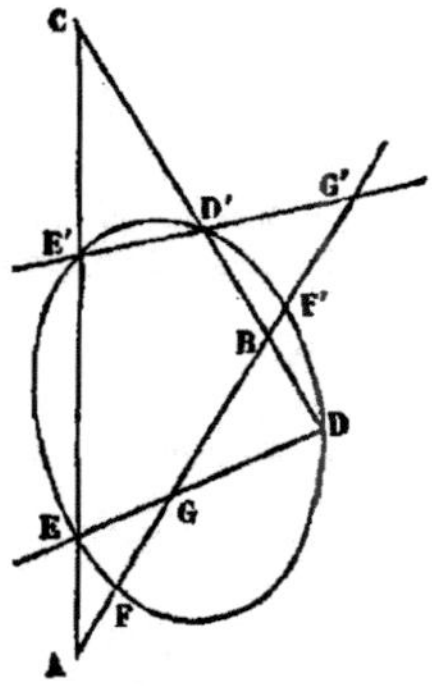

Fig. 231.

cut by the transversals DE and $D'E'$ in the points D and D', E and E', G and G'; by the theorem of Menelaus (Art. 139)

$$\frac{BD}{CD}\cdot\frac{CE}{AE}\cdot\frac{AG}{BG}=1, \quad . \quad . \quad . \quad . \quad . \quad . \quad (2)$$

and

$$\frac{BD'}{CD'}\cdot\frac{CE'}{AE'}\cdot\frac{AG'}{BG'}=1. \quad . \quad . \quad . \quad . \quad . \quad . \quad (3)$$

Again, $DEE'D'$ is a quadrangle inscribed in the conic, and by Desargues' theorem (Art. 183) the transversal AB meets the opposite sides and the conic in three pairs of points in involution; therefore (Art. 130) the anharmonic ratios $(ABFG)$ and $(BAF'G')$ are equal; thus (Art. 45) $(ABFG)=(ABG'F')$, or $(ABFG):(ABG'F')=1$, which gives

$$\frac{AF.AF'}{BF.BF'}:\frac{AG.AG'}{BG.BG'}=1. \quad . \quad . \quad . \quad . \quad . \quad . \quad (4)$$

* DE LA HIRE, *loc. cit.*, lib. iii. prop. 21. † *Géométrie de position*, p. 437.

Multiplying together (2), (3), and (4), we obtain the relation stated in the enunciation *.

386. Conversely, *if on the sides BC, CA, AB respectively of a triangle ABC there be taken three pairs of points D and D', E and E', F and F' such that the segments determined by them and the vertices of the triangle satisfy the relation* (1) *of Art.* 385, *these six points lie on a conic.*

For let the conic be drawn which passes through the five points D, D', E, E', F, and let F'' be the point where it cuts AB again. By Carnot's theorem a relation holds which differs only from (1) in that it has F'' in the place of F'. This relation, combined with (1), gives

$$AF' : BF' = AF'' : BF'',$$

whence $$(ABF'F'') = 1;$$

and therefore (Art. 72, VII) F'' coincides with F'.

* CARNOT'S theorem, being evidently true for the circle (since in this case $BD \cdot BD' = CD \cdot CD'$, &c.), may be proved without making use of involution properties as follows:

Let I, J, K be the points at infinity on BC, CA, AB respectively, and suppose Fig. 231 to have been derived by projecting from any vertex on any plane a triangle $A_1B_1C_1$ whose sides are cut by a circle in D_1 and D_1', E_1 and E_1', F_1 and F_1' respectively. Let I_1, J_1, K_1 be the points on the sides B_1C_1, C_1A_1, A_1B_1 which project into I, J, K respectively; they will of course be collinear. Then

$$\frac{BD}{CD} = (BCDI) \quad \text{(Art. 64)}$$
$$= (B_1C_1D_1I_1) \text{ (Art. 63)}$$
$$= \frac{B_1D_1}{C_1D_1} : \frac{B_1I_1}{C_1I_1}.$$

So $$\frac{BD'}{CD'} = \frac{B_1D_1'}{C_1D_1'} : \frac{B_1I_1}{C_1I_1};$$

$$\therefore \frac{BD \cdot BD'}{CD \cdot CD'} = \frac{B_1D_1 \cdot B_1D_1'}{C_1D_1 \cdot C_1D_1'} : \frac{B_1I_1^2}{C_1I_1^2}$$
$$= \frac{C_1I_1^2}{B_1I_1^2}. \quad \text{(Euc. iii. 35, 36.)}$$

Similarly, $$\frac{CE \cdot CE'}{AE \cdot AE'} = \frac{A_1J_1^2}{C_1J_1^2},$$

and $$\frac{AF \cdot AF'}{BF \cdot BF'} = \frac{B_1K_1^2}{A_1K_1^2}.$$

Multiplying these three equations together, and remembering that by the theorem of Menelaus the product on the right-hand side is equal to unity, we have the result required.

Carnot's theorem is true not only for a triangle but for a polygon of any number of sides; the proof just given can clearly be extended so as to show this, the theorem of Menelaus being capable of extension to the case of a polygon.

Menelaus' theorem is included in that of Carnot. It is what the latter reduces to when the conic degenerates into two straight lines of which one lies at infinity.

387. If the point A pass off to infinity (Fig. 232) the ratios $AF : AE$ and $AF' : AE'$ become in the limit each equal to unity, and the equation (1) of Art. 385 accordingly reduces to

$$\frac{BD \, . \, BD'}{CD \, . \, CD'} \cdot \frac{CE \, . \, CE'}{BF \, . \, BF'} = 1. \quad . \quad . \quad . \quad . \quad . \quad (5)$$

Draw parallel to BC a straight line to cut CEE' in Q and the conic in P and P'; the preceding equation, applied to the triangle whose vertices are C, Q, and the point at infinity where PP' and BC meet, gives

$$\frac{QE \, . \, QE'}{CE \, . \, CE'} \cdot \frac{CD \, . \, CD'}{QP \, . \, QP'} = 1.$$

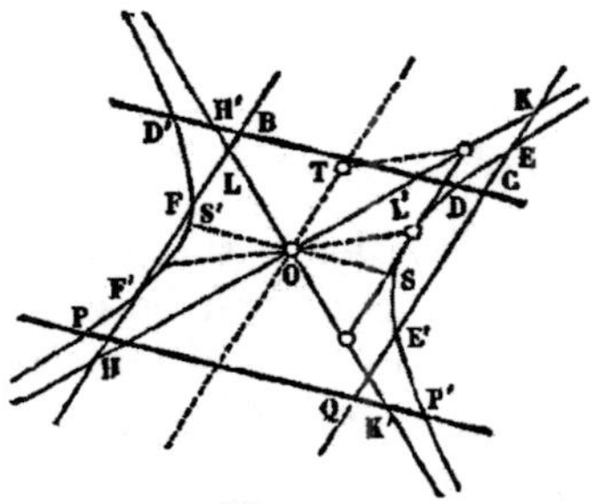

Fig. 232.

Multiplying together these last two equations, we obtain

$$\frac{BD \, . \, BD'}{BF \, . \, BF'} = \frac{QP \, . \, QP'}{QE \, . \, QE'};$$

that is to say:

If through any point Q there be drawn in given directions two transversals to cut a conic in P, P' and E, E' respectively, then the rectangles $QP \, . \, QP'$ and $QE \, . \, QE'$ are to one another in a constant ratio *†.

* APOLLONIUS, *loc. cit.*, lib. iii. 16–23; DESARGUES, *loc. cit.*, p. 202; DE LA HIRE, *loc. cit.*, bk. v. props. 10, 12.

† From this follows at once the result already proved in a different manner in Art. 316, viz. that *if a conic is cut by a circle, the chords of intersection make equal angles with the axes.*

For let P, P', E, E' be the points of intersection of a circle with the conic; then (Euc. iii. 35) $QP \, . \, QP' = QE \, . \, QE'$. But if MCM', NCN' be the diameters of the conic parallel respectively to QPP' and QEE', we have, by the theorem in the text,

$$QP \, . \, QP' : QE \, . \, QE' = CM \, . \, CM' : CN \, . \, CN'$$
$$= CM^2 : CN^2.$$

Therefore $CM = CN$, and consequently CM and CN (and therefore also QPP' and QEE') make equal angles with the axes.

388. Suppose in equation (5) of Art. 387 that the conic is a hyperbola and that in place of BC is taken an asymptote HK of the curve; then the ratio $HD.HD' : KD.KD'$ becomes equal to unity, and therefore

$$HF.HF' = KE.KE',$$

that is to say:

If through any point H (or H') lying on an asymptote there be drawn, parallel to a given straight line, a transversal to cut a hyperbola in two points F and F' (D and D'), then the rectangle $HF.HF'$ ($H'D.H'D'$) contained by the intercepts will be constant.

If the diameter parallel to the given direction $H'D$ meets the curve, then if S and S' are the points where it meets it, and if O is the centre,

$$H'D.H'D' = OS.OS' = -OS^2.$$

If the diameter OT parallel to the given direction HF does not meet the curve, a tangent can be drawn which shall be parallel to it. The square on the portion of this tangent intercepted between its point of contact and the asymptote will be equal to the rectangle $HF.HF'$ by the theorem now under consideration; but this portion is (Art. 303) equal to the parallel semidiameter OT; therefore $HF.HF' = OT^2$, or:

If a transversal cut a hyperbola in F and F' (in D and D') and an asymptote in H (in H'), the rectangle $HF.HF'$ ($H'D.H'D'$) is equal to $\pm$ the square on the parallel semidiameter OT (OS); the positive or negative sign being taken according as the curve has or has not tangents parallel to the transversal.

389. If the transversal cuts the other asymptote in L (in L'), then by Art. 193

$$HF' = FL \text{ or } H'D' = DL',$$

and consequently

$$FH.FL = -OT^2 \text{ or } DH'.DL' = OS^2;$$

therefore:

If a transversal drawn from any point F (D) on a hyperbola cut the asymptotes in H and L (in H' and L'), the rectangle $FH.FL$ ($DH'.DL'$) is equal to $\mp$ the square on the parallel semidiameter; the negative or positive sign being taken according as the curve has or has not tangents parallel to the transversal.

390. From the proposition of the last Article may be deduced a *construction for the axes of a hyperbola, having given a pair of conjugate semidiameters OF and OT in magnitude and direction* (Fig. 233).

We first construct the asymptotes. Of the two given semidiameters, let OF be the one which cuts the curve. Draw through F a parallel to OT; this will be the tangent at F. Take on this parallel FP and FQ each equal to OT; then OP and OQ will be the asymptotes (Art. 304). In order now to obtain the directions of the axes, we have only to find the bisectors of the angle included by the asymptotes, or, in other words, the two perpendicular rays OX, OY which are conjugate to one another in the involution of which OP and OQ are the double rays (Arts. 296, 297).

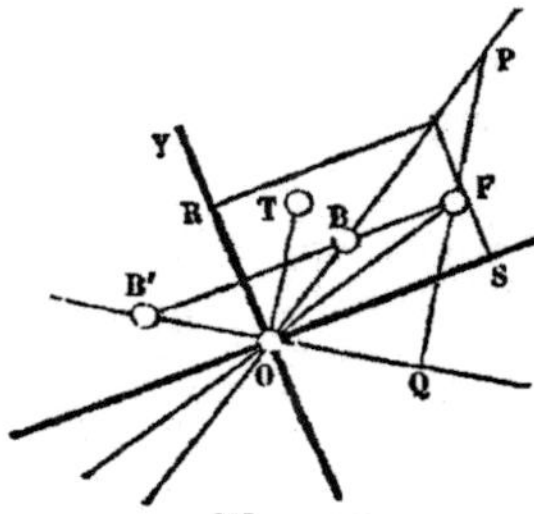

Fig. 233.

To determine the lengths of the axes, draw through F a parallel to OX, and let it cut the asymptotes in B and B'; and on OX take OS the mean proportional between FB and FB'. Then will OS be the length of the semiaxis in the direction OX; and OX will or will not cut the curve according as the segments FB, FB' have or have not the same direction. Again, construct the parallelogram of which OS is one side, which has an adjacent side along OY, and one diagonal along an asymptote; its side OR will be the length of the semiaxis in the direction OY (Art. 304).

391. In the plane of a triangle ABC take any two points O and O'; if OA, OB, OC meet the respectively opposite sides BC, CA, AB of the triangle in D, E, F, Ceva's theorem (Art. 137) gives

$$\frac{BD}{CD}\cdot\frac{CE}{AE}\cdot\frac{AF}{BF}=-1.$$

Similarly, if $O'A$, $O'B$, $O'C$ meet the respectively opposite sides in D', E', F', then

$$\frac{BD'}{CD'}\cdot\frac{CE'}{AE'}\cdot\frac{AF'}{BF'}=-1.$$

If these equations be multiplied together, equation (1) of Art. 385 is obtained; therefore:

If from any two points the vertices of a triangle are projected upon the respectively opposite sides, the six points so obtained lie on a conic.

For example, the middle points of the sides of a triangle and the feet of the perpendiculars from the vertices on the opposite sides are six points on a conic *.

* This conic is a circle (the nine-point circle). See STEINER, *Annales de Mathématiques* (Montpellier, 1828), vol. xix. p. 42; or his Collected Works, vol. i. p. 195.

392. Problem. *To construct a conic which shall pass through three given points A, B, C, and with regard to which the pairs of corresponding points of an involution lying on a given straight line u shall be conjugate points.*

Let AB and AC (Fig. 234) be joined, and let them meet u in D and E. Let the points corresponding in the involution to D and E respectively be D' and E'; let D'' be the harmonic conjugate of D

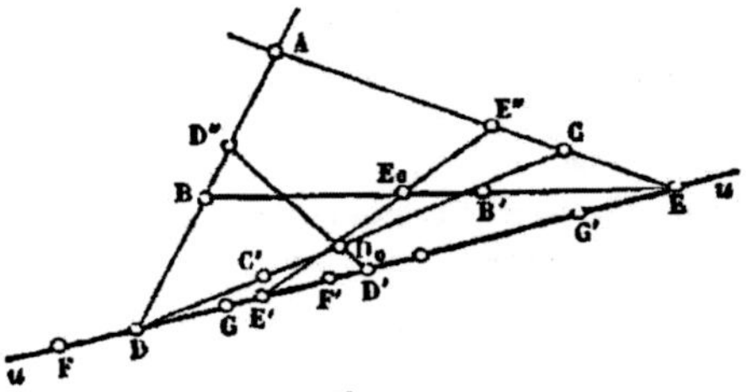

Fig. 234.

with respect to A and B, and let E'' be the harmonic conjugate of E with respect to A and C. Thus D will be conjugate (with respect to the required conic) both to D' and to D'', and therefore $D'D''$ will be the polar of D. So too $E'E''$ will be the polar of E.

Join BE, CD, and let them cut $E'E''$ and $D'D''$ in E_0 and D_0 respectively; then E_0 will be conjugate to E and D_0 to D. If then two points B', C' be found such that the ranges $BB'EE_0$ and $CC'DD_0$ are harmonic, they will both belong to the required conic.

In the figure, F and F', G and G' are the pairs of points which determine on u the involution of conjugate points.

393. Problem. *To construct a conic which shall pass through four given points Q, R, S, T and shall divide harmonically a given segment MN* (Fig. 235).

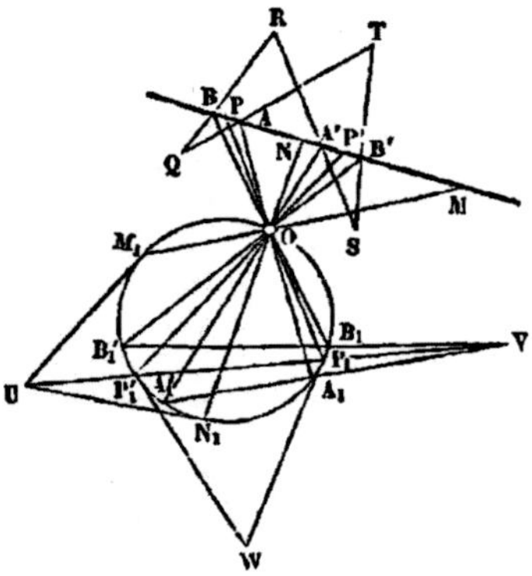

Fig. 235.

Let the pairs of opposite sides of the quadrangle $QRST$ meet the straight line MN in A and A', B and B'. If the required conic cuts MN, the two points of intersection will be a pair of the involution determined by A and A', B and B' (Art. 183). If then the involution of which M and N are the double points and the involution determined by the pairs of points A and A', B and B' have a pair P and P' in common, the required conic will pass through each of the points P and P' (Arts. 125, 208).

In order to construct these points, describe any circle (Art. 208) and from any point O on it project the points A, A', B, B', M, N upon the circumference, and let $A_1, A_1', B_1, B_1', M_1, N_1$ be their respective projections. If the chords A_1A_1' and B_1B_1' meet in V, and the tangents at M_1 and N_1 meet in U, all straight lines passing through U determine on the circumference, and consequently (by projection from O) on the straight line MN, pairs of conjugate points of the first involution, and the same is true, with regard to the second involution, of straight lines passing through V. If the straight line UV meets the circle in two points P_1 and P_1', let these be joined to O; the joining lines will cut MN in the required points P and P'.

Let W be the pole of UV with respect to the circle. Every straight line passing through W and cutting the circle determines on it two points which are harmonically conjugate with regard to P_1 and P_1'; and these points, when projected from O on MN, will give two points which are harmonically conjugate with regard to P and P', and which are therefore conjugate to one another with respect to the required conic. If then UV does not cut the circle, so that the points P and P' cannot be constructed, draw through W two straight lines cutting the circle, and project the points of intersection from the centre O upon the straight line MN; this will give two pairs of points which will determine the involution on MN of conjugate points with respect to the conic. The problem therefore reduces to that treated of in the preceding Article.

394. Problem. *To construct a conic which shall pass through four given points Q, R, S, T, and through two conjugate points (which are not given) of a known involution lying on a straight line u.*

This problem is similar to the preceding one; since it amounts to constructing the pair of conjugate points common to the given involution and to that determined on u by the pairs of opposite sides of the quadrangle $QRST$ (Art. 183).

Such a common pair will always exist when the given involution has no double points; and the two points composing it will both lie on the required conic. If the given involution has two double points M and N, the present problem becomes identical with that of Art. 393.

The problem clearly admits of only one solution, and the same is the case with regard to those of the two preceding Articles.

395. Consider a hyperbola whose asymptotes are perpendicular to one another, and to which, on this account, is given the name of *rectangular* hyperbola (Fig. 236). Since the asymptotes are harmonically conjugate with regard to any pair of conjugate diameters (Art. 296), they will in this case be the bisectors of the angle included between any such pair (Art. 60). But the parallelogram described on two conjugate semidiameters as adjacent sides has its diagonals parallel to the asymptotes (Art. 304); in this case therefore every such parallelogram is a rhombus; that is, *every diameter is equal in length to its conjugate.* On account of this property the rectangular hyperbola is also called *equilateral**.

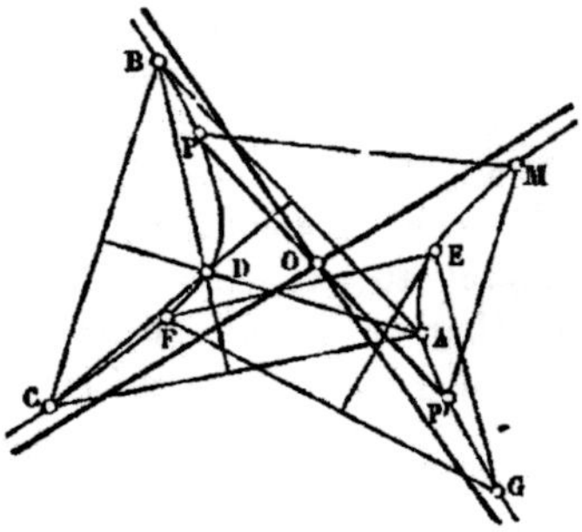

Fig. 236.

I. Since the chords joining the extremities P and P' of any diameter to any point M on the curve are parallel to a pair of conjugate diameters (Art. 287), the angles made by PM and $P'M$ with either asymptote are equal in magnitude and of opposite sign. If the points P and P' remain fixed, while M moves along the curve, the rays PM and $P'M$ trace out two pencils which are oppositely equal to one another (Art. 106).

II. Conversely, *the locus of the points of intersection of pairs of corresponding rays of two oppositely equal pencils is an equilateral hyperbola.*

For, in the first place, the locus is a conic, since the two pencils are projective (Art. 150). Further, the two pencils have each a pair of rays which include a right angle, and which are parallel respectively to the corresponding rays of the other pencil (Art. 106); the conic has thus two points at infinity lying in directions at right angles to one another, and is therefore an equilateral hyperbola. It will be seen moreover that the centres P and P' of the two pencils are the extremities of a diameter. For the tangent p at P is the ray corresponding to $P'P$ regarded as a ray p' of the second pencil, and the tangent q' at P' is the ray corresponding to PP' regarded as a ray q of the first pencil (Art. 150); but the angles pq and $p'q'$ must be equal and opposite; therefore, since p' and q coincide, p and q' must be parallel to one another.

III. The angular points of a triangle ABC and its orthocentre (centre of perpendiculars) D are the vertices of a

* APOLLONIUS, *loc. cit.*, vii. 21; DE LA HIRE, *loc. cit.*, book v. prop. 13.

complete quadrangle in which each side is perpendicular to the one opposite to it, and whose six sides determine on the straight line at infinity three pairs of points subtending each a right angle at any arbitrary point S. The three pairs of rays formed by joining these points to S belong therefore to an involution in which every ray is perpendicular to its conjugate (Arts. 131 left, 124, 207).

But this involution of rays projects from S the involution of points which, in accordance with Desargues' theorem, is determined on the straight line at infinity by the pairs of opposite sides of the quadrangle and by the conics (hyperbolas *) circumscribed about it. The pairs of conjugate rays therefore of the first involution give the directions of the asymptotes of these conics; thus:

If a conic pass through the angular points of a triangle and through the orthocentre, it must be an equilateral hyperbola †.

IV. Conversely, if an equilateral hyperbola be drawn to pass through the vertices A, B, C of a triangle, it will pass also through the orthocentre D. For imagine another hyperbola which is determined (Art. 162, I) by the four points A, B, C, D and by one of the points at infinity on the given hyperbola. This new hyperbola will be an equilateral one by the foregoing theorem, and will consequently pass through the second point at infinity on the given curve; and since the two hyperbolas thus have five points in common (A, B, C, and two at infinity) they must be identical; which proves the proposition. Therefore:

If a triangle be inscribed in an equilateral hyperbola, its orthocentre is a point on the curve.

V. If the point D approach indefinitely near to A, *i.e.* if BAC becomes a right angle, we have the following proposition:

If EFG (Fig. 236) *is a triangle, right-angled at E, which is*

* No ellipse or parabola can be circumscribed about the quadrangle here considered (Art. 219).

† This may be deduced directly from Pascal's theorem. For let a conic be drawn through A, B, C, D, and let I_1 and I_2 be the points where it meets the line at infinity. Since $ABCDI_1I_2$ is a hexagon inscribed in a conic, the intersections of AB and DI_1, of BC and I_1I_2, and of CD and I_2A, are three collinear points. Therefore the straight line joining the point in which DI_1 meets AB to that in which AI_2 meets CD must be parallel to BC. Thus AI_2 must be at right angles to DI_1, and as these lines are parallel to the asymptotes of the conic the latter is a rectangular hyperbola.

inscribed in an equilateral hyperbola, the tangent at E is perpendicular to the hypotenuse FG.

VI. Through four given points Q, R, S, T can be drawn only one equilateral hyperbola (Art. 394). The orthocentre of each of the triangles QRS, RST, STQ, QRT lies on the curve *.

VII. *Given four tangents to an equilateral hyperbola, to construct the curve.*

Since the diagonal triangle of the quadrilateral formed by the four tangents is self-conjugate with respect to the hyperbola, the centre of the latter will lie on the circle circumscribing this triangle (Art. 370, II). But the centre of the hyperbola lies also on the straight line which joins the middle points of the diagonals of the quadrilateral (Art. 318, II). Either of the points of intersection of this straight line with the circle will therefore give the centre of an equilateral hyperbola satisfying the problem; there are therefore two solutions. For another method of solution see Art. 372.

VIII. *The polar reciprocal of any conic with respect to a circle* K *having its centre on the director circle is an equilateral hyperbola.*

For since the tangents to the conic from the centre O of the circle K are mutually perpendicular, the conic which is the polar reciprocal of the given one must cut the straight line at infinity in two points subtending a right angle at O. That is to say, it must be an equilateral hyperbola.

396. Suppose given a conic, a point S, and its polar s; and let a straight line passing through S cut the conic in A and A'. Let the figure be constructed which is homological with the given conic, S being taken as centre of homology, s as axis of homology, and A, A' as a pair of corresponding points. Then every other point B' which corresponds to a point B on the conic will lie on the conic itself. For if AB meets the axis s in P, then B', the point of intersection of SB and $A'P$, is likewise a point on the conic (Art. 250). The curve homological with the given conic will therefore be the conic itself. Any two corresponding points (or straight lines) are separated harmonically by S and s; this is, in fact, the case of *harmonic homology* (Arts. 76, 298).

To the straight line at infinity will therefore correspond the

* These theorems are due to BRIANCHON and PONCELET; they were enunciated in a memoir published in vol. xi. of the *Annales de Mathématiques* (Montpellier, 1821), and were given again in vol. ii. (p. 504) of PONCELET'S *Applications d'Analyse et de Géométrie* (Paris, 1864).

straight line j which is parallel to s and which lies midway between S and s; and the points in which j meets the conic will correspond to the points at infinity on the same conic.

From this may be derived a very simple *method of determining whether a given arc of a conic*, however small, *belongs to an ellipse, a parabola, or a hyperbola.*

Draw a chord s joining any two points in the arc; construct its pole S, and draw a straight line j parallel to s and equidistant from S and s. If j does not cut the arc, the latter is part of an ellipse (Fig. 237 *a*). If j touches the arc at a point J, the arc belongs to a parabola of which SJ is a diameter (Fig. 237 *b*). If, finally, j cuts the arc in two points J_1, J_2 (Fig. 237 *c*), the arc will be part of a hyperbola whose asymptotes are parallel to SJ_1 and SJ_2*.

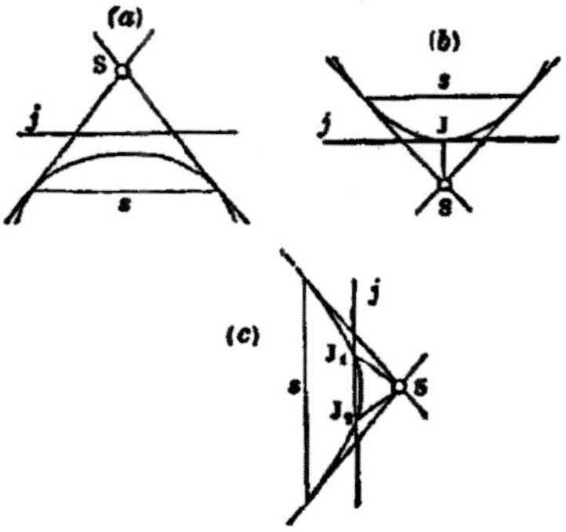

Fig. 237.

397. Problem. *Given a tangent to a conic, its point of contact, and the position (but not the magnitude) of a pair of conjugate diameters; to construct the conic* (Fig. 238).

Suppose O the point of intersection of the given diameters, and P and Q the points in which they are cut by the given tangent. Through the point of contact M of this tangent draw parallels to OQ, OP to meet OP, OQ in P' and Q' respectively. Since the polar of M (the tangent) passes through P, the polar of P will pass

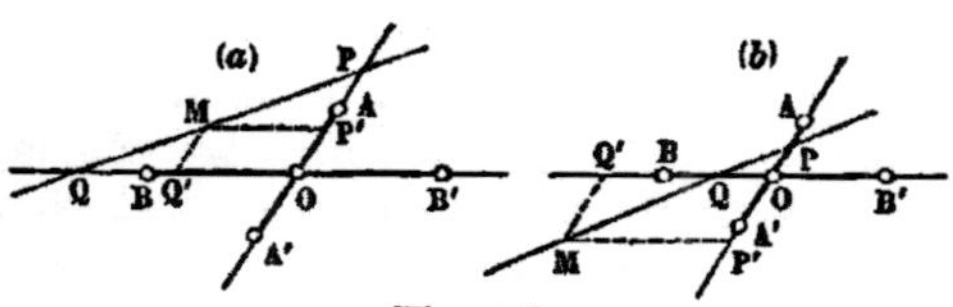

Fig. 238.

through M; and since the polar of P is parallel to OQ, it must be MP'; therefore P and P' are conjugate points.

If now points A and A' be taken on OP such that OA and OA' may each be equal to the mean proportional between OP and OP', then AA' will be equal in length to the diameter in the direction OP (Art. 290). In the same way the length of the other diameter BB' will be found by making OB and OB' each equal to the mean proportional between OQ and OQ'.

* Poncelet, *loc. cit.*, Arts. 225, 226.

If the points P and P' fall on the same side of O, the involution of conjugate points has a pair of double points A and A' (Art. 128); that is to say, the diameter OP meets the curve. If, on the other hand, P and P' lie on opposite sides of O, the involution has no double points, and the diameter OP does not meet the curve. In this case A and A' are two conjugate points lying at equal distances from O. The figure shows two cases: that of the ellipse (a) and that of the hyperbola (b).

398. Problem. *Given a point M on a conic and the positions of two pairs of conjugate diameters a and a', b and b', to construct the conic.*

I. *First solution* (Fig. 239). Through M draw chords parallel to each diameter, and such that their middle points lie on the respectively conjugate diameters. The other extremities A, A', B, B' of

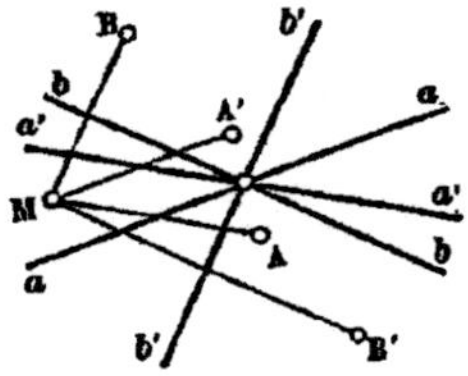

Fig. 239.

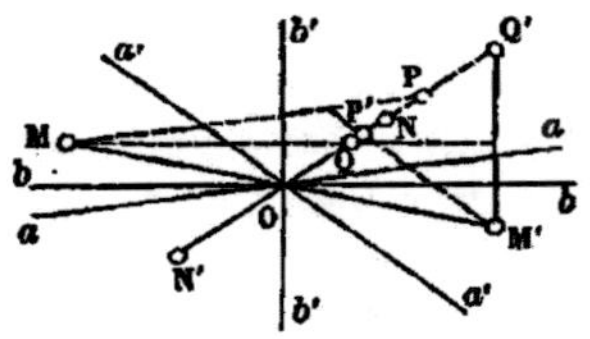

Fig. 240.

the four chords so drawn will be four points all of which lie on the required conic.

II. *Second solution* (Fig. 240). Denoting the diameter MOM' by c, if the ray c' be constructed which is conjugate to c in the involution determined by the pairs of rays a and a', b and b', then c' will be the diameter conjugate to c (Art. 296). Through M draw MP parallel to a, and through M' draw $M'P'$ parallel to a'; these parallels will intersect on the conic (Art. 288); let them cut c' in P and P' respectively. These last two points are conjugate with respect to the conic (Art. 299); thus if on c' two other points be found which correspond to one another in the involution determined by the pair P, P' and the central point O, then MQ and $M'Q'$ will intersect on the conic. If then on c' two points N and N' be taken such that the distance of either of them from O is a mean proportional between OP and OP', they will be the extremities of the diameter c' (Art. 290).

III. *Third solution.* Through the extremities M and M' of the diameter which passes through the given point draw parallels to a and a'; they will meet in a point A lying on the conic. Through the same points draw parallels to b and b'; these will meet in another point B also lying on the conic (Art. 288). Produce AO to A', making OA' equal to AO; and similarly BO to B', making OB' equal to BO; then will A' and B' be points also lying on the required conic (Art. 281).

399. Problem. *Given in position two pairs of conjugate diameters a and a', b and b' of a conic, and a tangent t, to construct the conic.*

I. *First solution* (Fig. 241). Let O be the point of intersection of the given diameters, that is, the centre of the conic. Draw parallel to t and at a distance from O equal to that at which t lies, a straight line t'; this will be the tangent parallel to t. Let the points of intersection of t and t' with a and a' be joined; this will give two other parallel tangents u and u' (Art. 288). Another pair of parallel tangents v and v' will be obtained by joining the points where t and t' meet b and b'.

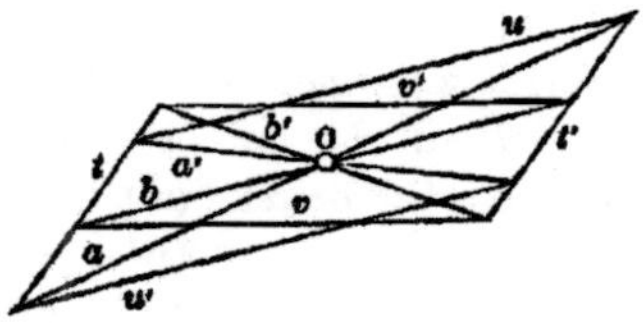

Fig. 241.

II. *Second solution.* The conjugate diameters a and a', b and b', will meet t in two pairs of points A and A', B and B' which determine an involution whose centre is the point of contact of t (Art. 302). The problem therefore reduces to one already solved (Art. 397). If the involution has double points, the straight lines joining these points to O will be the asymptotes.

400. Problem. *Given two points M and N on a conic and the position of a pair of conjugate diameters a and a', to construct the conic* (Fig. 242).

Let M' and N' be the other extremities of the diameters passing through M and N. Through M and M' draw MH, $M'H$ parallel to a and a' respectively; similarly, through N and N' draw NK, $N'K$ parallel to a and a' respectively. The points H and K will lie on the required conic.

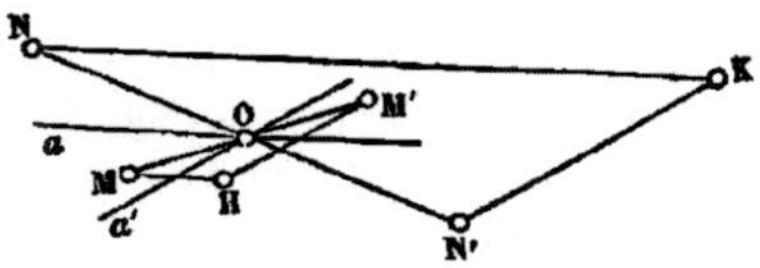

Fig. 242.

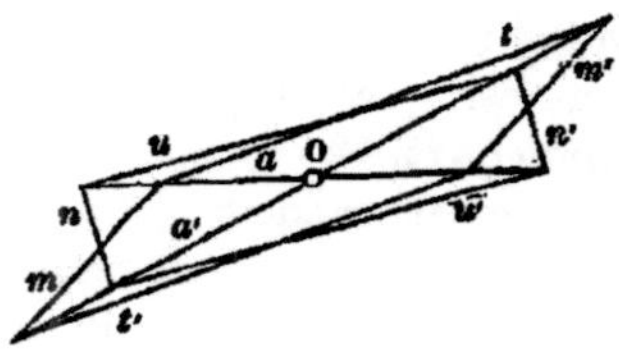

Fig. 243.

401. Problem. *Given two tangents m and n to a conic, and the position of a pair of conjugate diameters a and a', to construct the conic* (Fig. 243).

Draw the straight lines m' and n' parallel respectively to m and n, and at distances from the centre O equal respectively to those at which m and n lie; then m' will be the tangent parallel to m, and n'

the tangent parallel to n. Join the points where m and m' meet a and a' by the straight lines t and t', and the points where n and n' meet a and a' by the straight lines u and u'. The four straight lines t, t', u, u' will all be tangents to the required conic (Art. 288).

402. PROBLEM. *Given five points on a conic, to construct a pair of conjugate diameters which shall make with one another a given angle* *.

Construct first a diameter AA' of the conic (Art. 285); and on it describe a segment of a circle containing an angle equal to the given one. Find the points in which the circle of which this segment is a part cuts the conic again (Art. 227); if M is one of these points, AM and $A'M$ will be parallel to a pair of conjugate diameters. Since then AMA' is equal to the given angle, the problem will be solved by drawing the diameters parallel to AM and $A'M$.

If the segment described is a semicircle, this construction gives the axes.

403. PROBLEM. *To construct a conic with respect to which a given triangle EFG shall be self-conjugate, and a given point P shall be the pole of a given straight line p* †.

Let p meet FG in A. The polar of A will pass through E the pole of FG, and through P the pole of p, and will therefore be EP. Similarly FP, GP will be the polars of the points B, C in which p is cut by GE, EF respectively. Let A' be the point in which FG intersects EP; then F and G, A and A', are two pairs of conjugate points with respect to the conic, and if the involution which they determine has a pair of double points L and L', these points will lie on the required conic (Art. 264). The same construction may be repeated in the case of the other two sides of the triangle EFG.

If the point P lies within the triangle EFG, the points A', B', C' lie upon the sides FG, GE, EF respectively (not produced ‡). The straight line p may cut two of the sides of the triangle, or it may lie entirely outside the triangle. In the first case the involutions lying on the two sides of the triangle which are cut by p are both of the non-overlapping (hyperbolic) kind, and therefore each possesses double points (Art. 128); these give four points of the required curve, and the problem reduces to that of describing a conic passing through four given points and with respect to which two other given points are conjugates (Art. 393). In the second case, on the other hand, the pairs of conjugate points on each of the sides of the triangle EFG overlap, and the involutions have no double points (Art. 128); in

* DE LA HIRE, *loc. cit.*, book ii. prop. 38.

† STAUDT, *Geometrie der Lage*, Art. 237.

‡ We shall say that a point A' lies *on* the side FG of the triangle, when it lies between F and G; and that a straight line *cuts* the side FG, when its point of intersection with FG lies between F and G.

this case the conic does not cut any of the sides of the self-conjugate triangle; therefore (Art. 262) it does not exist.

If the point P lies outside the triangle, one only of the three points A', B', C' lies on the corresponding side; the two others lie on the respective sides produced. If these two other sides are cut by p, none of the involutions possesses double points, and the conic does not exist. If, on the other hand, p cuts the first side, or if p lies entirely outside the triangle, the conic exists, and may be constructed as above.

In all cases, whether the conic has a real existence or not, the *polar system* (Art. 339) exists. It is determined by the self-conjugate triangle EFG, the point P, and the straight line p. To construct this system is a problem of the first degree, while the construction of the conic is a problem of the second degree.

404. PROBLEM. *Given a pentagon $ABCDE$, to describe a conic with regard to which each vertex shall be the pole of the opposite side* *.

Let F be the point of intersection of AB and CD. If the conic **K** be constructed (Art. 403) with regard to which ADF is a self-conjugate triangle and E the pole of BC, then the points B and C in which BC is cut by AF and DF respectively will be the poles of ED and EA, the straight lines which join E to the points D and A respectively. Every vertex of the pentagon will therefore be the pole of the opposite side; that is, **K** will be the conic required.

If the conic **C** be constructed which passes through the five vertices of the pentagon, and also the conic **C**$'$ which touches the five sides of the pentagon (Art. 152), these two conics will be polar reciprocals one of the other with respect to **K** (Art. 322).

405. PROBLEM. *Given five points A, B, C, D, E (no three of which are collinear), to determine a point M such that the pencil $M(ABCDE)$ shall be projective with a given pencil $abcde$* (Fig. 244).

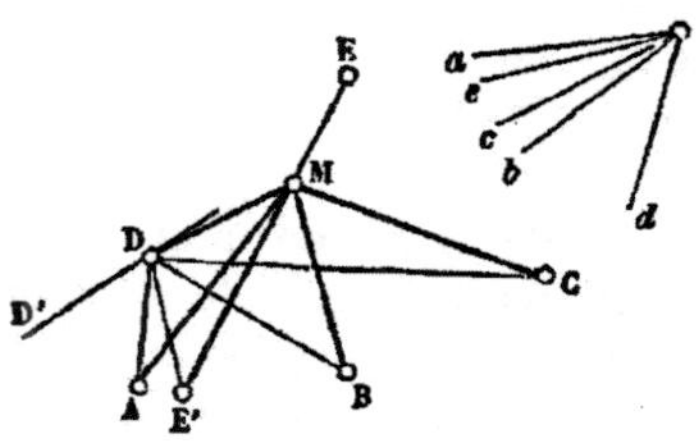

Fig. 244.

Through D draw two straight lines DD', DE' such that the pencil $D(ABCD'E')$ is projective with $abcde$ (Art. 84, right). Construct the point E' in which DE' meets the conic which passes through the four points $ABCD$ and touches DD' at D (Art. 165); then construct the point M in which the same conic meets EE'. M will be the point required. For since M, A, B, C, D, E' lie on the same conic, the pencil $M(ABCDE')$ is projective with the pencil $D(ABCD'E')$,

* STAUDT, *loc. cit.*, Arts. 238, 258.

which by construction is projective with the given pencil *abcde*. Since then *ME'* and *ME* are the same ray, the problem is solved.

As an exercise may be solved the correlative problem, viz.

Given five straight lines a, b, c, d, e, no three of which are concurrent, to draw a straight line m to meet them in five points forming a range projective with a given range ABCDE *.

406. Problem. *To trisect a given arc AB of a circle* †.

On the given arc take (Fig. 245) a point *N*, and from *B* measure in the opposite direction to *AN* an arc *BN'* equal to twice the arc *AN*. If *BT* be the tangent at *B*, and if *O* be the centre of the circle

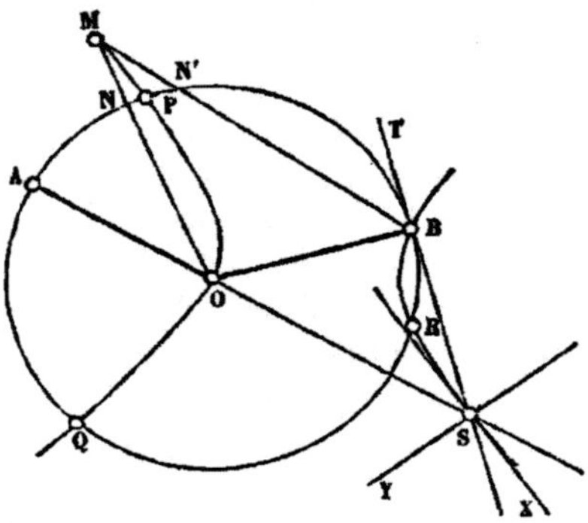

Fig. 245.

of which the arc *AB* is a part, the angles *AON* and *TBN'* are equal and opposite. If *N* and *N'* vary their positions simultaneously, the rays *ON* and *BN'* will describe two oppositely equal pencils, and the locus of their point of intersection *M* will therefore (Art. 395, II) be an equilateral hyperbola passing through *O* and *B*. The asymptotes of this hyperbola are parallel to the bisectors of the angle made by *AO* and *BT* with one another; for these straight lines are corresponding rays (being the positions of the variable rays *ON* and *BN'* for which the arcs *AN* and *BN'* are each zero). The centre of the hyperbola is the middle point of the straight line *OB* which joins the centres of the two pencils.

The hyperbola having been constructed by help of Pascal's theorem, the point *P* will have been found in which it cuts the arc *AB*. Two corresponding points *N* and *N'* coalesce in this point; therefore the arc *AP* is half of the arc *PB*, and *P* is that point of trisection of the arc *AB* which is the nearer to *A*.

The hyperbola meets the circle in two other points *R* and *Q*. The point *R* is one of the points of trisection of the arc which together

* Staudt, *loc. cit.* Art. 263.

† Staudt, *Beiträge*, Art. 432; Chasles, *Sections coniques*, Art. 37.

with AB makes up a semicircle; and the point Q is one of the points of trisection of the arc which together with AB makes up the circumference of the circle.

407. It has been seen (Art. 191) that if P', P'', Q', Q'' (Fig. 246) are four given collinear points, and if any conic be described to pass through P' and P'', and then a tangent be drawn to this conic from Q' and another from Q'', the chord joining the points of contact of these tangents passes through one of the double points M', N' of the involution which is determined by the two pairs of points P' and P'', Q' and Q''. The two tangents which can be drawn from Q', combined with the two from Q'', give four such chords of contact, of which two pass through M' and two through N'. From this may

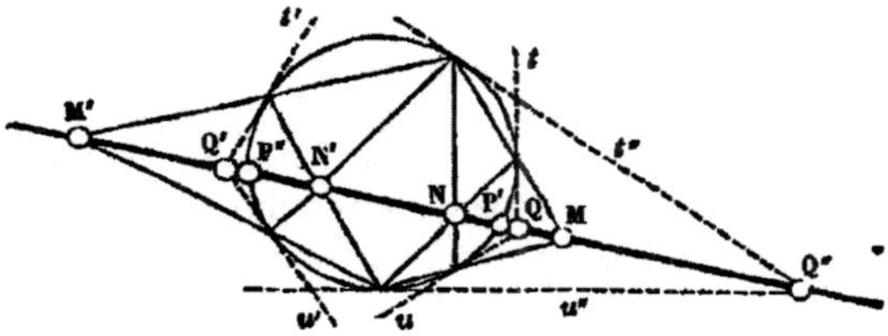

Fig. 246.

be deduced *a construction for the double points of the involution* $P'P''$, $Q'Q''$, or, what is the same thing (Art. 125), *for the two points* M', N' *which divide each of the two given segments* $P'P''$ *and* $Q'Q''$ *harmonically*.

Describe any circle to pass through P' and P'', and draw to it from Q' the tangents t' and u', and from Q'' the tangents t'' and u''. The chord of contact of the tangents t' and t'' and that of the tangents u' and u'' will cut the straight line $P'P''$ in the two required points M' and N'.

408. This construction has been applied by Brianchon * to the solution of the two problems considered in Art. 221, viz.

I. *To construct a conic of which two points* P', P'' *and three tangents* q, q', q'' *are given.*

Join $P'P''$, and let it cut the three given tangents in Q, Q', Q'' respectively (Fig. 246). Describe any circle through P', P'' and draw to it tangents from Q, Q', Q''. The chords which join the points of contact of the tangents from Q'' to the points of contact of the tangents from Q meet $P'P''$ in two points M and N; and similarly the tangents from Q'' combined with those from Q' determine two points M' and N'.

The chord of contact of the tangents q', q'' to the required conic will therefore pass through one of the points M, N, and that of the

* Brianchon, *loc. cit.*, pp. 47, 51.

tangents q', q'' will pass through one of the points M', N'. The four combinations MM', MN', NM', NN' give the four solutions of the problem.

The problem therefore reduces to the following: *To describe a conic which shall touch three given straight lines q, q', q'' in such a way that the chords of contact of the two pairs of tangents q, q'' and q', q'' shall pass respectively through two given points M and M'.* Let $QQ'Q''$ (Fig. 247) denote the triangle formed by the three given tangents, and let A, A', A'' be the points of contact to be determined. By a corollary to Desargues' theorem (Art. 194), the side $q \equiv Q'Q''$ is divided harmonically at the point of contact A and at the point where it is cut by the chord $A'A''$. If these four harmonic points be projected on MQ'' from A'' as centre, it follows that the segment RQ'' intercepted on MQ'' between q'' and q' is divided harmonically by M and the chord $A'A''$.

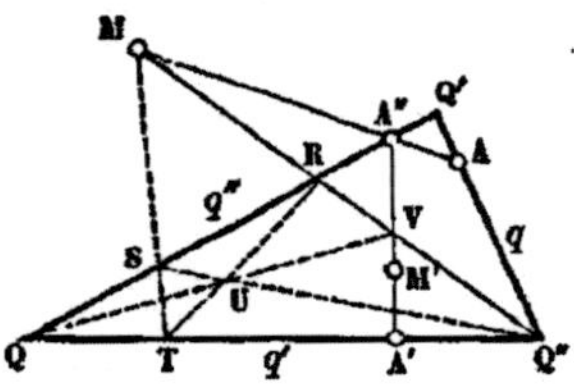

Fig. 247.

Let then MQ'' be joined; it will cut q'' in some point R; and let the point V be determined which is harmonically conjugate to M with regard to R and Q''. In order to do this, draw through M any straight line to cut q'' and q' in S and T respectively; join SQ'' and TR, meeting in U; and join QU, meeting RQ'' in V. Join VM'; it will meet q' and q'' in A' and A''; and finally if MA'' be joined, it will cut $Q'Q''$ in A.

II. *To construct a conic of which three points P, P', P'' and two tangents q, q' are given.*

Join PP', and let it meet q and q' in Q and Q' respectively; join PP'', and let it meet q and q' in R and R' respectively. Describe a circle round $PP'P''$, and to it draw tangents from Q and Q'; the chords of contact will meet PP' in two points M and N. Similarly draw the tangents from R and R'; the chords of contact will meet PP'' in two other points M' and N'. Then each of the straight lines MN', NN', $M'N$, MM' will meet the tangents q and q' in two of the points of contact of these two tangents with a conic circumscribing the triangle $PP'P''$.

This construction differs from that given in Art. 221 (left) only in the method of finding the double points M and N, M' and N'.

409. THEOREM. *If two angles AOS and $AO'S$ of given magnitude turn about their respective vertices O and O' in such a way that the point of intersection S of one pair of arms lies always on a fixed straight line u, the point of intersection of the other pair of arms will describe a conic* (Fig. 248).

The proof follows at once from the property that the pencils traced out by the variable rays OA and OS, OS and $O'S$, $O'S$ and $O'A$ are projective two and two (Arts. 42, 108), and that consequently the pencils traced out by OA and $O'A$ are projective. This theorem is due to NEWTON, and was given by him under the title of *The Organic Description* of a conic *.

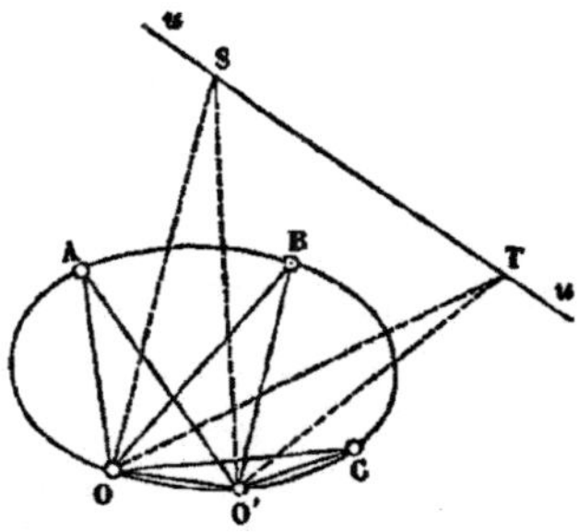

Fig. 248.

410. The following, which depend on the foregoing theorem, may serve as exercises to the student:—

1. Deduce a construction for a conic passing through five given points O, O', A, B, C.

2. Given these five points, determine the magnitude of the angles AOS, $AO'S$ and the position of the straight line u in order that the conic generated may pass through the five given points.

3. On the straight line OO' which joins the vertices of the two given angles a segment of a circle is described containing an angle equal to the difference between four right angles and the sum of the given angles. Show that according as the circle of which this segment is a part cuts, does not cut, or touches the straight line u, so the conic generated will be a hyperbola, an ellipse, or a parabola.

4. Determine the asymptotes of the conic, supposing it to be a hyperbola; or its axis, in the case where it is a parabola.

5. When is the conic (*a*) a circle, (*b*) an equilateral hyperbola, (*c*) a pair of straight lines?

6. Examine the cases in which the two given angles are directly equal, or oppositely equal, or supplementary †.

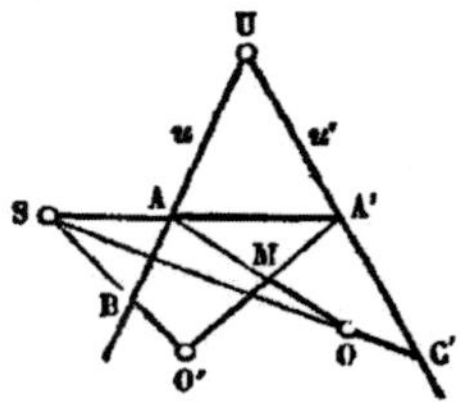

Fig. 249.

411. THEOREM. *If a variable triangle AMA' move in such a way that its sides turn severally round three given points O, O', S* (Fig. 249) *while two of its vertices A, A' slide along two fixed straight lines u, u' respectively, the locus of the third vertex M is a conic passing through the following five points, viz. O, O', uu', and the intersections B and C' of u and u' with $O'S$ and OS respectively* ‡.

* *Principia*, lib. i. lemma xxi; *Enumeratio linearum tertii ordinis* (*Opticks*, 1704), p. 158, § xxxi.

† MACLAURIN, *Geometria Organica* (London, 1720), sect. i. prop. 2.

‡ See Art. 156.

412. Theorem. (The theorem of Art. 411 is a particular case of this). *If a variable polygon move in such a way that its n sides turn severally round n fixed points $O_1, O_2, \ldots O_n$* (Fig. 250) *while $n-1$ of its vertices slide respectively along $n-1$ fixed straight lines $u_1, u_2, \ldots u_{n-1}$, then the last vertex will describe a conic; and the locus of the point of intersection of any pair of non-adjacent sides will also be a conic**.

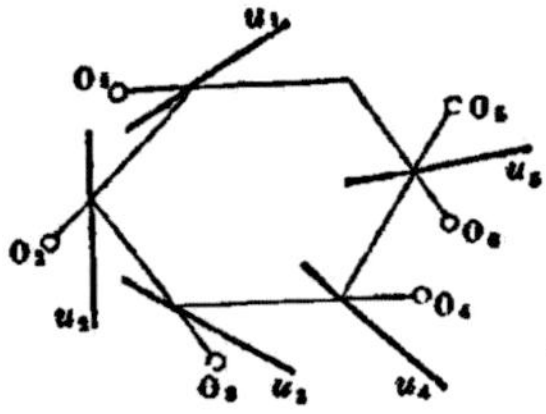

Fig. 250.

The proof of this theorem and its correlative is left to the student †.

413. Theorem. *From two given points A and A′ tangents AB, AC and A′B′, A′C′ are drawn to a conic; then will the four points of contact B, C, B′, C′, and the two given points A, A′ all lie on a conic* (Fig. 251 ‡).

Let $A'C'$, $A'B'$ meet BC in D and E respectively; these points will evidently be the poles of AC', AB' respectively. The pencil $A(BCB'C')$ is projective with the range of poles $BCED$ (Art. 291), and therefore with the pencil $A'(BCED)$ or $A'(BCB'C')$; which proves the theorem.

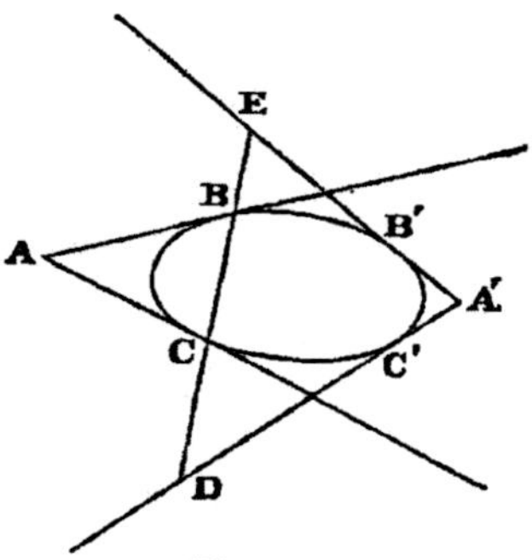

Fig. 251.

414. Theorem (correlative to that of Art. 413). *From two given points A and A′ tangents AB, AC and A′B′, A′C′ are drawn to a conic; then will the four tangents and the two chords of contact all touch a conic* ‡.

For (Fig. 251) the range of points $BC(AB, AC, A'B', A'C')$ or $BCED$ is projective with the pencil $A(BC\,B'C')$ formed by their polars; but this pencil is projective with the range $B'C'(AB, AC, A'B', A'C')$; therefore the six lines AB, AC, $A'B'$, $A'C'$, BC, $B'C'$ all touch a conic.

415. Theorem. *On each diagonal of a complete quadrilateral is taken a pair of points dividing it harmonically; if of these six points three (one from each diagonal) lie in a straight line, the other three will also lie in a straight line.*

Corollary. *The middle points of the three diagonals of a complete quadrilateral are collinear.*

* This theorem is due to Maclaurin and Braikenridge (*Phil. Trans.*, London, 1735).

† Poncelet, *loc. cit.*, Art. 502.

‡ Chasles, *Sections coniques*, Arts. 213, 214.

416. THEOREM. *If from any point O on the circle circumscribing a triangle ABC straight lines OA', OB', OC' be inflected to meet the sides BC, CA, AB in A', B', C' respectively, and to make with them equal angles (both as regards sign and magnitude); then the three points A', B', C' will be collinear* (Fig. 252).

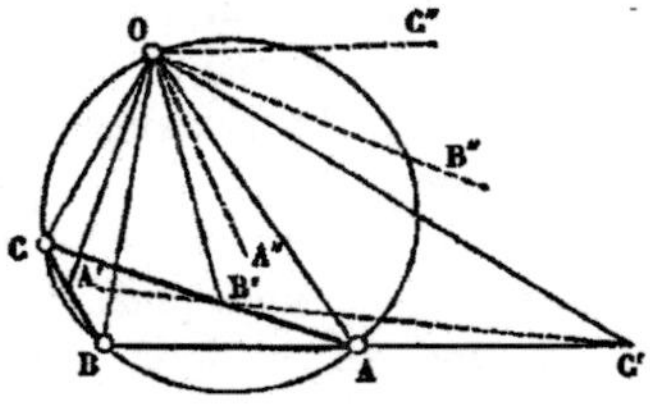

Fig. 252.

Through O draw OA'', OB'', OC'' parallel to BC, CA, AB respectively; then it is easily seen that the angles AOA'', BOB'', COC'' have the same bisectors. The same will therefore be true with regard to the angles AOA', BOB', COC'; consequently (Art. 142) the arms of these last three angles will form an involution, and therefore (Art. 135) the points A', B', C' will be collinear * †.

417. THEOREM. *If from the vertices of a triangle circumscribed about a circle straight lines be inflected to meet any tangent to the circle, so that the angles they subtend at the centre may be equal (in sign and magnitude), then the three straight lines will meet in a point* ‡.

The proof is similar to that of the theorem in the preceding Article.

418. PROBLEMS. (1). Given three collinear segments AA', BB', CC'; to find a point at which they all subtend equal angles (Art. 109).

In what case can these angles be right angles? (See Art. 128).

(2). Given two projective ranges lying on the same straight line; to find a point which is harmonically conjugate to a given point on the line, with respect to the two self-corresponding points of the two ranges (which last two points are not given) §.

(3). Given two pairs of points lying on a straight line; to determine on the line a fifth point such that the rectangle contained by its distances from the points of the first pair shall be to that contained

* CHASLES, *loc. cit.*, Art. 386.

† *Otherwise:* Since the triangles BOC', COB' are similar,

$$BC' : CB' = OB : OC.$$

So also
$$CA' : AC' = OC : OA,$$

and
$$AB' : BA' = OA : OB;$$

whence by multiplication, paying attention to the signs of the segments,

$$BC' . CA' . AB' = -C'A . B'C . A'B,$$

which shows (Art. 139) that A', B', C' are collinear.

‡ CHASLES, *loc. cit.*, Art. 387.

§ CHASLES, *Géom. sup.*, Art. 269.

by its distances from the points of the second pair in a given ratio *.

(4). Through a given point to draw a transversal which shall cut off from two given straight lines two segments (measured from a fixed point on each line) which shall have a given ratio to one another; *or*, the rectangle contained by which shall be equal to a given one †.

419. It will be a useful exercise for the student to apply the theory of pole and polar to the solution of problems of the first and second degree, supposing given a ruler, and a fixed circle and its centre. We give some examples of problems treated in this manner:

I. *To draw through a given point P a straight line parallel to a given straight line q.*

The pole Q of q and the polar p of P (with respect to the given circle) must be found; if A be the point where p is cut by the straight line OQ joining Q to the centre of the circle, then the polar a of A will be the straight line required.

II. *To draw from a given point P a perpendicular to a given straight line q.*

Draw through P a straight line parallel to OQ; it will be the perpendicular required.

III. *To bisect a given segment AB.*

Let a and b be the polars of A and B respectively, and c that diameter of the given circle which passes through ab; if d be the harmonic conjugate of c with respect to a and b, the pole of d will be the middle point of AB.

IV. *To bisect a given arc MN of a circle.*

Construct the pole S of the chord MN; the diameter passing through S will cut MN in the middle point of the latter.

V. *To bisect a given angle.*

If from a point on the circle parallels be drawn to the arms of the given angle, the problem reduces to the preceding one.

VI. *Given a segment AC; to produce it to B so that AB may be double of AC.*

Let a and c be the polars of A and C respectively, d the diameter of the given circle which passes through ac, and b the ray which makes the pencil $abcd$ harmonic; the pole of b will be the required point B.

* This is the problem 'de sectione determinata' of APOLLONIUS. See CHASLES, *Géom. sup.*, Art. 281.

† These are the problems 'de sectione rationis' and 'de sectione spatii' of APOLLONIUS. See CHASLES, *Géom. sup.*, Arts. 296, 298.

VII. *To construct the circle whose centre is at a given point U and whose radius is equal to a given straight line UA.*

Produce AU to B, making UB equal to AU (by VI), and draw perpendiculars at A and B to AB (by II). Bisect the right angles at A and B (by V); and let the bisecting lines meet in C and D. We have then only to construct the conic of which AB and CD are a pair of conjugate diameters (Art. 301).

420. The following problems* depend for their solution on the theorem of Art. 376.

I. *Given three points A, B, C on a conic and one focus F, to construct the conic.*

With centre F and any radius describe a circle **K**, and let the polars of A, B, C with respect to this circle be a, b, c respectively. Describe a circle touching a, b, c and take its polar reciprocal with respect to **K**; this will be the conic required.

Since there can be drawn four circles touching a, b, c (the inscribed circle of the triangle abc and the three escribed circles), there are four conics which satisfy the problem.

II. *Given two points A, B on a conic, one tangent t, and a focus F, to construct the conic.*

Describe a circle **K** as in the last problem, and let a, b be the polars of A, B, and T the pole of t, with respect to **K**. Draw a circle to pass through T and to touch a and b; the polar reciprocal of this circle with respect to **K** will be the conic required.

Since four circles can be drawn to pass through a given point and touch two given straight lines, this problem also admits of four solutions.

III. *Given one point A on a conic, two tangents b, c, and a focus F, to construct the conic.*

Describe a circle **K** as in the last two problems; let a be the polar of A, and let B, C be the poles of b, c respectively with regard to this circle. Draw a circle to pass through B and C and to touch a; its polar reciprocal with respect to **K** will be the conic required.

Since two circles can be described through two given points to touch a given straight line, this problem admits of two solutions.

IV. *Given three tangents a, b, c to a conic and one focus F, to construct the conic.*

Describe a circle **K** as in the last three problems, and let A, B, C be the poles of a, b, c respectively with regard to this circle. Draw the circle through A, B, C and take its polar reciprocal with respect to **K**; this will be the conic required.

This problem clearly admits of only one solution.

* Solutions of these problems were given by De la Hire (see Chasles, *Aperçu historique*, p. 125), and by Newton (*Principia*, lib. i. props. 19, 20, 21).

421. Problem. *Given the axes of a conic in position (not in magnitude) and a pair of conjugate straight lines which cut one another orthogonally, to construct the foci.*

If O be the centre of the conic, and P, P' and Q, Q' the points in which the two conjugate lines respectively cut the axes, then of the two products $OP \,.\, OP'$ and $OQ \,.\, OQ'$, one will be positive and the other negative. This determines which of the two given axes is the one containing the foci. If now a circle be circumscribed about the triangle formed by the two given conjugate lines and the non-focal axis, it will cut the focal axis at the foci (Art. 343).

422. The following are left as exercises to the student.

1. Given the axes of a conic in position, and also a tangent and its point of contact, construct the foci, and determine the lengths of the axes (Art. 344).

2. Given the focal axis of a conic, the vertices, and one tangent, construct the foci (Art. 360).

3. Given the tangent at the vertex of a parabola, and two other tangents, find the focus (Art. 358).

4. Given the axis of a parabola, and a tangent and its point of contact, find the focus (Art. 346).

5. Given the axis and the focus of a parabola, and one tangent, construct the parabola by tangents (Arts. 346, 349, 358).

6. The locus of the pole of a given straight line r with respect to any conic having its foci at two given points is a straight line r' perpendicular to r. The two lines r, r' are harmonically separated by the two foci.

7. The locus of the centre of a circle touching two given circles consists of two conics having the centres of the given circles for foci.

8. The locus of a point whose distance from a given straight line is equal to its tangential distance from a given circle consists of two parabolas.

9. In a central conic any focal chord is proportional to the square of the parallel diameter.

10. In a parabola, twice the distance of any focal chord from its pole is a mean proportional between the chord and the parameter.

INDEX.

Affinity, pp. 18, 19.
Angle of constant magnitude turning round its vertex traces out two directly equal pencils, 91.
bisection of an, 300.
trisection of an, 294.
Angles, two, of given magnitude; generation of a conic by means of, 297.
Anharmonic ratio defined, 54, 57.
unaltered by projection, 54.
of a harmonic form is − 1, 57.
cannot have the values + 1, 0, or ∞, 62.
of four points or tangents of a conic, 122.
Anharmonic ratios, the six, 60, 61.
Apollonius, x, xi, xii.
on the parabola, 127, 218.
on the hyperbola, 130, 142, 156, 158, 286.
on the diameters of a conic, 217, 223, 230, 232, 234, 235.
on focal properties of a conic, 253, 258, 259, 262.
section-problems, 300.
Arc of a conic, determination of kind of conic to which it belongs, 289.
of a circle, trisection of, 294.
of a circle, bisection of, 300.
Asymptotes, tangents at infinity, 16, 129.
meet in the centre of the conic, 219.
determination of the, given five points on the conic, 178, 179.
Auxiliary conic, 203, 239, 240.
circle of a conic, 260.
Axes of a conic defined, 227, 228.
case of the parabola, 228.
focal and non-focal, 252.
bisectors of the angle between its chords of intersection with any circle, 236, 281.
Axes of a conic, construction of the, given a pair of conjugate diameters, 232, 283.
given five points, 236, 292.
Axis of perspective or homology, 10.
of affinity, 18.
of symmetry, 64.

Bellavitis, xi, 64, 161.
Bisection of a given segment or angle by means of the ruler only, 300.
Brianchon, x, xi, xii, 124, 125.
Brianchon's theorem, xi, 124.
points, the sixty, 126.

Carnot's theorem, xi, 279, 280.
Centre of projection, 1, 3.
of perspective or homology, 10, 12, 98.
of similitude, 18.
of symmetry, 64.
of an involution, 102.
Centre of a conic, the pole of the line at infinity, 218.
bisects all chords, 219.
the point of intersection of the asymptotes, 219.
when external and when internal to the conic, 219.
locus of, given four tangents, 237.
construction of the, given five points, 220.
construction of the, given five tangents, 238.
Ceva, theorem of six segments, 111.
Chasles, xi, xii.
on homography, 34.
method of generating conics, 127.
correlative to the theorem 'ad quatuor lineas,' 159.
on the geometric method of false position, 194.
solutions of problems of the second degree, 200.
Circle, curve homological with a, 14, 15.
generated by the intersection of two directly equal pencils, 114.
harmonic points and tangents of a, 115, 116.
fundamental projective properties of points and tangents of a, 115.
of curvature at a point on a conic, 190.
cutting a conic; the chords of intersection make equal angles with the axes, 236, 281.
circumscribing triangle formed by three tangents to a parabola, 253.
auxiliary, of a conic, 260.
Class of a curve, 4.
is equal to the degree of its polar reciprocal with regard to a conic, 240.

Coefficient of homology, 63.
Collinear projective ranges, 68.
their self-corresponding points, 78, 91, 92, 93.
construction for these, 170.
Complementary operations, 33.
Concentric pencils, 69.
construction for their self-corresponding rays, 169.
Cone, sections of the, 14, 18.
Confocal conics, 266.
Congruent figures, 64.
Conic, homological with a circle, 15, 16.
generated by two projective pencils, 119.
generated as an envelope from two projective ranges, 120.
determined by five points or five tangents, 123.
fundamental projective property of points and tangents, 118.
projective ranges of points and series of tangents of a, 161.
homological with itself, 228, 288.
polar reciprocal of a, 240.
homological with a given conic, and having its centre at a given point, 249.
confocal with a given conic, and passing through a given point, 266.
Conic, construction of a, having given five points or tangents, 131, 149, 176, 179, 180, 297.
four points and the tangent at one of them, 137, 177.
three points and the tangents at two of them, 139, 177.
three tangents and the points of contact of two of them, 143, 177.
four tangents and the point of contact of one of them, 146, 177.
four points and a tangent, 180.
four tangents and a point, 180.
three points and two tangents, 182, 296.
three tangents and two points, 182, 295.
the asymptotes and one point or tangent, 277.
the two foci and one tangent, 264.
the two foci and one point, 265.
one focus and three tangents, 268, 301.
one focus and three points, 301.
one focus, two points and a tangent, 301.
one focus, two tangents and a point, 301.
a pair of conjugate diameters, 229, 231.
a pair of conjugate diameters in position, and two points or tangents, 291.
a pair of conjugate diameters in position, and a tangent and its point of contact, 289.
two pairs of conjugate diameters in position, and one point or tangent, 290, 291.
two reciprocal triangles, 247.
a self-conjugate triangle, and a point and its polar, 292.
a self-conjugate pentagon, 293.
three points and the osculating circle at one of them, 190.
Conic, construction of a, homological with itself, 228, 288.
passing through three points and determining a known involution on a given line, 284.
passing through four points and dividing a given segment harmonically, 284.
passing through four points and through a pair of conjugate points of a given involution, 285.
Conics, osculating, 189.
having a common self-conjugate triangle, 213, 214.
circumscribing the same quadrangle, 150, 214, 237.
inscribed in the same quadrilateral, 150, 213, 214, 237.
Conjugate axis of a hyperbola, 228.
Conjugate diameters, defined, 219.
of a circle cut orthogonally, 222.
form an involution, 227.
parallelogram described on a pair as adjacent sides is of constant area, 234.
sum or difference of squares is constant, 235.
construction of, given two pairs, 232.
construction of, given five points on the conic, 236.
including a given angle, construction of, 292.
Conjugate lines meeting in a point, one orthogonal pair can be drawn, 227.
orthogonal, the involution determined by them on an axis of the conic, 251.
orthogonal, with respect to a parabola, 253.
Conjugate points and lines with regard to a conic, 204.
involution-properties of, 209.
Conjugates, harmonic, 46.
in an involution, 101.
Construction of a figure homological with a given one, 13.
for the fourth element of a harmonic form, 47.
for the fourth point of a range whose anharmonic ratio is given, 55.

of pairs of corresponding elements of two projective forms, when three are given, 70.
for the self-corresponding elements of two superposed projective forms, 169.
for the sixth element of an involution, 109.
of pairs of elements of an involution, given two, 104.
for the centre of an involution, 109.
for the double elements of an involution, 169, 175, 295.
for the common pair of two superposed involutions, 173.
for the pole of a line or polar of a point, 205, 206.
of a triangle self-conjugate to a conic, 207.
of the centre and axes of a conic, 220, 236, 238, 283, 292.
of conjugate diameters, 232, 236, 292.
for diameters of a parabola, having given four tangents, 238.
for the focus of a parabola, given four tangents, 254.
for the foci of a conic, given the axes and a pair of orthogonal conjugate lines, 302.
Copolar and coaxial triangles, 7, 8.
Correlative figures, 26, 85, 241.
Curvature, circle of, 190.

Degree of a curve, 4.
is equal to the class of its polar reciprocal with respect to a conic, 240.
De la Hire, x, xii.
Desargues, ix, x, xii, 101, 102, 107, 148.
Desargues' theorem, 148.
Descriptive, the term, as distinguished from metrical, 50.
Diagonal triangle, of a quadrangle or quadrilateral, 30.
common to the complete quadrilateral formed by four tangents to a conic, and the complete quadrangle formed by their points of contact, 140.
Diagonals of a complete quadrilateral, each is cut harmonically by the other two, 46.
their middle points are collinear, 109, 299.
form a triangle self-conjugate to any conic inscribed in the quadrilateral, 208.
if the extremities of two are conjugate points with regard to a conic, those of the third are so too, 245.
Diameters of a conic defined, 217.
of a parabola, 218.
conjugate, 219.
ideal, 223.
of a parabola, construction for, given four tangents, 238.
Dimension of a geometric form, 25.
Directly equal ranges, defined, 88.
generated by the motion of a segment of constant length, 89.
Directly equal pencils, defined, 90.
two, the projection of two concentric projective pencils, 89.
two, generate a circle by their intersection, 114.
subtended at a focus of a conic by the points in which a variable tangent cuts two fixed ones, 264.
Director circle, defined, 269.
the locus of the intersection of orthogonal tangents, 269.
cuts orthogonally the circumscribing circle of any self-conjugate triangle, 270.
Directrix, defined, 254.
property of focus and, 256.
Directrix of a parabola, the locus of the intersection of orthogonal tangents, 270.
the locus of the centre of the circumscribing circle of a self-conjugate triangle, 271.
the locus of the orthocentre of a circumscribing triangle, 273.
Division of a given bisected segment into *n* equal parts, by means of the ruler only, 97.
Double elements of an involution, 102.
they separate harmonically any pair of conjugates, 103.
construction for the, 169, 295.
Duality, the principle of, 26–32.

Eccentricity, 259.
of the polar reciprocal of a circle with respect to another circle, 274.
Ellipse, 16.
its centre an internal point, 219.
is cut by all its diameters, 220.
is symmetrical in figure, 228.
Envelope of connectors of corresponding points of two projective ranges is a conic, 120.
if the ranges are similar, it is a parabola, 128.
of a straight line the product of whose distances from two given points is constant, 260.
Equal ranges and pencils, 86–90.
Equianharmonic forms and figures are projective, and vice versa, 54, 56, 62, 66.
Equilateral hyperbola, why so called, 286.
triangles self-conjugate with regard to a, 271.
inscribed in a quadrilateral, 272.

circumscribing a triangle passes through the orthocentre, 287.
is the polar reciprocal of a conic with regard to a point on the director circle, 288.
construction of, given four tangents, 272, 288.
Euclid, porisms of, x, 96.
External and internal points with regard to a conic, 203.

False position, geometrical method of, 194.
Focal axis of a conic, 252.
radii of a point on a conic, 253.
radii, their sum or difference is constant, 258.
Foci, defined, 250.
are points such that conjugate lines meeting in them cut orthogonally, 250.
are internal points lying on an axis, 250.
are the double points of the involution determined on an axis by pairs of orthogonal conjugate lines, 251.
of a parabola, one at infinity, 253.
of parabolas inscribed in a given triangle, locus of, 254.
properties of, with regard to tangent and normal, 259–264.
reciprocation with respect to the, 274, 275.
construction of, under various conditions, 302.
Focus of a parabola, 253.
inscribed in a given triangle, locus of, 254.
reciprocal of the curve with regard to, 275.
Forms, geometric, defined, 22, 164.
elements of, 23, 164.
prime, of one, two, three dimensions, 24.
dual generation of, 23, 24, 26.
projective, 34–38.
harmonic, 39 49.
projective, when in perspective, 67.
projective, superposed, 68, 69.

Gaskin, 189, 269, 271.
Gergonne, x.

Harmonic forms defined, 39, 40.
forms are projective, 41, 43.
pairs of points necessarily alternate, 45.
conjugates, 46.
point or ray, construction for the fourth, 47.
forms, metrical relations, 57, 58.
homology, 64, 228, 288.
points and tangents of a circle, 115, 116, 169.
and of a conic, 122, 157, 168.
Hesse, theorem relating to the extremities of the diagonals of a complete quadrilateral, 245.
Hexagon, inscribed in a line-pair, 76.
circumscribed to a point-pair, 76.
inscribed in a conic, 124.
circumscribed to a conic, 124.
complete, contains sixty simple hexagons, 125.
Homographic, the term, 34.
figures, construction of, 81.
figures may be placed in homology, 84.
Homological figures, construction of, 13–20.
metrical relations between, 63–65.
Homology, defined, 9, 10.
in space, 20.
plane of, 20.
coefficient or parameter of, 63.
harmonic, 64, 228, 288.
Homothetic figures, 18.
Hyperbola, tangent-properties of a, 129, 130.
and asymptotes cut by a transversal, 156, 282.
tangent cut off by the asymptotes is bisected at the point of contact, 158.
centre is an external point, 219.
is cut by one only of every pair of conjugate diameters, 220.
is symmetrical in figure, 228.
properties of the asymptotes and conjugate points and lines, 277.
equilateral, 285.

Ideal diameters and chords, 223, 226.
Infinity, points and line at, 5.
line at, a tangent to the parabola, 16.
plane at, 21.
Internal and external points with regard to a conic, 203.
Intersection of a conic with a straight line; constructions, 176, 177, 180, 226.
of two conics; constructions, 189.
Involution, defined, 101.
the two kinds, elliptic and hyperbolic, 105, 168.
construction for the sixth element of an, 109.
determined by two pairs of conjugates, 104, 165.
of points or tangents of a conic, 165.
construction for the double elements of an, 169, 295.
formed by cutting a conic by a pencil, 166.
of conjugate points or lines with regard to a conic, 209.
of conjugate diameters of a conic, 227.

Involution-properties of the complete quadrangle and quadrilateral, 107.
of a conic and an inscribed or circumscribed quadrangle, 148, 225.
of a conic and an inscribed or circumscribed triangle, 152, 157.
of a conic, two tangents, and their chord of contact, 154.
of conjugate points and lines with regard to a conic, 209.

Lambert, ix, xi, 96–98.
Latus rectum, 257, 258.
Locus of the centre of perspective of two figures when one is turned round the axis of perspective, 12, 98.
of the intersection of corresponding rays of two projective pencils is a conic, 119.
ad quatuor lineas, 158.
of middle points of parallel chords of a conic, 217.
of poles of a straight line with regard to conics inscribed in a quadrilateral, 237.
of the centre of a conic, given four tangents, 237.
of foot of perpendicular from the focus of a conic on a tangent, 260.
of the intersection of orthogonal tangents to a conic, 269.

Maclaurin, xi, 127, 141, 185, 297, 298.
Major and minor axes of an ellipse, 228.
Menelaus, theorem on triangle cut by a transversal, 112, 280.
Metrical, the term, distinguished from descriptive, 50.
Möbius, theorem on figures in perspective, 12.
on anharmonic ratio, x, 56, 61.
Monge, xii.

Newton, locus of centre of a conic inscribed in a quadrilateral, 238.
organic description of a conic, xi, 297.
Nine-point circle, 283.
Normal, 252.

Oppositely equal pencils, 90.
they generate an equilateral hyperbola by their intersection, 286.
Oppositely equal ranges, 88.
Organic description of a conic, 297.
Orthocentre of a triangle circumscribing a parabola lies on the directrix, 273.
of a triangle inscribed in an equilateral hyperbola lies on the curve, 287.
Orthogonal projection, 19.
pair of rays in a pencil in involution, 172.
pair of conjugate diameters of a conic, 227.
conjugate lines with respect to a conic, 251, 252.
Osculating conics, 189.
circle of a conic, 190.

Pappus, x, xii.
on a hexagon inscribed in a line-pair, 76.
porisms of, 95, 96.
fundamental property of the anharmonic ratios, 54.
problem 'ad quatuor lineas,' 158.
on the focus and directrix property of a conic, 257.
Parabola, touches the line at infinity, 16.
is determined by four points or tangents, 127.
two fixed tangents are cut proportionally by the other tangents, 128.
generated as an envelope from two similar ranges, 128.
diameters of a, 218.
construction of the diameters, having given four tangents, 238.
focal properties of the, 253, 254.
focus and directrix property, 257.
self-conjugate triangle, property of, 271.
inscribed in a triangle, its directrix passes through the orthocentre, 273.
Parabola, construction of a, given four points, 181.
given four tangents, 135.
given three tangents and a point, 182.
under various conditions, 138, 139, 143, 146.
given the axis, the focus, and one point, 266.
given two tangents, the point of contact of one of them, and the direction of the axis, 278.
given two tangents and their points of contact, 279.
Parallel lines meet at infinity, 5.
projection, 19.
lines, construction of, with the ruler only, 96, 300.
Parallelogram, inscribed in or circumscribed about a conic, 219, 221.
described on a pair of conjugate semi-diameters of a conic is of constant area, 234.
Parameter of homology, 63.
Pascal's theorem, xi, 124.
lines, the sixty, 125.
Pencil, flat, defined, 22.
axial, 22.

harmonic, 40, 42.
in involution, 101.
in involution, orthogonal pair of rays of a, 172.
cut by a conic in pairs of points forming an involution, 166.
Pentagon, inscribed in a conic, 136.
circumscribed to a conic, 145.
self-conjugate with regard to a conic, 293.
Perpendiculars, centre of, see Orthocentre.
from a focus on tangents to a conic, the locus of their feet a circle, 259.
from the foci of a conic on a tangent, their product constant, 260.
from any point of the circumscribing circle of a triangle to the sides, their feet collinear, 261, 299.
construction of, with the ruler only, 97, 300.
Perspective, figures in, 3.
triangles in, 7, 8, 246.
forms in, 35.
plane, 10.
relief, 20.
Plane of points or lines, 22.
Planes, harmonic, 42.
involution of, 101.
Points, harmonic, on a straight line, 40.
harmonic, on a circle, 116.
harmonic, on a conic, 122, 157.
projective ranges of, on a conic, 161.
Polar reciprocal curves and figures, 240, 241.
of a conic with respect to a conic is a conic, 240.
of a circle with respect to a circle, 274.
of a conic with respect to a focus, 274, 275.
of a conic with respect to a point on the director circle, 288.
Polar system, defined, 248.
determined by two triangles in perspective, 248.
determined by a self-conjugate triangle and a point and its polar, 293.
Pole and polar, defined, 201, 202.
reciprocal property of, 204.
theory of, applied to the solution of problems, 300.
construction of, 205, 206, 248.
Poles, range of, projective with the pencil formed by their polars, 209, 224.
of a straight line with regard to all conics inscribed in the same quadrilateral lie on a fixed straight line, 237.
Polygon, inscribed in a conic, whose sides pass through fixed points, 151, 185, 187.
circumscribed to a conic, whose vertices slide on fixed lines, 152, 186.
whose sides pass through fixed points and whose vertices lie on fixed lines, 184.
Poncelet, ix, x, xii.
on variable polygons inscribed in or circumscribed to a conic, 151, 184–187.
on ideal chords, 226.
on polar reciprocal figures, 240.
on triangles inscribed in one conic and circumscribed about another, 244.
Porisms, of Euclid and Pappus, 95, 96.
of in- and circumscribed triangle, 94, 244.
of the inscribed and self-conjugate triangle, 243.
of the circumscribed and self-conjugate triangle, 243.
Power of a point with respect to a circle, 58.
Prime-forms, the six, 24.
Problems, solved with ruler only, 96–98.
of the second degree, 176–200.
solved by means of the ruler and a fixed circle, 194, 300.
solved by polar reciprocation, 301.
Projection, operation of, 2, 22, 164.
central, 3.
orthogonal, 19.
parallel, 19.
of a triad of elements into any other given triad, 36.
of a quadrangle into any given quadrangle, 80.
of a plane figure into another plane figure, 81.
Projective forms and figures, 34.
forms, when in perspective, 67.
forms, when harmonic, 69.
ranges, metrical relations of, 62.
forms, construction of, 70–74.
figures, construction of, 81–84.
plane figures can be put into homology, 84.
properties of points and tangents of a circle, 114–117.
properties of points and tangents of a conic, 118–130.
Projectivity of any two forms ABC and $A'B'C'$, 36.
of two forms $ABCD$ and $BADC$, 38.
of harmonic forms, 41, 43.
of the anharmonic ratio, 54.
of any two plane quadrangles, 80.
of a range of poles and the pencil formed by their polars, 209, 224.

Quadrangle, complete, defined, 29.
two plane quadrangles always projective, 80.

harmonic properties, 39, 47.
involution properties, 107, 225.
inscribed in a conic, 138, 140, 208, 225.
if two pairs of opposite sides are conjugate lines with regard to a conic, the third pair is so too, 246.
Quadrangles having the same diagonal points; their eight vertices lie on a conic or a line-pair, 210.
Quadrilateral, complete, defined, 29.
harmonic properties, 39, 46.
involution properties, 107, 225.
middle points of diagonals are collinear, 109, 299.
circumscribed to a conic, 142, 208, 225, 272.
locus of centres of inscribed conics, 237.
theorem of Hesse relating to the extremities of the three diagonals, 245.
Quadrilaterals having the same diagonals; their eight sides touch a conic or a point-pair, 212.

Range, defined, 22.
harmonic, 40.
Ranges, projective, on a conic, 161.
Ratio, of similitude, 18.
harmonic, 57.
anharmonic, 54–62.
Reciprocal figures, 85.
points and lines with regard to a conic, 204.
triangles, two, are in perspective, 246.
Reciprocation, polar, 241.
with respect to a circle, 274, 275.
applied to solution of problems, 301.
Rectangular hyperbola, see Equilateral.
Ruler only, problems solved with, 96–98.
Ruler and fixed circle, problems solved by help of the, 194, 300.

Section, operation of, 2, 22, 164.
of a cone, 14, 18.
of a cylinder, 19.
Segment, dividing two given ones harmonically, 58, 103, 295.
of constant magnitude sliding along a line generates two directly equal ranges, 89.
bisected, its division into n equal parts by aid of the ruler only, 97.
Segments of a straight line, metrical relations between, 51, 52.
Self-conjugate pentagon with regard to a conic, 293.
Self-conjugate triangle, 207–209.
circumscribing circle of a, its properties, 271.
Self-conjugate triangles with regard to a conic, two; properties of, 242.
Self-corresponding elements, defined, 67.
of two superposed projective forms, 68, 69, 78, 91–93.
general construction for these, 169.
of two coplanar projective figures, 79.
of two projective ranges on or series of tangents to a conic, 162, 163.
Sheaf, defined, 22.
Signs, rule of, 51.
Similar ranges and pencils, 86, 87, 128.
and similarly placed figures, 18.
Staudt, vi, vii.
on the geometric prime-forms, 24.
on the principle of duality, 26.
on harmonic forms, 39.
on the construction of two projective figures, 81.
on the polar system, 248.
on an involution of points on a conic, 165.
Steiner, vii, x, xii.
on the sixty Pascal lines and Brianchon points, 125.
on the solution of problems of the second degree by means of a ruler and a fixed circle, 194.
Superposed geometric forms, 68, 69.
construction of their self-corresponding elements, 169.
plane figures, if projective, cannot have more than three self-corresponding elements, 79.
Supplemental chords, 221.
Symmetry, a special case of homology, 64.

Tangents, harmonic, of a circle, 116, 117.
harmonic, of a conic, 168.
to a conic, series of projective, 163, 164.
orthogonal, to a conic, 269.
to a conic from a given point; constructions, 176, 177, 179, 226.
common, to two conics; constructions, 190.
Tetragram and Tetrastigm, 29.
Townsend, 200.
Transversal, cut by the sides of a triangle, 112.
cutting a quadrangle or a quadrilateral, 107, 108.
cutting a conic and an inscribed quadrangle, 150.
drawn through a point to cut a conic; property of the product of the segments, 281.
cutting a hyperbola and its asymptotes, 156, 282.
Transverse axis of a hyperbola, 228.

Triangle, inscribed in one triangle and circumscribed about another, 94.
inscribed in a conic, 143, 216.
circumscribed to a conic, 144, 216.
inscribed or circumscribed, involution-properties, 152, 157.
self-conjugate with regard to a conic, 207, 270.
circumscribed to a parabola, 253, 273.
self-conjugate with regard to a parabola, 271.
self-conjugate with regard to an equilateral hyperbola, 271.
cut by a conic, Carnot's theorem, 279.
inscribed in an equilateral hyperbola, 287.

Triangles, two, self-conjugate with regard to a conic; properties of, 242.
inscribed in one conic and self-conjugate to another, 243.
circumscribed to one conic and self-conjugate to another, 243.
inscribed in one conic and circumscribed to another, 244.
reciprocal, are in perspective, 246.
formed by two pairs of tangents to a conic and their chords of contact, 298.
Trisection of an arc of a circle, 294.

Vanishing points and lines, 5.
plane, 21.
Vertex of a conic, 228, 256.
circle of curvature at a, 190.

THE END.

Select Works

PUBLISHED BY THE CLARENDON PRESS.

ALDIS. A Text-Book of Algebra: with Answers to the Examples. By W. S. ALDIS, M.A. Crown 8vo, 7*s*. 6*d*.

BAYNES. Lessons on Thermodynamics. By R. E. BAYNES, M.A. Crown 8vo, 7*s*. 6*d*.

CHAMBERS. A Handbook of Descriptive Astronomy. By G. F. CHAMBERS, F.R.A.S. *Fourth Edition.*

Vol. I. The Sun, Planets, and Comets. 8vo, 21*s*.
Vol. II. Instruments and Practical Astronomy. 8vo, 21*s*.
Vol. III. The Starry Heavens. 8vo, 14*s*.

CLARKE. Geodesy. By Col. A. R. CLARKE, C.B., R.E. 12*s*. 6*d*.

CREMONA. Graphical Statics. Two Treatises on the Graphical Calculus and Reciprocal Figures in Graphical Statics. By LUIGI CREMONA. Translated by T. HUDSON BEARE. Demy 8vo, 8*s*. 6*d*.

DONKIN. Acoustics. By W. F. DONKIN, M.A., F.R.S. *Second Edition.* Crown 8vo, 7*s*. 6*d*.

EMTAGE. An Introduction to the Mathematical Theory of Electricity and Magnetism. By W. T. A. EMTAGE, M.A. 7*s*. 6*d*.

JOHNSTON. A Text-Book of Analytical Geometry. By W. J. JOHNSTON, M.A. Crown 8vo, 10*s*. 6*d*.

MINCHIN. A Treatise on Statics, with Applications to Physics. By G. M. MINCHIN, M.A. *Fourth Edition.*

Vol. I. Equilibrium of Coplanar Forces. 8vo, 10*s*. 6*d*.

Vol. II. Non-Coplanar Forces. 8vo, 16*s*.

—— **Uniplanar Kinematics of Solids and Fluids.** Crown 8vo, 7*s*. 6*d*.

—— **Hydrostatics and Elementary Hydrokinetics.** Crown 8vo, 10*s*. 6*d*.

PRICE. Treatise on Infinitesimal Calculus. By BARTHOLOMEW PRICE, D.D., F.R.S.

Vol. I. Differential Calculus. *Second Edition.* 8vo, 14*s.* 6*d.*
Vol. II. Integral Calculus, Calculus of Variations, and Differential Equations. *Second Edition.* 8vo, 18*s.*
Vol. III. Statics, including Attractions; Dynamics of a Material Particle. *Second Edition.* 8vo, 16*s.*
Vol. IV. Dynamics of Material Systems. *Second Edition.* 8vo, 18*s.*

RUSSELL. An Elementary Treatise on Pure Geometry, with numerous Examples. By J. WELLESLEY RUSSELL, M.A. Crown 8vo, 10*s.* 6*d.*

SELBY. Elementary Mechanics of Solids and Fluids. By A. L. SELBY, M.A. Crown 8vo, 7*s.* 6*d.*

SMYTH. A Cycle of Celestial Objects. Observed, Reduced, and Discussed by Admiral W. H. SMYTH, R.N. Revised, condensed, and greatly enlarged by G. F. CHAMBERS, F.R.A.S. 8vo, 12*s.*

STEWART. An Elementary Treatise on Heat. With numerous Woodcuts and Diagrams. By BALFOUR STEWART, LL.D., F.R.S. *Fifth Edition.* Extra fcap. 8vo, 7*s.* 6*d.*

THOMSON. Recent Researches in Electricity and Magnetism. By J. J. THOMSON, M.A., D.Sc., F.R.S., Editor of Maxwell's 'Electricity and Magnetism.' 8vo, 18*s.* 6*d.*

VAN 'T HOFF. Chemistry in Space. Translated and Edited by J. E. MARSH, B.A. Crown 8vo, 4*s.* 6*d.*

WALKER. The Theory of a Physical Balance. By JAMES WALKER, M.A. 8vo, stiff cover, 3*s.* 6*d.*

WATSON and BURBURY.

I. A Treatise on the Application of Generalised Coordinates to the Kinetics of a Material System. By H. W. WATSON, D.Sc., and S. H. BURBURY, M.A. 8vo, 6*s.*

II. The Mathematical Theory of Electricity and Magnetism.

Vol. I. Electrostatics. 8vo, 10*s.* 6*d.*
Vol. II. Magnetism and Electrodynamics. 8vo, 10*s.* 6*d.*

Oxford:

AT THE CLARENDON PRESS.

LONDON: HENRY FROWDE,

OXFORD UNIVERSITY PRESS WAREHOUSE, AMEN CORNER, E.C.

Clarendon Press, Oxford.

SELECT LIST OF STANDARD WORKS.

DICTIONARIES	page 1
LAW	„ 2
HISTORY, BIOGRAPHY, ETC.	„ 3
PHILOSOPHY, LOGIC, ETC.	„ 6
PHYSICAL SCIENCE	„ 7

1. DICTIONARIES.

A New English Dictionary on Historical Principles, founded mainly on the materials collected by the Philological Society. Edited by James A. H. Murray, LL.D., and H. Bradley, M.A. Imperial 4to.

Vol. I, A and B, and Vol. II, C, half-morocco, 2*l.* 12*s.* 6*d.* each.
Vol. III, D and E.
D. Edited by Dr. Murray. [*In the Press.*]
E. Edited by Henry Bradley, M.A.
E—EVERY, 12*s.* 6*d.* [*Published.*]
EVERYBODY—EZOD, 5*s.* [*Published.*]

An Etymological Dictionary of the English Language, arranged on an Historical Basis. By W. W. Skeat, Litt.D. *Second Edition.* 4to. 2*l.* 4*s.*

A Middle-English Dictionary. By F. H. Stratmann. A new edition, by H. Bradley, M.A. 4to, half-bound, 1*l.* 11*s.* 6*d.*

An Anglo-Saxon Dictionary, based on the MS. collections of the late Joseph Bosworth, D.D. Edited and enlarged by Prof. T. N. Toller, M.A. Parts I–III. A–SÁR. 4to, stiff covers, 15*s.* each. Part IV, § 1, SÁR–SWÍÐRIAN. Stiff covers, 8*s.* 6*d.*

An Icelandic-English Dictionary, based on the MS. collections of the late Richard Cleasby. Enlarged and completed by G. Vigfússon, M.A. 4to. 3*l.* 7*s.*

A Greek-English Lexicon, by H. G. Liddell, D.D., and Robert Scott, D.D. *Seventh Edition, Revised and Augmented.* 4to. 1*l.* 16*s.*

A Latin Dictionary. By Charlton T. Lewis, Ph.D., and Charles Short, LL.D. 4to. 1*l.* 5*s.*

A Sanskrit-English Dictionary. Etymologically and Philologically arranged. By Sir M. Monier-Williams, D.C.L. 4to. 4*l.* 14*s.* 6*d.*

A Hebrew and English Lexicon of the Old Testament, with an Appendix containing the Biblical Aramaic, based on the Thesaurus and Lexicon of Gesenius, by Francis Brown, D.D., S. R. Driver, D.D., and C. A. Briggs, D.D. Parts I–III. Small 4to, 2*s.* 6*d.* each.

Thesaurus Syriacus: collegerunt Quatremère, Bernstein, Lorsbach, Arnoldi, Agrell, Field, Roediger: edidit R. Payne Smith, S.T.P. Vol. I, containing Fasc. I–V, sm. fol. 5*l.* 5*s.*

Fasc. VI. 1*l.* 1*s.*; VII. 1*l.* 11*s.* 6*d.*; VIII. 1*l.* 16*s.*; IX. 1*l.* 5*s.*

2. LAW.

Anson. *Principles of the English Law of Contract, and of Agency in its Relation to Contract.* By Sir W. R. Anson, D.C.L. *Seventh Edition.* 8vo. 10*s.* 6*d.*

—— *Law and Custom of the Constitution.* 2 vols. 8vo.

Part I. Parliament. *Second Edition.* 12*s.* 6*d.*

Part II. The Crown. 14*s.*

Baden-Powell. *Land-Systems of British India;* being a Manual of the Land-Tenures, and of the Systems of Land-Revenue Administration prevalent in the several Provinces. By B. H. Baden-Powell, C.I.E. 3 vols. 8vo. 3*l.* 3*s.*

—— *Land-Revenue and Tenure in British India.* By the same Author. With Map. Crown 8vo, 5*s.*

Digby. *An Introduction to the History of the Law of Real Property.* By E ›lm E. Digby, M.A. *Fourth Edition.* 8vo. 12*s.* 6*d.*

Greenidge. *Infamia; its place in Roman Public and Private Law.* By A. H. J. Greenidge, M.A. 8vo. 10*s.* 6*d.*

Grueber. *Lex Aquilia.* The Roman Law of Damage to Property: being a Commentary on the Title of the Digest 'Ad Legem Aquiliam' (ix. 2). By Erwin Grueber, Dr. Jur., M.A. 8vo. 10*s.* 6*d.*

Hall. *International Law.* By W. E. Hall, M.A. *Third Edition.* 8vo. 22*s.* 6*d.*

—— *A Treatise on the Foreign Powers and Jurisdiction of the British Crown.* By W. E. Hall, M.A. 8vo. 10*s.* 6*d.*

Holland and Shadwell. *Select Titles from the Digest of Justinian.* By T. E. Holland, D.C.L., and C. L. Shadwell, B.C.L. 8vo. 14*s.*

Also sold in Parts, in paper covers:—

Part I. Introductory Titles. 2*s.* 6*d.*

Part II. Family Law. 1*s.*

Part III. Property Law. 2*s.* 6*d.*

Part IV. Law of Obligations (No. 1). 3*s.* 6*d.* (No. 2). 4*s.* 6*d.*

Holland. *Elements of Jurisprudence.* By T. E. Holland, D.C.L. *Sixth Edition.* 8vo. 10*s.* 6*d.*

Holland. *The European Concert in the Eastern Question*; a Collection of Treaties and other Public Acts. Edited, with Introductions and Notes, by T. E. Holland, D.C.L. 8vo. 12s. 6d.

—— *Gentilis, Alberici, De Iure Belli Libri Tres.* Edidit T. E. Holland, I.C.D. Small 4to, half-morocco, 21s.

—— *The Institutes of Justinian*, edited as a recension of the Institutes of Gaius, by T. E. Holland, D.C.L. *Second Edition.* Extra fcap. 8vo. 5s.

Markby. *Elements of Law considered with reference to Principles of General Jurisprudence.* By Sir William Markby, D.C.L. *Fourth Edition.* 8vo. 12s. 6d.

Moyle. *Imperatoris Iustiniani Institutionum Libri Quattuor*; with Introductions, Commentary, Excursus and Translation. By J. B. Moyle, D.C.L. *Second Edition.* 2 vols. 8vo. Vol. I. 16s. Vol. II. 6s.

—— *Contract of Sale in the Civil Law.* By J. B. Moyle, D.C.L. 8vo. 10s. 6d.

Pollock and Wright. *An Essay on Possession in the Common Law.* By F. Pollock, M.A., and R. S. Wright, B.C.L. 8vo. 8s. 6d.

Poste. *Gaii Institutionum Juris Civilis Commentarii Quattuor*; or, Elements of Roman Law by Gaius. With a Translation and Commentary by Edward Poste, M.A. *Third Edition.* 8vo. 18s.

Raleigh. *An Outline of the Law of Property.* By Thos. Raleigh, M.A. 8vo. 7s. 6d.

Sohm. *Institutes of Roman Law.* By Rudolph Sohm, Professor in the University of Leipzig. Translated by J. C. Ledlie, B.C.L. With an Introductory Essay by Erwin Grueber, Dr. Jur., M.A. 8vo. 18s.

Stokes. *The Anglo-Indian Codes.* By Whitley Stokes, LL.D.

Vol. I. Substantive Law. 8vo. 30s.

Vol. II. Adjective Law. 8vo. 35s.

First and Second Supplements to the above, 1887–1891. 8vo. 6s. 6d.

Separately, No. 1, 2s. 6d.; No. 2, 4s. 6d.

3. HISTORY, BIOGRAPHY, ETC.

Arbuthnot. *The Life and Works of John Arbuthnot, M.D.* By George A. Aitken. 8vo, cloth, with portrait, 16s.

Bentham. *A Fragment on Government.* By Jeremy Bentham. Edited with an Introduction by F. C. Montague, M.A. 8vo. 7s. 6d.

Boswell's *Life of Samuel Johnson, LL.D.* Edited by G. Birkbeck Hill, D.C.L. In six volumes, medium 8vo. With Portraits and Facsimiles. Half-bound, 3l. 3s.

Carte's *Life of James Duke of Ormond.* 6 vols. 8vo. 1l. 5s.

Casaubon (Isaac). 1559–1614. By Mark Pattison. 8vo. 16s.

Clarendon's *History of the Rebellion and Civil Wars in England.* Re-edited from a fresh collation of the original MS. in the Bodleian Library, with marginal dates and occasional notes, by W. Dunn Macray, M.A., F.S.A. 6 vols. Crown 8vo. 2l. 5s.

Earle. *Handbook to the Land-Charters, and other Saxonic Documents.* By John Earle, M.A. Crown 8vo. 16s.

Finlay. *A History of Greece from its Conquest by the Romans to the present time,* B.C. 146 to A.D. 1864. By George Finlay, LL.D. A new Edition, edited by H. F. Tozer, M.A. 7 vols. 8vo. 3l. 10s.

Fortescue. *The Governance of England.* By Sir John Fortescue, Kt. A Revised Text. Edited, with Introduction, Notes, &c., by Charles Plummer, M.A. 8vo, half-bound, 12s. 6d.

Freeman. *The History of Sicily from the Earliest Times.*

Vols. I. and II. 8vo, cloth, 2l. 2s.

Vol. III. The Athenian and Carthaginian Invasions. 8vo, cloth, 24s.

Vol. IV. From the Tyranny of Dionysios to the Death of Agathoklês. Edited by Arthur J. Evans, M.A. 21s.

—— *History of the Norman Conquest of England; its Causes and Results.* By E. A. Freeman, D.C.L. In Six Volumes. 8vo. 5l. 9s. 6d.

—— *The Reign of William Rufus and the Accession of Henry the First.* 2 vols. 8vo. 1l. 16s.

French Revolutionary Speeches. (See Stephens, H. Morse.)

Gardiner. *The Constitutional Documents of the Puritan Revolution,* 1628–1660. Selected and Edited by Samuel Rawson Gardiner, M.A. Crown 8vo. 9s.

Greswell. *History of the Dominion of Canada.* By W. Parr Greswell, M.A. Crown 8vo. With Eleven Maps. 7s. 6d.

—— *Geography of the Dominion of Canada and Newfoundland.* Crown 8vo. With Ten Maps. 6s.

—— *Geography of Africa South of the Zambesi.* With Maps. Crown 8vo. 7s. 6d.

Gross. *The Gild Merchant;* a Contribution to British Municipal History. By Charles Gross, Ph.D. 2 vols. 8vo. 24s.

Hastings. *Hastings and the Rohilla War.* By Sir John Strachey, G.C.S.I. 8vo, cloth, 10s. 6d.

Hodgkin. *Italy and her Invaders.* With Plates and Maps. By T. Hodgkin, D.C.L. Vols. I–IV, A.D. 376–553. 8vo.

Vols. I. and II. *Second Edition.* 2l. 2s.

Vols. III. and IV. 1l. 16s.

Vols. V. and VI. *In the Press.*

—— *The Dynasty of Theodosius; or, Seventy Years' Struggle with the Barbarians.* By the same Author. Crown 8vo. 6s.

Hume. *Letters of David Hume to William Strahan.* Edited with Notes, Index, &c., by G. Birkbeck Hill, D.C.L. 8vo. 12s. 6d.

Johnson. *Letters of Samuel Johnson, LL.D.* Collected and edited by G. Birkbeck Hill, D.C.L., Editor of Boswell's 'Life of Johnson' (see Boswell). 2 vols. half-roan, 28*s.*

Kitchin. *A History of France.* With Numerous Maps, Plans, and Tables. By G. W. Kitchin, D.D. In three Volumes. *Second Edition.* Crown 8vo, each 10*s.* 6*d.*

Vol. I. to 1453. Vol. II. 1453–1624. Vol. III. 1624–1793.

Ludlow. *The Memoirs of Edmund Ludlow, Lieutenant-General of the Horse in the Army of the Commonwealth of England*, 1625–1672. Edited, with Appendices and Illustrative Documents, by C. H. Firth, M.A. 2 vols. 8vo. 1*l.* 16*s.*

Luttrell's (*Narcissus*) *Diary.* A Brief Historical Relation of State Affairs, 1678–1714. 6 vols. 1*l.* 4*s.*

Lucas. *Introduction to a Historical Geography of the British Colonies.* By C. P. Lucas, B.A. With Eight Maps. Crown 8vo. 4*s.* 6*d.*

—— *Historical Geography of the British Colonies:*

Vol. I. The Mediterranean and Eastern Colonies (exclusive of India). With Eleven Maps. Crown 8vo. 5*s.*

Vol. II. The West Indian Colonies. With Twelve Maps. Crown 8vo. 7*s.* 6*d.*

Vol. III. West Africa. With Five Maps. Crown 8vo. 7*s.* 6*d.*

Machiavelli. *Il Principe.* Edited by L. Arthur Burd, M.A. With an Introduction by Lord Acton. 8vo. Cloth, 14*s.*

Prothero. *Select Statutes and other Constitutional Documents, illustrative of the Reigns of Elizabeth and James I.* Edited by G. W. Prothero, Fellow of King's College, Cambridge. Crown 8vo. 10*s.* 6*d.*

Ralegh. *Sir Walter Ralegh.* A Biography. By W. Stebbing, M.A. 8vo. 10*s.* 6*d.*

Ramsay (Sir J. H.). *Lancaster and York.* A Century of English History (A.D. 1399–1485). By Sir J. H. Ramsay of Bamff, Bart., M.A. With Maps, Pedigrees, and Illustrations. 2 vols. 8vo. 36*s.*

Ranke. *A History of England, principally in the Seventeenth Century.* By L. von Ranke. Translated under the superintendence of G. W. Kitchin, D.D., and C. W. Boase, M.A. 6 vols. 8vo. 3*l.* 3*s.*

Rawlinson. *A Manual of Ancient History.* By George Rawlinson, M.A. *Second Edition.* 8vo. 14*s.*

Rhŷs. *Studies in the Arthurian Legend.* By John Rhŷs, M.A. 8vo. 12*s.* 6*d.*

Ricardo. *Letters of David Ricardo to T. R. Malthus* (1810–1823). Edited by James Bonar, M.A. 8vo. 10*s.* 6*d.*

Rogers. *History of Agriculture and Prices in England,* A.D. 1259–1702. By James E. Thorold Rogers, M.A. 6 vols., 8vo. 7*l.* 2*s.*

—— *First Nine Years of the Bank of England.* 8vo. 8*s.* 6*d.*

—— *Protests of the Lords, including those which have been expunged, from 1624 to 1874;* with Historical Introductions. In three volumes. 8vo. 2*l.* 2*s.*

Smith's *Wealth of Nations.* With Notes, by J. E. Thorold Rogers, M.A. 2 vols. 8vo. 21*s.*

Stephens. *The Principal Speeches of the Statesmen and Orators of the French Revolution*, 1789–1795. With Historical Introductions, Notes, and Index. By H. Morse Stephens. 2 vols. Crown 8vo. 21s.

Stubbs. *Select Charters and other Illustrations of English Constitutional History, from the Earliest Times to the Reign of Edward I.* Arranged and edited by W. Stubbs, D.D., Lord Bishop of Oxford. *Seventh Edition.* Crown 8vo. 8s. 6d.

—— *The Constitutional History of England, in its Origin and Development. Library Edition.* 3 vols. Demy 8vo. 2l. 8s.

Also in 3 vols. crown 8vo. price 12s. each.

Stubbs. *Seventeen Lectures on the Study of Medieval and Modern History.* Crown 8vo. 8s. 6d.

—— *Registrum Sacrum Anglicanum.* An attempt to exhibit the course of Episcopal Succession in England. By W. Stubbs, D.D. Small 4to. 8s. 6d.

Swift (F. D.). *The Life and Times of James the First of Aragon.* By F. D. Swift, B.A. 8vo. 12s. 6d.

Vinogradoff. *Villainage in England.* Essays in English Mediaeval History. By Paul Vinogradoff, Professor in the University of Moscow. 8vo, half-bound. 16s.

4. PHILOSOPHY, LOGIC, ETC.

Bacon. *The Essays.* With Introduction and Illustrative Notes. By S. H. Reynolds, M.A. 8vo, half-bound. 12s. 6d.

—— *Novum Organum.* Edited, with Introduction, Notes, &c., by T. Fowler, D.D. *Second Edition.* 8vo. 15s.

—— *Novum Organum.* Edited, with English Notes, by G. W. Kitchin, D.D. 8vo. 9s. 6d.

Berkeley. *The Works of George Berkeley, D.D., formerly Bishop of Cloyne; including many of his writings hitherto unpublished.* With Prefaces, Annotations, and an Account of his Life and Philosophy. By A. Campbell Fraser, Hon. D.C.L., LL.D. 4 vols. 8vo. 2l. 18s.

The Life, Letters, &c., separately, 16s.

Bosanquet. *Logic; or, the Morphology of Knowledge.* By B. Bosanquet, M.A. 8vo. 21s.

Butler's *Works, with Index to the Analogy.* 2 vols. 8vo. 11s.

Fowler. *The Elements of Deductive Logic, designed mainly for the use of Junior Students in the Universities.* By T. Fowler, D.D. *Ninth Edition,* with a Collection of Examples. Extra fcap. 8vo. 3s. 6d.

—— *The Elements of Inductive Logic, designed mainly for the use of Students in the Universities.* By the same Author. *Sixth Edition.* Extra fcap. 8vo. 6s.

Fowler and Wilson. *The Principles of Morals.* By T. Fowler, D.D., and J. M. Wilson, B.D. 8vo, cloth, 14s.

Green. *Prolegomena to Ethics.* By T. H. Green, M.A. Edited by A. C. Bradley, M.A. 8vo. 12s. 6d.

Hegel. *The Logic of Hegel.* Translated from the Encyclopaedia of the Philosophical Sciences. With Prolegomena to the Study of Hegel's Logic and Philosophy. By W. Wallace, M.A. *Second Edition, Revised and Augmented.* 2 vols. Crown 8vo. 10s. 6d. each.

Hegel's *Philosophy of Mind.* Translated from the Encyclopaedia of the Philosophical Sciences. With Five Introductory Essays. By William Wallace, M.A., LL.D. Crown 8vo. 10s. 6d.

Hume's *Treatise of Human Nature.* Edited, with Analytical Index, by L. A. Selby-Bigge, M.A. Crown 8vo. 9s.

Hume's *Enquiry concerning the Human Understanding, and an Enquiry concerning the Principles of Morals.* Edited by L. A. Selby-Bigge, M.A. Crown 8vo. 7s. 6d.

Locke. *An Essay Concerning Human Understanding.* By John Locke. Collated and Annotated, with Prolegomena, Biographical, Critical, and Historic, by A. Campbell Fraser, Hon. D.C.L., LL.D. 2 vols. 8vo. 1*l.* 12s.

Lotze's *Logic*, **in Three Books;** of Thought, of Investigation, and of Knowledge. English Translation; Edited by B. Bosanquet, M.A. *Second Edition.* 2 vols. Cr. 8vo. 12s.

—— *Metaphysic*, in Three Books; Ontology, Cosmology, and Psychology. English Translation; Edited by B. Bosanquet, M.A. *Second Edition.* 2 vols. Cr. 8vo. 12s.

Martineau. *Types of Ethical Theory.* By James Martineau, D.D. *Third Edition.* 2 vols. Cr. 8vo. 15s.

—— *A Study of Religion: its Sources and Contents. Second Edition.* 2 vols. Cr. 8vo. 15s.

5. PHYSICAL SCIENCE.

Chambers. *A Handbook of Descriptive and Practical Astronomy.* By G. F. Chambers, F.R.A.S. *Fourth Edition*, in 3 vols. Demy 8vo.

Vol. I. The Sun, Planets, and Comets. 21s.

Vol. II. Instruments and Practical Astronomy. 21s.

Vol. III. The Starry Heavens. 14s.

De Bary. *Comparative Anatomy of the Vegetative Organs of the Phanerogams and Ferns.* By Dr. A. de Bary. Translated by F. O. Bower, M.A., and D. H. Scott, M.A. Royal 8vo. 1*l.* 2s. 6d.

—— *Comparative Morphology and Biology of Fungi, Mycetozoa and Bacteria.* By Dr. A. de Bary. Translated by H. E. F. Garnsey, M.A. Revised by Isaac Bayley Balfour, M.A., M.D., F.R.S. Royal 8vo, half-morocco, 1*l.* 2s. 6d.

De Bary. *Lectures on Bacteria.* By Dr. A. de Bary. *Second Improved Edition.* Translated by H. E. F. Garnsey, M.A. Revised by Isaac Bayley Balfour, M.A., M.D., F.R.S. Crown 8vo. 6s.

Goebel. *Outlines of Classification and Special Morphology of Plants.* By Dr. K. Goebel. Translated by H. E. F. Garnsey, M.A. Revised by Isaac Bayley Balfour, M.A., M.D., F.R.S. Royal 8vo, half-morocco, 1*l.* 1s.

Sachs. *Lectures on the Physiology of Plants.* By Julius von Sachs. Translated by H. Marshall Ward, M.A., F.L.S. Royal 8vo, half-morocco, 1*l.* 11*s.* 6*d.*

—— *A History of Botany.* Translated by H. E. F. Garnsey, M.A. Revised by I. Bayley Balfour, M.A., M.D., F.R.S. Crown 8vo. 10*s.*

Fossil Botany. *Being an Introduction to Palaeophytology from the Standpoint of the Botanist.* By H. Graf zu Solms-Laubach. Translated by H. E. F. Garnsey, M.A. Revised by I. Bayley Balfour, M.A., M.D., F.R.S. Royal 8vo, half-morocco, 18*s.*

Annals of Botany. Edited by Isaac Bayley Balfour, M.A., M.D., F.R.S., Sydney H. Vines, D.Sc., F.R.S., D. H. Scott, M.A., Ph.D., F.L.S., and W. G. Farlow, M.D.; assisted by other Botanists. Royal 8vo, half-morocco, gilt top.

Vol. I. Parts I–IV. 1*l.* 16*s.*
Vol. II. Parts V–VIII. 2*l.* 2*s.*
Vol. III. Parts IX–XII. 2*l.* 12*s.* 6*d.*
Vol. IV. Parts XIII–XVI. 2*l.* 5*s.*
Vol. V. Parts XVII–XX. 2*l.* 10*s.*
Vol. VI. Parts XXI–XXIV. 2*l.* 4*s.*
Vol. VII. Parts XXV–XXVIII. 2*l.* 10*s.*
Vol. VIII. Parts XXIX, XXX. 12*s.* each.

Biological Series.

I. *The Physiology of Nerve, of Muscle, and of the Electrical Organ.* Edited by J. Burdon Sanderson, M.D., F.R.SS. L.&E. Medium 8vo. 1*l.* 1*s.*

II. *The Anatomy of the Frog.* By Dr. Alexander Ecker, Professor in the University of Freiburg. Translated, with numerous Annotations and Additions, by G. Haslam, M.D. Med. 8vo. 21*s.*

IV. *Essays upon Heredity and Kindred Biological Problems.* By Dr. A. Weismann. Vol. I. Translated and Edited by E. B. Poulton, M.A., S. Schönland, Ph.D., and A. E. Shipley, M.A. *Second Edition.* Crown 8vo. 7*s.* 6*d.*

——Vol. II. Edited by E. B. Poulton, and A. E. Shipley. Crown 8vo. 5*s.*

Prestwich. *Geology, Chemical, Physical, and Stratigraphical.* By Joseph Prestwich, M.A., F.R.S. In two Volumes.

Vol. I. Chemical and Physical. Royal 8vo. 1*l.* 5*s.*

Vol. II. Stratigraphical and Physical. With a new Geological Map of Europe. Royal 8vo. 1*l.* 16*s.*

Price. *A Treatise on the Measurement of Electrical Resistance.* By W. A. Price, M.A., A.M.I.C.E. 8vo. 14*s.*

Smith. *Collected Mathematical Papers of the late Henry J. S. Smith, M.A., F.R.S.* Edited by J. W. L. Glaisher, Sc.D., F.R.S. 2 vols. 4to 3*l.* 3*s.*

Oxford
AT THE CLARENDON PRESS
LONDON: HENRY FROWDE
OXFORD UNIVERSITY PRESS WAREHOUSE, AMEN CORNER, E.C.

Published by

Facsimile Publisher
Ashok Vihar, Delhi-110052 (India)
Website: www.facsimilepublishers.com

Distributed by

Gyan Books Pvt. Ltd.
5, Ansari Road, Daryaganj, New Delhi-110002 (India)
E-mail: books@gyanbooks.com, Website: www.gyanbooks.com

8 888000 686439